Forecasting in Financial and Sports Gambling Markets

Forecasting in Financial and Sports Gambling Markets

Adaptive Drift Modeling

William S. Mallios
Craig School of Busines
California State University, Fresno
Fresno, California

A JOHN WILEY & SONS, INC., PUBLICATION

Published by John Wiley & Sons, Inc., Hoboken, New Jersey.
Published simultaneously in Canada.

Library of Congress Cataloging-in-Publication Data:

Mallios, William S.

Forecasting in financial and sports gambling markets: adaptive drift modeling / William S. Mallios.

Includes bibliographical references and index.

ISBN 978-0-470-48452-4

Printed in Singapore

10 9 8 7 6 5 4 3 2 1

Contents

Preface

Adam Smith's last words in 1790: *I believe we should adjourn this meeting to another place*. The meeting is re-adjourned—as it has been so many times. The focus is on adaptive drift modeling—models that adapt to the specific evolving market and then drift over time. Forecasting procedures are for purposes of active trading in finance and betting against the line in sports. Such modeling has relevance to other evolving markets, including the influenza markets.

There is ample motivation for this book: the recent financial market abyss, the proliferation of sports gambling, the metastasis of lotteries, the ongoing epidemics of financial and mathematical illiteracy—epidemics that are allied with the emerging epidemic of adolescent problem gambling—and the assortment of *sausage legislation* proposed for financial market reform and regulation and legalized online gambling. All of these topics are driven by a common denominator: human behavior—the type of behavior that has been described as *the madness of crowds* combined with *the cunning of the few*.

Thomas Huxley, the 19th-century biologist and defender of Darwin, said that the great tragedy of science was the slaying of a beautiful hypothesis by an ugly fact. A counter opinion is that the beauty of science is the eventual slaying of outdated hypotheses and dogma through evolving and enlightened inquiry.

For over 50 years, the prevailing investment wisdom was *buy and hold for the long term*. Indeed, markets were said to be efficient in which case active or short-term trading would inevitably result in portfolio losses. Researchers were quick to apply efficient market dogma to the sports gambling markets: that is, that over the longer term, the bookmakers' lines can't be beaten. Popular books such as *A Random Walk Down Wall Street* (Malkiel, 1985) portrayed active traders as inevitable losers. Even the *Wall Street Journal* carried a series that compared dart throwing in selecting equities with selections by expert traders.

However, with financial market deregulation and the entry of hedge funds, particularly during the latter stages of the Clinton Administration, active trading strategies began to dominate the buy-and-hold strategies. Financial innovations began to sprout under the cover of efficient market dogma. Finally, the innovations bubbled and the dogma crumbled. In testimony before Congress in 2006, former Fed Chairman Alan Greenspan humbly admitted: *I made a mistake in presuming that the self interests of organizations, specifically banks and others, were such that they were best capable of protecting their own shareholders and their equity in the firms*.

In lay terms, hedge funds are investment vehicles limited by law to the very rich. In contrast to mutual funds, they are largely unregulated and invest opaquely. They *hedge* their investment monies, not so much in the sense of *hedging or protecting against risk*, but rather for purposes of maximizing profits. Hedge funds are known as *quant funds* when they employ quantitative (statistical) modeling in forecasting short-term price movements. Quant funds were subjected to severe criticism when the subprime mortgage crisis spilled over to other financial markets—at which point the hedge funds were affected adversely. In many quarters, quant modeling was condemned, sometimes in buffoonish fashion, along with the efficient market hypothesis.

> *When you see a quantitative "expert," shout for help, call for his disgrace, make him accountable. Ask for the drastic overhaul of business schools.... Ask for the Nobel Prize in Economics to be withdrawn from the authors of these theories, as the Nobel's credibility can be extremely harmful.* (Taleb and Triana, 12/8/08)

A number of explanations were given for the forecasting failures of quant modeling—failures that also apply to forecasting in the sports gambling markets.

1. It is typically the case that modeling complexities induce less capable analysts to impose invalid or oversimplified modeling assumptions, which usually lead to invalid forecasts, especially during periods of unexpected volatility.
2. Quant models have short shelf lives and tend to be of limited value when they are not updated on a continuing basis to accommodate changing market dynamics.

3. Recent relevant and vital information, often from allied sciences, is not incorporated in the model-building procedure. (For example, the weather determines price differences in the agricultural commodities market, which has led hedge funds to hire meteorologists to interact with their traders.)
4. With so few qualified modelers, the best modelers are lured away by competing funds. The modelers then use the same models to chase the same money.
5. Market shocks (i.e., unexpected, often unpredictable events) are either not incorporated or are incorporated inappropriately in forecasting models. Moreover, there has been a failure to recognize that the volatility associated with sufficiently large shocks may destabilize model structure, at least temporarily, to the extent that model forecasts become unreliable.
6. In situation 5, there is typically a failure to reconstruct and adapt the forecasting model so that it applies to evolving market conditions, which includes adapting the model to incremental changes during periods when markets are relatively stable.

Given the highly competitive and risky environments of current-day financial and sports gambling markets, the focus is on the dynamic process of constructing effective forecasting rules that are based on both graphical patterns and adaptive drift modeling of *cointegrated* time series. The graphical patterns are in terms of *candlestick charts* and their variants, a well-known charting procedure dating back to feudal Japan. Charting objectives are to identify optimal time periods in financial markets and optimal games in sports gambling markets for which forecasting rules and models are likely to provide profitable trading or wagering outcomes.

The modeling of *cointegrated* time series means that forecasts are with reference to a system of simultaneous time series wherein long-term relations exist between the individual series comprising the system. Disequilibria between such relations are known to affect subsequent outcomes within individual relations. As such, estimates of the between-series disequilibria can be used in forecasting subsequent movements within the individual time series. For example, consider a time-varying, emotional attachment variable for each of two lovers. The two variables are clearly related, but the relation between the two is subject to disequilibria over time. When, at any point in time, a major disequilibrium occurs—in the sense of, say, a temporarily strained personal relationship—the tendency is for the relation to return subsequently to normal. In this case, between-relation disequilibria can be used to predict subsequent outcomes for each of the individual variables. On the other hand, the disequilibria can become sufficiently large—analogous to periods of extreme volatility in financial markets—to the extent that the lovers may split (temporarily or permanently) and their responses may no longer be cointegrated.

Optimal profit-making situations in financial markets occur when markets are *inefficient*, in which case short-term price movements are more likely to be predictable. In the sports gambling markets, periods of *market inefficiency* are in terms of forthcoming games where outcomes are likely to differ considerably from the bookmakers' lines.

Shocks, defined as unexpected deviations from the norm (or from *what is expected*), may or may not be predictable. However, once they occur and are known or estimated, their effects are often highly consequential in effectively forecasting subsequent outcomes. In fact, shocks are the key to successful forecasting in the markets under study.

Shocks are best illustrated in modeling National Football League or National Basketball Association game outcomes. A bookmaker's line on a game is based on the gambling public's expectation of what the game outcome will be. Specifically, the bookmaker's job is to determine that line (or spread) which evenly divides the money wagered on the game. Since the parties covering the bets charge a commission (usually, 10 percent) on each bet that is made, it is irrelevant whether the line is realistic or not as long as the payouts to the winners are covered by the losers' losses.

To illustrate the effects of gambling shocks, suppose that a heavily favored team is upset by an underdog, such as having the 2008–2009 Los Angeles Lakers, an 11.5-point favorite, lose to the Sacramento Kings in midseason. The likely Laker team reaction to the loss is to *reevaluate game strategies, identify mental lapses, elevate testosterone levels, and then make up for the miserable performance* not only in their next game or games but also in their next meeting with the Kings. In this context, the gambling shocks are reflections of physiological–psychological–sociological variables that affect player and team personnel. As such, shocks tend to be determining factors in subsequent game outcomes. Discussions of shock effects in financial markets, termed moving average effects, are presented in Chapter 4.

The creation of sports hedge funds appears inevitable—if they do not already exist in the opaque and ill-regulated world of hedge funds. A bet on the favored 2009 New York Yankees in October carried less risk than an active trader's long or short position on Bank of America during the same time period—at least for bettors without access to insider information. In a similar vein, online sports gambling will eventually be legalized for purposes of enriching government coffers—in the same way that Prohibition was repealed to provide lucrative tax revenues. Concurrently, the lottery markets will continue to flourish in the form of stupidity taxes that prey on those who are infected by the raging epidemics of mathematical and financial illiteracy and the related epidemic of adolescent problem gambling.

The great economist Woody Allen once said: *More than any time in our history, mankind faces a crossroads. One path leads to despair and utter hopelessness, the other to total extinction. Let us pray we have the wisdom to choose correctly*. Financial and sports gambling markets will continue to be an inevitable part of the economic and social fabric for unforeseeable future generations. Reasonable courses must be chartered. This meeting will be readjourned again and again and again.

Updates of adaptive drift modeling forecasts in sports gambling and financial markets are available at *www.MalliosAssociates.com*.

Acknowledgments

Throughout the writing of this book, Ronna Mallios provided both expertise and critiques. The association between gradual and abrupt drift in modeling and the Darwin–Gould–Eldredge theories resulted from communications with Seth Mallios. Peter Mallios contributed literary criticisms. Bo Hatfield provided the means of converting sports and financial data bases into formats that allowed applications of adaptive drift forecasting.

WILLIAM S. MALLIOS

1

Introduction

1.1 FAVORABLE BETTING SCENARIOS

The buy-and-hold strategies under efficient market dogma have shifted toward active trading strategies under adaptive market alternatives. Microeconomics appears to be back. It would be better if, as Keynes said, markets were not the by-product of a casino. But, in fact, they are.

In light of the greatest downturn since the Great Depression, the shift to active trading is not without critics. Under Saint Joan's banner, French President Sarkozy has taken steps to instill moral values in the global market economy by urging policymakers to consider fresh ways of combating financial *short-termism*[1] (Hall, 1/3/09). Perhaps Mr. Sarkozy has taken a perverse view of Keynes' dictum that *economics is a moral and not a natural science*.

The recessionary angst of late 2008 saw many favorable betting scenarios in financial and sports gambling markets. Attractive bets included:

[1]In an effort to add intellectual glamor and impetus to his presidency, Sarkozy proposed that Albert Camus' remains be moved from his simple village grave in Lourmarin to the Pantheon in Paris, burial place of France's greatest heroes and intellectual leaders. *Sarkozy has made an art of snatching high-profile figures from the left for his government to offset his occasional foray farther right on issues such as immigration and security* (Thompson and Hollinger, 12/02/09).

Forecasting in Financial and Sports Gambling Markets: Adaptive Drift Modeling, By William S. Mallios

establishing short positions on Goldman Sachs shares during November and betting on the Los Angeles Lakers (favored by 3 points) in their Christmas Day rematch with the Boston Celtics. (The Celtics embarrassed the Lakers in the previous National Basketball Association championship series). The attractiveness of each bet depended on the effectiveness of the gambler's forecasting models—models that are assumed based on public information.

It has been argued that profitable modeling forecasts tend to favor the sports gambling markets since they are accommodated by greater regulation and surveillance, considerably less opacity, and public point spreads that reflect the gambling public's expectations. For example, the New England Patriots' loss to the New York Giants in the 2008 Super Bowl was an outcome that superseded the New York Jets' upset win over the Baltimore Colts in the 1969 Super Bowl. The Patriots were prohibitive 12-point favorites; the bookmakers' line on total points scored was 53.5. Relative to the lines on the difference and total points scored, the Patriots had vastly overperformed throughout the first half of the season, then underperformed but kept winning until the finale (see Figure 1.2.2). New England had clearly peaked by midseason.

In contrast, the Giants jelled in the second half of the season and peaked during the play-offs (see Figure 1.2.3). In the finale, the Giants won 17–14, an outcome that was easily amenable to effective forecasting; see Table 1.1.1 and the modeling procedure described in Section 10.2.

Relative to the line, the Giants' expected winning margin of 3.4 points was a far more realistic estimate of the outcome. (See Section 11.2 for the calculation of the expected winning margin.) However, whether or not the line is realistic, there are always two groups of winners—those covering[2] the bets and those betting on the winning side of the line—and one group of losers—those betting on the losing side of the line. Those covering the bets charge a commission per bet and are always the winners as long as

TABLE 1.1.1 Super Bowl 2008: NE Patriots vs. NY Giants +12[a]

Outcome	Odds to $1	Probability
Patriots to win by more than 7 points	$2.01 to 1	0.33
Game decided by at most 7 points	$13.3 to 1	0.11
Giants to win by more than 7 points	$0.79 to 1	0.56

[a] NYG expected winning margin: 3.4 points. Outcome: NYG won by 3 points.

[2] In the United States, the service of accepting or covering a bet is typically provided by the sports book at the casinos. In Europe the service is also provided by online bookmakers such as Ladebrokes.

the line splits the money wagered (i.e., losing bets pay off the winning bets after commissions). Thus, a bookmaker's line is simply a measure of the gambling public's expectation of a game's outcome—regardless of whether or not that expectation is realistic.

A financial market analogy to Table 1.1.1 is illustrated in terms of Microsoft's (stock symbol: MSFT) price movements during 1999–2000, a volatile period during final inflation and deflation of the NASDAQ bubble. From 3/27/00 to 4/3/00, the MSFT closing price dropped from \$53.13/share to \$44.53/share. Figure 1.1.1 presents weekly price changes and volumes through the January–June 2000 period, and Figure 1.1.3 presents these changes in terms of a *candlestick chart* (see Section 5.1 for detailed discussions).

In Figure 1.1.3, each week in Figure 1.1.1 is represented by a candlestick that depicts four summary prices for MSFT: the opening price (O), the high (H), the low (L), and the closing price (C) for the week. A candlestick is composed of a body and a wick that extends above and below the body. The body is white if $C > O$ and dark if $O > C$. The maximum (minimum) of the wick is the high (low) for the week; see Figure 1.1.2 for an illustration of three hypothetical candlesticks. The lower portion of Figure 1.1.3 presents the 25- and 100-day moving averages for C (where five trading days correspond to one week). The moving averages are based on successive days prior to each weekly candlestick.

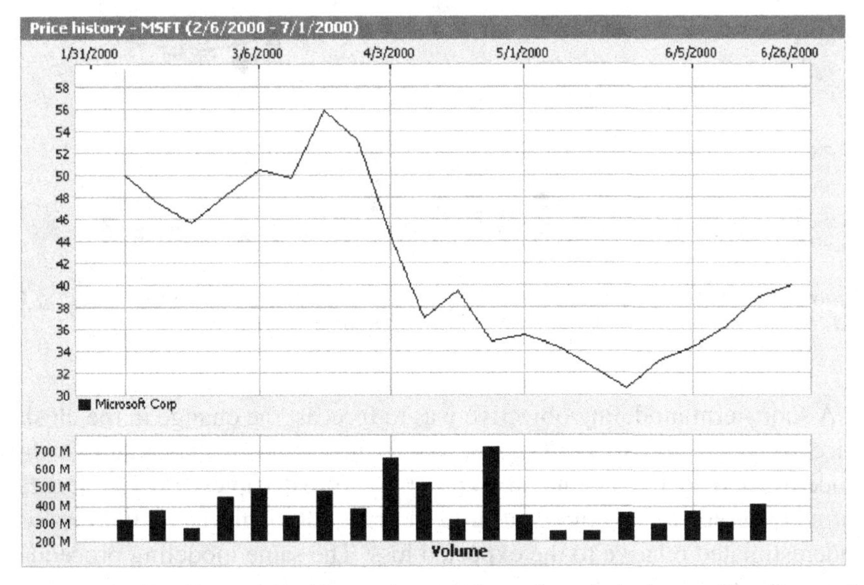

Figure 1.1.1 *Line chart of weekly per share closing prices, including weekly volumes, for Microsoft (MSFT).*

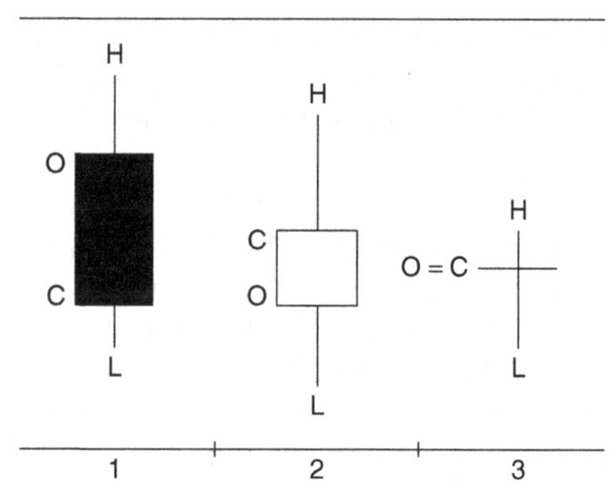

Figure 1.1.2 Three candlesticks with (1) $H > O > C > L$, (2) $H > C > O > L$, and (3) $H > O = C > L$.

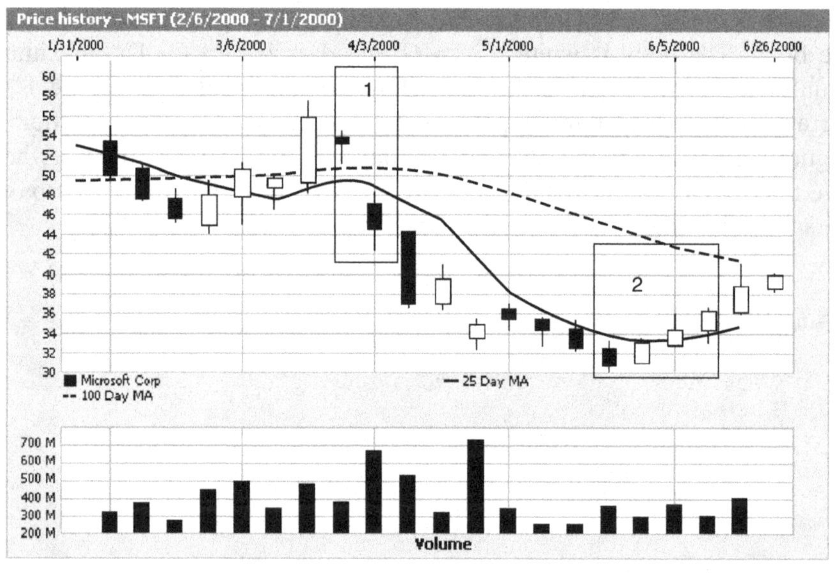

Figure 1.1.3 Candlestick chart for weekly per share Microsoft price changes including 25- and 100-day moving averages and weekly volumes. (Source: MSN Money)

A short-term modeling objective was to forecast the change in the closing price from 3/27/00 to 4/3/00; see box 1 in Figure 1.1.1. Adaptive drift modeling led to the results in Table 1.1.2 (see Chapter 9). The forecast correctly projected a significant drop in price, although the actual loss was underestimated relative to the expected loss. The same modeling procedures were used to forecast losses correctly through mid-May and gains in the rebound that followed; see box 2 in Figure 1.1.1.

TABLE 1.1.2 Odds on D(MSFT, t), the Change in the Closing Price per Share of Microsoft from 3/27/00 to 4/5/00[a]

Outcomes for D(MSFT, t)	Odds to $1	Probability
> $6	$19.0 to 1	0.05
[$6, $2)	$11.5 to 1	0.08
[−$2, $2]	$7.33 to 1	0.12
(−$2, −$6]	$3.00 to 1	0.25
< −$6	$1.04 to 1	0.49

[a] Expected gain/loss for D(MSFT, t) = −$4.24. Observed gain/loss for D(MSFT, t) = −$9.60.

1.2 GAMBLING SHOCKS

A *gambling shock* (GS) is defined as the difference between the game outcome and the line. For example, if the line on the difference favors the Patriots by 12 points and they lose by 3 points, $GS\,\text{difference(NE)} = GSD\text{(NE)} = -3 - (12) = -15$. If the line on the total points scored in the Giants–Patriots game is 51 and the total points scored is 31, then $GS\,\text{total(NE)} = GS\,\text{total(NYG)} = GST(*) = 31 - 51 = -20$. Larger values of $|GSD|$ and/or $|GST|$ for a particular team generally affect that team's subsequent performance or performances in that they may reflect the effects of motivation, injuries, personnel problems, and so on—all of which translate into physiological, psychological, and sociological variables.

When, for example, the Giants suffered through two embarrassing losses to the Dallas Cowboys during the 2007–2008 regular season, the likelihood of a Giants' upset win against Dallas in the play-offs was exceptionally high (especially in view of the Giants late-season performances). In fact, when the Giants lost a game throughout the regular season, they usually won their next game (as shown in Figure 1.2.3).

When there are marked differences in player talent between opposing teams, the GS may act as a surrogate for fans and teams in the evaluation of team and player performances. The home team fans may take consolation when their underdog team loses by less than the spread—especially if they've bet on their team.

> *We play hard and cover. We lead the league in covering the point spread.* (Hubie Brown, coach of the last-place New York Knicks, *Sports Illustrated, 1986)*

Figure 1.2.1 depicts game outcomes and accompanying gambling shocks for the 23 Los Angeles Lakers' play-off games leading to their 2008–2009 National Basketball Association (NBA) title. The Lakers won

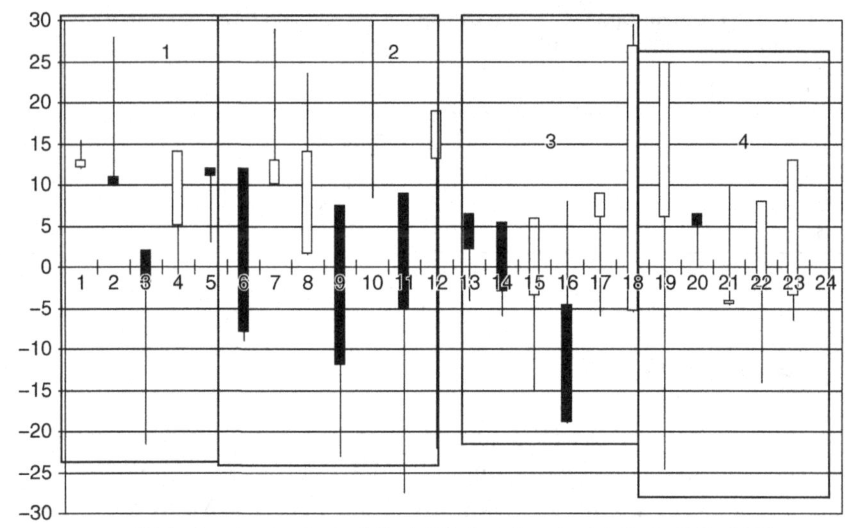

White Candlestick Body: Diff < LDiff, Dark Candlestick Body: Diff > LDiff
Candlestick Wick: GST > 0 above body and GST < 0 below body

Figure 1.2.1 *2008–2009 NBA play-off games for Los Angeles Lakers with wins over Utah (5 games), Houston (7 games), Denver (6 games), and Oriando (won NBA title in 5 games).*

in five games against Utah, seven against Houston, six against Denver, and then five in the finale against Orlando. White bodies denote games in which the Lakers beat the line on the difference. The minimum value of the white body is the line on the difference for Lakers, and the maximum of a white body is the Lakers' winning/losing margin. Dark bodies denote games in which the Lakers did not beat the line; that is, the maximum value of a dark body is the line on the difference, and the minimum value of a dark body is the winning or losing margin. A white (dark) body indicates that the Lakers overperformed (underperformed) relative to the line on the difference. The size (magnitude) of the body reflects (equals) the size of the gambling shock on the difference for the Lakers. An observed difference above (below) zero signifies a Lakers' win (loss).

The size of the gambling shock for the total points scored is given by the wick (or stick) that extends either above or below each body. When the wick extends above (below) the body, $GS\text{total} > 0$ ($GS\text{total} < 0$). For example, in the first play-off game against the Utah Jazz, the Lakers were favored by 12 points and won 113–100; the line on the total was 210.5. Thus, $GS\text{difference(LAL)} = 13 - 12 = 1$ (a small white body) and $GS\text{total(LAL)} = 213 - 210.5 = 2.5$ (a short wick extending above the white body). In this game, the gambling public's expectations were on the rational side.

Several predictive indicators are apparent in Figure 1.2.1—indicators that are revealed in analyses of Lakers' regular-season games. (*Note:* The obvious purpose of adaptive drift modeling is to uncover such indicators so that they can be applied in successful forecasting of subsequent game outcomes relative to the lines.)

Predictive Indicator 1: The Revenge Factor A Lakers' (LAL) dark or white body loss is followed by an LAL white body win. (*Note:* There are no white body losses for LAL, but there is one for the New York Giants in Figure 1.2.3.)

Predictive Indicator 2: The Complacency Factor An LAL dark body win is followed by an LAL loss.

Predictive Indicator 3: The Exhaustion Factor An LAL larger-than-average white body with a large lower stick tends to be followed by an LAL red body (except for the last game).

Figure 1.2.1 Candlestick chart for the Los Angeles Lakers during the 2008–2009 play-offs, including the first round versus Utah (won in five games; see box 1), the second round versus Houston (won in seven; see box 2), the third round versus Denver (won in six; see box 3) and the finals versus Orlando (won the NBA championship in five; see box 4). Candlesticks are defined as follows: white body, LineDiff(LAL) > 0; dark body, LineDiff(LAL) < 0; upper wick, *GS*Total(LAL) > 0; lower wick, *GS*Total(LAL) < 0.

Predictive indicator 1, the revenge factor, reflects the motivation to win convincingly in the game following a loss. Predictive indicator 2 may reflect complacency on the part of the Lakers—in the sense that they won their previous game but by less than the expected margin and that this same complacency will characterize their performance in their subsequent game. Predictive indicator 3 represents a larger than expected win and a lower than expected total points scored—an outcome that tends to describe exceptional offense and defense effort on the part of the Lakers. To achieve both in a single game may presage fatigue in the following game. The exception to predictive indicator 3 is the sixth and last game of the play-offs—a point in the series where Orlando appeared to have thrown in the towel (analogous to the LAL loss in the sixth and last game against the Celtics in the finale of the 2007–2008 play-offs).

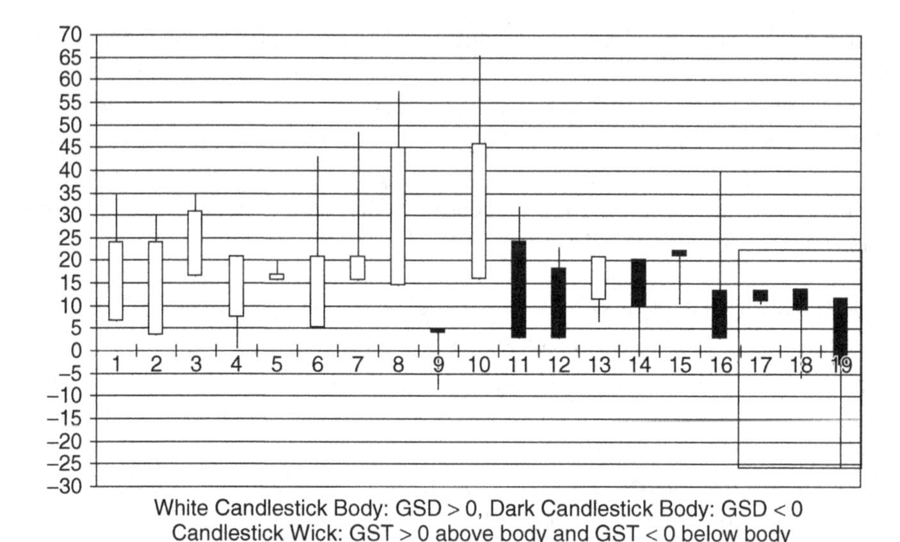

Figure 1.2.2 *New England Patriots: 2007–2008 regular season and three play-off games, ending with Super Bowl loss to New York Giants.*

The three indicators frequently characterize exceptional NBA teams such as the 2008–2009 Lakers, the 2007–2008 Celtics, and the San Antonio Spurs during their recent championship seasons. For teams of lesser talent, predictive indicators tend to interact with a variety of other variables and often require more complicated explanations. The key point to be emphasized is that each team tends to be unique and that forecasting models should be team specific. Universal models that are said to apply to all teams are as useful as deterministic models of human behavior.

For the Giants' 2008 Super Bowl win over the Patriots, the candlestick charts in Figures 1.2.2 and 1.2.3 depict, respectively, all regular and postseason games for the Patriots and Giants in 2007–2008 which culminated in the Giants' Super Bowl win and the Patriots' only loss of the season.

The Patriot chart in Figure 1.2.2 depicts a classic example of a team that had peaked by midseason. Specifically, white bodies dominate the first half of the season and dark bodies dominate the last half. Moreover, $GST > 0$ dominates in the first half and $GST < 0$ is overrepresented in the second half.

The chart for the Giants in Figure 1.2.3 contrasts sharply with that for the Patriots. The Giants began the season poorly, losing to the Dallas Cowboys

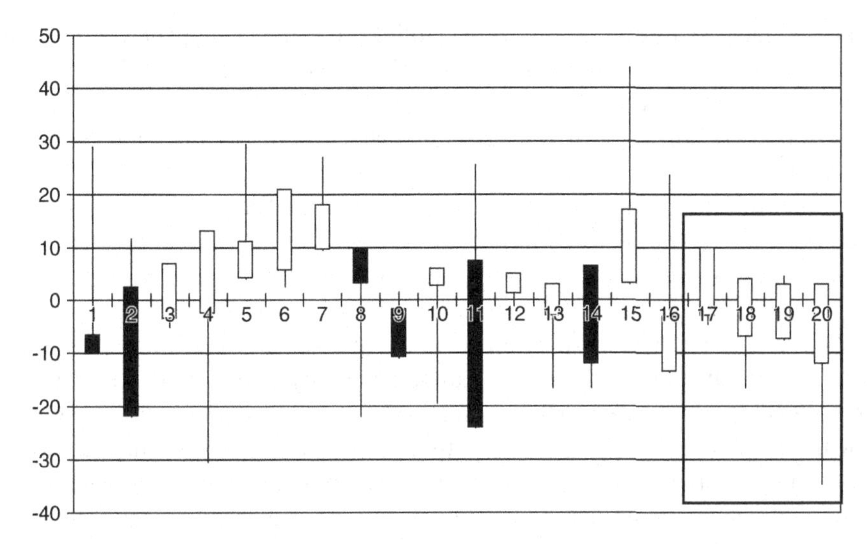

Figure 1.2.3 *New York Giants: 2007–2008 regular season and four play-off games, ending with Super Bowl win over the New England Patriots.*

35–45 in week 1 and to the Green Bay Packers 13–35 in week 2. These losses were followed by a succession of five white body wins, followed by a seven-week period of mediocrity, and ending in a succession of six white bodies.

With the exception of the first week, a Giants' loss was always followed by a white body win, as was the case for the Lakers in Figure 1.2.1. Also similar to the Lakers, the single dark body win is followed by a dark body loss. Although not shown in the chart, a loss by the Giants was followed by a win against that team if they met for the second time during the season—with the exception that the Giants lost to Dallas twice during the regular season prior to beating the Cowboys in the play-offs (see the forecasts for the Giants' play-off games in Section 9.1).

1.3 THE DARK SIDE OF SPORTS: THE FIXES

The only good thing that can be said about the dark side of sports is that it is pale in comparison to the dark side of finance.

Regarding the line on any given game, profits for those covering the bets are assured from rules such as the *11 for 10 rule*: the gambler puts up

\$11 for each \$10 bet. Once posted publicly, the line is adjusted whenever necessary to balance the total amount that was bet. Kenny White of the Las Vegas Sports Consultants has been quoted as saying that the movement of the line by half a point typically indicates betting on one side of about \$50,000.

When the line is rational, as in the rational expectations hypothesis (see Section 2.1), the market is efficient. Irrational lines are indicative of market inefficiency. There is no comparable line in financial markets[3] for the simple reason that prices could easily be manipulated to beat such a line.[4]

While involvement of players in fixing games in the U.S. professional ranks is highly unlikely, there are understandably concerns by the NCAA in college sports. College athletes are not salaried, and many are in school for the express purpose of participating in their athletic specialty. In addition, many of these athletes are from low-income areas of inner cities. As such, college sports are susceptible to violations and scandals and fixes that include point shaving.

Regarding NCAA violations, the following allegations are said to represent the tip of the iceberg:

1. The Memphis men's basketball program was charged by the NCAA with major violations during the 2007–2008 season.
2. Wolfers (2006) suggests that point-spread shaving may have influenced outcomes in nearly 500 games over a 16-year period. *Wolfers looked at point-spread favorites of 12 or more points and*

[3]In proposing changes in compensation incentives for business executives, Martin argues that a stock price in business is the moral equivalent of a point spread in football betting. However, *while National Football League players are forbidden from betting on the spread and are compensated, not on the line, but on their field performances, business executives are compensated primarily on increases in stock prices... instead of real-market measures such as revenue growth, market share, profits and book equity return* (Martin, 5/12/09).

[4]In sports gambling markets, there are strong arguments for discounting the possibility of a fix. *To assure a "fix" one would have to bribe or threaten a player who plays the majority of the games and* [whose earnings are in the millions]. *The size of the bribe required to induce a player to forgo this salary for the rest of his career if caught therefore would be very large. This, in turn, would require that an extremely large amount of money be bet on the game in question to cover the bribe and make a profit...It is very unlikely that any bookmaker would take such a large bet and if one attempted to break up the bet into a more reasonable size...the level of activity on a single game clearly would be noticed by the bookmakers and probably would lead them to call off all bets on the game* (Dobra et al., 1990). Although there have been no recent allegations of fixes by players in the professional ranks, the same is not true for referees and umpires. This topic is discussed in Section 11.4, where surveillance procedures are proposed for detecting possible referee and umpire irregularities in Major League Baseball (MLB), National Basketball Association (NBA), and National Football League (NFL) games.

concluded those teams tend to cover the point spread at a lower rate than they should. While that one statistic isn't enough to conclude there is something unusual going on, Wolfers also mentions the NCAA survey where eight out of 388 basketball players admitted to taking money in exchange for playing poorly or at least had knowledge of teammates who did (Alan Moody, About.com).

3. In 1951, City College of New York, the collegiate basketball champion, became the symbol of corruption when its players were among 33 players from seven universities arrested for fixing 86 games.
4. In 1961, North Carolina State was involved in point-shaving scandals. Everett Case, then N.C. State coach, blamed it on players from the north that he had recruited. Sport writer Red Smith wrote that Case *got integrity confused with geography*. Congress then passed the Interstate Wire Act in 1961 to prevent sports gambling. In simple terms, the Wire Act makes it illegal to book bets on the phone or through the Internet. *Whittier law professor Nelson Rose said: The Wire Act was designed to cut "the wire" which was the telegraph that every illegal bookie had to have to know who won a horse race before his patrons* (Bowe, 7/24/06).

NBA referee Tim Donaghy officiated in 772 regular-season games and 20 play-off games from 1994 to 2007. Donaghy was alledged to have bet on games that he officiated during his last two seasons and to have made calls affecting point spreads on those games. Donaghy pleaded guilty to two federal charges related to the investigation and was sentenced to prison. However, he could face more charges at the state level if it is determined that he deliberately miscalled individual games. *According to data obtained from a Las Vegas company, Donaghy refereed in 11 games after 1/1/2007 in which the consensus Las Vegas line moved two points or more. The team on which bettors wagered heavily enough to move the line that far won seven of those 11 games. ... A 7–4 record would not be compelling to a statistician, who would consider the good possibility of that happening randomly. But Jimmy Vaccaro, the chief oddsmaker for American Wagering, which runs 60 sports books across Nevada, said that such performance would leave any gambler giddy. ... If you win seven out of 11, more than 60%, you'd be a billionaire in about a year*[5] (Schwarz, 7/21/07).

[5]Graphical methods for monitoring individual referee performances in terms of game outcomes relative to point spreads on those games are presented in Section 10.8.

More recently, in November 2009, match fixing in European football was faced with a startling revelation. The Union of European Football Associations, soccer's governing body in Europe, reported *without a doubt the biggest fraud scandal to ever hit European football*. The game had seen bribery scandals almost since the rules of the game were codified in 1863. Investigations identified approximately 200 games in which there was suspected fraud. *Sports authorities have been slow to realize the scale of the problem. If fans come to believe that the product they are watching is rigged, then they may turn off. It happened to other once-popular sports: rowing was a phenomenon in Britain in the 19th century before it became tainted by match-fixing* (Kuper and Blitz, 11/21/09).

2

Market Perspectives: Through a Glass Darkly

2.1 CHANGING PARADIGMS

> *October is one of the peculiarly dangerous months to speculate on stocks. The others are July, January, September, April, November, May, March, June, December, August and February.* (Mark Twain)

In 1841, Mackay wrote: *Money has often been a cause of the delusion of multitudes. Sober nations have all at once become desperate gamblers, and risked almost their existence upon the turn of a piece of paper.... Men, it has been said, think in herds; it will be seen that they go mad in herds, while they recover their senses slowly, and one by one*. A century later, Keynes was to view stock markets as casinos guided by animal spirits that react in herds. By the early 21st-century finance was viewed as *the mirror of mankind, revealing every hour of every working day, the way we value ourselves and the resources of the world around us. It is not the fault of the mirror if it reflects our blemishes as clearly as our beauty* (Fergusen, 2008).

Markets reflect the human condition as it oscillates between rational and irrational states (Simon, 1984). Interactively, the human condition reflects the dynamic nature of markets where periods of market efficiency are dominated by rational-type behaviors and periods of market inefficiency by irrational-type behaviors. The greater (less) the market inefficiency, the greater (less) the likelihood of profitable trading rules and forecasts.

Market inefficiency refutes the *efficient market hypothesis* (EMH), developed independently by Fama and Samuelson in the 1960s. The EMH assumes that current prices reflect all available information rationally and instantaneously—as does a bookmaker's line in sports gambling markets. Three forms of the EMH in the sports gambling markets are (1) the *weak form*, which assumes that the line incorporates all relevant information in past games; (2) the *semistrong form*, which assumes that the line incorporates all relevant public information as well as past game outcomes, and (3) the *strong form*, which assumes that some bettors have access to insider information (which is typically tied to knowledge of unpublicized injuries or personnel problems).

However, somewhere along the way, what started as a critique of the wrong way that people try to beat the markets turned into a source of new techniques for making money. Specifically, *the efficient market hypothesis, the Nicene Creed of the market rationalists, inspired a wave of innovative financial products, such as derivatives and securitized subprime mortgages, that believers claimed would allow users to exploit the wonders of the market. This gospel was embraced so enthusiastically by the markets that these products soon accounted for trillions of dollars of trades. Then it turned out that the market was not rational after all. Trillions were wiped out and, as one of the cheerleaders for rationality, Alan Greenspan, the former chairman of the Federal Reserve, put it, "the whole intellectual edifice collapsed"* (*The Economist*, 6/11/09).

Behavioral economists were highly critical of the EMH well before its fall. Irrational-type behavior by market participants—which, according to behavioral economists, typifies most participants most of the time—refuted the EMH and its companion, the *rational expectations hypothesis* (REH). The REH stipulates that people's expectations are informed predictions of future events and are essentially the same as predictions of *relevant theory* (Muth, 1981). Granger (1980) noted that REH proponents typically fail to specify a *relevant theory* that is generally accepted by the majority of economists.

EMH opponents have long argued that markets operate inefficiently when buyers and sellers have *unequal access to information* needed to make optimal choices. Apart from access to insider information, such informational asymmetry has been attributed to constraints on participants' decision-making abilities.[1] Under such constraints, participants possess only *bounded*

[1]Given the inability of most participants to extract information successfully from accessible data due to a lack of knowledge, time, or cost constraints, one cannot ignore the ongoing financial/mathematical illiteracy pandemic in the United States. Data per se are seemingly useless except for those well versed in information extraction or the limited few who can afford payment for such extraction (see Section 12.1).

rationality and must make decisions by *satisficing* or choosing that which may not be optimal but which they think will make them happy (Simon, 1984).

Dating back to the 1950s, assumptions regarding how people behave were at two poles: (1) rational behavior (as in the EMH) and (2) mindless behavior, resembling nuts and bolts in a large complex machine, as quantified in terms of multidimensional structural regression systems, which gained prominence in the 1950s (Marwah, 1997). In contrast, psychologists, led by A. Tversky (Tversky et al., 1981) and Nobel laureates D. Kahneman (Kahneman and Tversky, 1979) and V. Smith (2000), tested how people really do behave and, in doing so, created the discipline of behavioral economics and finance. In general, they found that people gather limited information, reason poorly, and act intuitively rather than rationally. Their conclusions were a devastating blow to postulates of rational decision making.[2]

In attempting to bridge differences between EMH and behavioral hypothesis alternatives, Lo (2004) applied principles of evolutionary biology—competition, adaptation, and natural selection—to the markets and formulated the *adaptive market hypothesis* (AMH).[3] Lo argued that much of what behavioralists cite as counterexamples to economic rationality—loss aversion, overconfidence, overreaction, and other behavioral biases—are consistent with an evolutionary model of people adapting to a changing environment and that market efficiency is highly context dependent and dynamic.

Specifically, the degree of market efficiency is related to market ecology factors such as the number and types of competitors in the market, the availability and magnitude of profit opportunities, and the adaptability of the market participants (Lo, 2005). *There is, however, one big difference between nature and finance. Whereas evolution in biology takes place in a natural environment, where change is essentially random,*[4] *evolution in*

[2]For five decades, the Chartered Financial Analyst Institute (CFAI) has been teaching the tenants of analysis based on the EM. Recently, the CFAI asked members whether they trusted *market efficiency—and discovered that more than two-thirds of respondents no longer believed market prices reflect all available information.... Although two-thirds of financial professionals surveyed said they regarded behavioral finance a useful addition to the EMH, just 14% thought it could alone become a new paradigm* (Tett, 6/16/09).

[3]Evolutionary arguments are also evident in recent explanations of the human *herd effect*. Burnham (2005) argues that humans have a *lizard brain* component that evolved when life was primitive. This part of the brain seeks clusters, which is useful when there is a need to avoid harm or find food instantly. However, for more complex adaptive systems such as financial markets, the danger is that investors get trapped into herd reasoning by buying at the top and selling at the bottom.

[4]Darwinian evolution proceeds through the natural selection of random hereditable changes or mutations. However, following the mutations, the survival of the fittest is not random.

financial services occurs within a regulatory framework where—to borrow a phrase from anti-Darwinian creationists— "intelligent design" play a part (Ferguson, 12/13/07).

2.2 MODELING COMMENTARIES

For the markets under study, *adaptive drift modeling* is developed as an alternative to existing versions of dynamic modeling based on Bayesian learning; (see Section 7.4). Given the notoriety associated with the failures of statistical (*quant*) modeling during the 2007–2008 financial crisis, we begin by distinguishing between adaptive statistical modeling and *expert-knowledge, nonstatistical modeling*[5] (even though the two can be integrated in terms of Bayesian analyses). Proponents of the latter include those who have done remarkably well in their long-term investments as well as those who view the future as mostly unknowable.[6]

Statistical modeling proponents view the near term as predictable during periods of market inefficiency if the modeling procedure adapts effectively to market dynamics and if it is recognized that sufficiently large shocks may disrupt model validity. Such sufficiently large shocks are analogous to *antigenic shifts* in molecular biology, or the Gould–Eldredge theses of *punctuated equilibrium* in evolution (see Section 4.3).

Commentaries by *statistical modeling* skeptics range from reasoned to unequivocal. Clausewitz, in his treatise *On War*, wrote: *It is simply not possible to construct a model for the art of war.... Such a faulty model creates an absurd difference between theory and practice which not only defies common sense, but, even worse, too often serves as a pretext by limited and ignorant minds to justify their congenital incompetence.*

Later, in 1869, General William Tecumseh Sherman wrote: *I know there exist many good men who honestly believe that one may, by the aid of modern science, sit in comfort and ease in his office chair and, with figures and algebraic symbols, master the great game of war. I think this is an insidious and most dangerous mistake.*

[5] Anthropologists use the neologisms *emic* and *etic* as terms for the insider's view (the expert opinion approach) and the outsider's view (e.g., the statistical modeling approach). During his stint on Monday Night Football, Howard Cosell provided a noninformative etic view. He had little knowledge of football but enhanced the network's coffers by mesmerizing the ranks of the viewing audience with his soap opera monologue and quips: *Sports is the toy department of human life.*

[6] As Karl Popper observed, to predict the creation of the wheel is to invent it. Paraphrasing Kay (11/26/08), *to anticipate a new market theory is to take the main step in bringing it into being.*

In a Ph.D. thesis written in 1900 at the University of Paris, Louis Bachelier (1964) was the first to state that stock prices follow a random walk model. Bachelier's work is said to have influenced Einstein in his 1905 work on Brownian motion—although Einstein makes no such acknowledgment in his book on the topic.

In the early 1950s, in a search for regular cycles, M. G. Kendall applied statistical methods, now largely outdated, in examining the behavior of stock and commodity prices. Instead of discovering regular cycles, he found that prices appeared to follow a random walk. Kendall's analysis results are a modeling consequence of the EMH. The random walk viewpoint persisted strongly, even into the early 21st century. From earlier days when the EMH held court, the following commentary prevailed: *The price-determining mechanism described as the random walk model is the only mechanism which is consistent with the unrestrained pursuit of profit motive by participants in markets of this type* (Granger, 1970).

Beginning in August 2007, dismal performances of *quant funds* prompted considerable commentary on the failures of statistical modeling. Modeling skeptics again came to the forefront. In *Black Swan*, Taleb (2007) describes history as opaque, essentially a black box of cause and effects, and his *Fooled by Randomness* (2005) has become an idiom used to describe when someone sees a pattern where there is just random noise.

Following publication of *Moneyball* by financial journalist Michael Lewis (2003), the baseball community reacted adversely to the notion that statistical modeling could beat trained professionals. Such reactions were somewhat marginalized by the success of Billy Bean, general manager of the Oakland Athletics. Bean, *Moneyball*'s central character, succeeded Sandy Anderson as general manager in 1998 and expanded on Anderson's application of *sabermetric* principles.

Sabermetrics, derived from the acronym SABR (Society for American Baseball Research), has been defined as *the search for objective knowledge about baseball. Sabermetrics* attempts to answer objective questions about baseball, such as *Which player on the Phillies contributed most to the team's offense?* Or, *how many RBI's will Alex Rodriguez hit next year?* It cannot deal with the subjective judgments which are also important to the game, such as *Who is your favorite player?*

In assessing risk and rewards, Bean viewed the search, identification, and purchase of undervalued baseball players, particularly in less well researched markets such as Asia and Central and South America, the same as a financial analyst's search, identification, and purchase of undervalued financial assets in emerging economies. In identifying hidden talents among prospective players, Bean utilized baseball statistics that were, at the time, unconventional. Bean coveted Kevin Youkilis because of one overlooked skill: his ability to draw walks. Bean said that Youkilis was the *Greek God of Walks*. Bean reasoned that the ability to take walks showed a good eye and great discipline—skills that are difficult to teach. Pitchers are obliged to throw such batters balls that they can hit. With the Boston Red Sox, Youkilis was to become one of baseball's most feared hitters (Authers, 5/23/09).

Theo Epstein, general manager of the Red Sox, was to follow Billy Bean in the use of baseball data mining techniques. The baseball modeling results for the Red Sox and Oakland Athletics in Sections 7.4 and 7.5 are intended to provide insights on forecasting team and pitcher performances.

2.3 SPORTS HEDGE FUNDS

> *When I was young, people called me a gambler. As the scale of my operations increased, I became known as a speculator. Now I am called a banker. But I have been doing the same thing all thee time.* (Quotation from Sir Ernest Cassel, private banker to King Edward VII)

Gambling is said to refer to games of pure chance (e.g., lotteries and roulette and slot machines) in which participants pursue monetary gains without using their skills and the odds of winning are independent (or nearly so) of the participants' skills. In contrast, monetary returns from speculation and investment depend on participants' skills. According to Keynes, *speculation... describes a situation where, instead of trying to make a forecast about the probable yield of an investment over* [the longer term], *people try to guess how the market, under the influence of mass psychology, will value it* [in the near term]. The Keynes-type speculator is the likely participant in sports gambling markets, although participants cover the spectrum from lottery and roulette mentalities to those possessing inside information.

There has been a public tendency to view sports gambling and poker as part of the same market. From a legal perspective, the regulation of gaming appears to be based on whether the preponderance for an outcome lies in pure chance or in some degree of skill. In terms of player skills, the poker player confronts observable cards and opposing players while the sports gambler confronts the line and observable data. The sports gambler usually has the advantage of greater time and information (unless there is a complete mismatch in talents among competing poker players).

Modeling poker outcomes is typically approached through Monte Carlo studies. Such methods were used to identify profitable card-counting strategies in blackjack; Thorp (1966) published one such strategy in *Beat the Dealer*. However, once these winning strategies were publicized, the casinos changed the game to their benefit by using multiple decks and/or having dealers shuffle after every hand. Knowledgeable blackjack gamblers were forced to devise new, less profitable betting schemes.

To evaluate poker skills, Fiedler and Rock (2009) developed the *critical repetition frequency* (CRF) which is defined as the threshold of repetitions at which a game becomes influenced predominantly by skill rather than chance. They show that poker can be classified as a game of skill, but emphasize that the results hold only for the sample under study.

In light of recent online poker scandals, there are indications that poker markets might not offer a level playing field. *It seems as though there is another major scandal on an almost weekly basis in the world of poker* (POKER-KING.com). As such, our commentaries on modeling in the sports gambling markets are divorced completely from commentaries on the poker and card game markets.

Given the evolutionary mechanisms that underlie the financial and sports gambling markets, sports hedge funds are likely in the offing—if they do not already exist opaquely. Dallas Mavericks' owner Mark Cuban proposed the establishment of hedge funds that would invest in the sports gambling markets with professional gamblers as portfolio managers. Although Cuban was apparently pressured to abandon his hedge fund idea, he remains no *less adamant about his view that investing in the stock market is riskier and less transparent than plunking down money on the Cowboys versus the Steelers* (Forbes.com). For societies that impose obstacles to gambling, financial markets serve the function of the gambling casino. As such, the entry of regulated sports hedge funds in the United States is seemingly inevitable.

> *Somewhere in a remote part of southern England, during a quiet Friday lunchtime, a man places a £30 bet on his computer on ice hockey matches in Switzerland, Slovenia, Russia and Germany. He bets that the matches will average more than five-and-a-half goals a game and stands to make a few hundred pounds. Moments later, in the London headquarters of Ladbrokes, the bet is processed. It is one of about 150,000 online transactions that Britain's biggest land-based bookmaker will undertake that day.... According to the Isle of Man–based Global Betting and Gaming Consultants, gambling around the world generates $370bn in annual gross win—the amount retained by operators after paying out winnings—with online gambling accounting for $17bn of that.* (Blitz, 2/4/09)

2.4 GAMBLING MARKETS: PROHIBITION, REPEAL, AND TAXATION

Prohibition by legislative fiat has an ignominious record. A late 19th-century German experience foretold the failure of the Volstead Act, the genesis of organized crime in the United States. In 1896, the Reichstag banned futures trading in response to the depressed prices of German commodities. The prohibition led to illegal speculation. Exports became excessive in order to capitalize on world prices. In turn, exports had to be made up by imports, at a heavy cost to all Germans. Futures trading resumed in 1900. *This German failure has been regarded as positive proof . . . that economic law [is] ultimately beyond legislative fiat* (Cowling, 1965). In a similar vein, repeal of the Volstead Act was for the purpose of gaining lucrative tax revenues in the Depression era.

> *In 1963 President Kennedy helped revive the London bond market by imposing a tax on US investment in foreign securities. That made the international bond market move to London, allowing the City to regain its 19th century status as Wall Street's rival in capital markets.*[7] *Now US regulators—through anti-gambling laws—are doing the same thing with the online betting and gambling markets, markets that are now flourishing in London. (Financial Times,* 8/23/05)

In October 2006, President Bush signed the *Safe Port Act*, a bill on coast guard security. The Act contained a controversial provision, engineered by then Senate majority leader Bill Frist, making it illegal for banks and credit card firms to make payments to online gaming sites. The Act did not affect Nevada's brick-and-mortar casinos. Just prior to passage of the Act, James Chanos (president of Kynikos Associates, the world's largest dedicated short-selling hedge fund, and best known for his prescient shorting of Enron) shorted shares of the online gambling businesses *Sportingbet* and *World Gaming*—backed by a large portion of his fund's $3 billion in assets. Hours following passage of the Act, shares in Sportingbet and World Gaming plunged dramatically by 58% and 76%, respectively (see Section 4.5).

[7] *The debate about New York's position relative to London has mostly been prompted by the shift in companies choosing to raise money in London rather than on the New York Stock Exchange or NASDAQ. The figures are stark: in 2006, $55bn was raised on the London Stock Exchange and the lightly-regulated Aim market. For the first time ever, this figure exceeded the amount raised on the NYSE and NASDAQ, where share issues reached $47bn* (*Financial Times*, 1/20/07).

The Kynikos Associates short selling turned into a financial killing comparable to that of George Soros, who, in 1992, shorted the British pound and was the primary beneficiary, to the tune of $1 billion.[8] Some analysts have speculated that Chanos anticipated that Frist would attach the controversial provision to the Safe Port Act and steer it successfully through the Senate. At the time, Frist had presidential ambitions and wanted to show his right-wing religious base that he opposed gambling.

The Frist provision, in effect, prohibited online gambling. Opponents argue that the provision *has little to do with gambling beliefs, but more with false arguments circulated by parties threatened by the online gambling industry... that it will be a costly mistake, politically and economically and... that online gaming companies should be required to incorporate and operate their businesses in the US.... Legalizing online gambling and regulating it would help establish an already thriving, high-tech global industry in the US, attracting investment, retaining entrepreneurs and increasing employment and tax revenues.... History has shown time and again that prohibitions, rather than solving problems, cause more* (Brenner and Turk, 10/5/06).

In 2007, Barney Frank, chairman of the House of Representatives financial services committee, announced plans to overturn the Frist amendment. Frank maintained that *attempts to make* [online gambling] *illegal smack of a rightwing puritanical zeal that led to Prohibition in the 1920s and 1930s*. Nearly two years later, on 6/6/09, Frank introduced bills that would allow licensed operators to run online poker games, casinos, lotteries, parimutuel betting, and bingo. However, sports gambling was excluded to prevent professional and college sports associations from opposing the bills.[9]

Given the need for Senate acquiescence—especially by Senate Majority Leader Harry Reid (Nevada), who faced a difficult reelection campaign in 2010—Frank designed his bills to allow Nevada casino companies to take bets from almost all other states while keeping out foreign competition (Rose, 2009).

Laws are like sausages. It's better not to see them made. (Otto von Bismark)

[8]Soros's *audacious bet against the British pound... stemmed from his savvy reading of Britain's economic malaise and a belief that, despite strong statement to the contrary, Britain would abandon the Exchange Rate Mechanism rather that continually defend the pound through repeated and expensive interventions in the currency markets* (*Business Week*, 8/23/93).

[9]The National Football League had teamed with the Christian right to oppose an earlier proposal by Frank to undo the Frist amendment. Former U.S. Senator Alfonse D'Amato stated that the NFL's lobbying shows that *they aren't going to want anyone betting on their games unless they can control it* (Kirchgaessner, 8/14/07).

Subsequently, NBA Commissioner David Stern bolstered both Senate approval of Frank's bills and Reid's reelection chances. In a December 2009 interview, Stern said that the NBA *may be nearing a point of considering supporting the legalization of gambling on sports. Stern said that evolving public perceptions of gambling and the potential to increase revenue could lead to a move to allow sports betting on league games. Stern contradicted the main argument used by the NFL in leading the fight to maintain the sports gambling ban set up by the 1992 Professional and Amateur Sports Protection Act.*[10] *While NFL representatives fiercely cling to the theory that legal sports betting will lead to fixed games, Stern said illegal gambling was just as dangerous.... The commissioner said that times have changed since gambling was seen as an immorality. Noting the proliferation of government involvement in lotteries and other generators of gaming revenue, Stern admitted the decision may become merely a business* (Burkhart, 12/11/09).

Amid the flux of proposed gaming legislation, state-owned lotteries and online gaming continue to be at loggerheads. In September 2009, state-owned lotteries across the European Union claimed "*a great victory*" *in their battle competition from Internet gaming sites at bay as European Court of Justice gave its first ruling in the case involving the Austrian Internet operator Bwi* [one of Europe's biggest Internet bookmakers] versus *the Portuguese state gambling monopoly. The Court said that EU states should be allowed to restrict freedom to provide services in the gambling area if there was an overwhelming public interest.... The EU Association of State Lotteries welcomed the ruling and concluded that "National governments can grant monopolies to state operations for gambling on the internet and... ban foreign online gambling operators such as Bwin even if they are based and licensed in another EU member state*" (Tait and Blitz, 2/4/09).

2.5 QUANTIFYING THE MADNESS OF CROWDS IN SPORTS GAMBLING MARKETS

> *We need a new science of macroeconomics. A science that starts from the assumption that individuals have severe cognitive limitations; that they do*

[10]*Plans for the state of Delaware to begin taking sports bets* (*on 9/1/09*) *took a hit today when a federal appeals court in Philadelphia ruled that the plans would violate the 1992 Sports and Amateur Protection Act that banned sports wagering. Delaware was one of four states specifically exempted by the law because it previously offered a state sports-based lottery, but the court apparently agreed with attorneys from the major sports leagues that the state's plan to accept single-game wagers on games was more than the exemption permitted* (*GGB News*, 8/24/09).

not understand much about the complexities of the world in which they live. This lack of understanding creates biased beliefs and collective movements of euphoria when agents underestimate risk, followed by collective depression in which perceptions of risk are dramatically increased. These collective movements turn uncorrelated risks into highly correlated ones. What Keynes called "animal spirits" are fundamental forces driving macroeconomic fluctuations. . . . The basic error of modern economics is the belief that the economy is simply the sum of microeconomic decisions of rational agents. But the economy is more than that. The interactions of these decisions create collective movements that are not visible at the micro level. It will remain difficult to model these collective movements. There is much resistance. Too many macro-economists are attached to their models because they want to live in the comfort of what they understand—the behavior of rational and superbly informed individuals. To paraphrase Isaac Newton, macroeconomics can calculate the motion of a lonely rational agent but not the madness of crowds. Yet if macroeconomics wants to become relevant again, its practitioners will have to start calculating this madness. It is going to be difficult, but that is no excuse not to try. (DeGrauwe, 7/22/09)

How does one begin to measure the *madness of crowds*? One way is to associate the crowd's level of madness or sanity with their level of irrationality or rationality (I/R) and then to measure I/R levels in terms of the crowd's expectations associated with a forthcoming micro event, such as the line on a game. The question, then, is whether specific expectations in the sports gambling markets are rational, irrational, or somewhere in between.

Since a bookmaker's lines are eventually determined collectively by those betting on a game, we propose evaluating the crowd's I/R level in terms of the difference between the actual game outcome and the bookmaker's line—a known public line as opposed to, say, an unknown rate of inflation that is expected by the public at some future time. We term this difference a *gambling shock* (GS). In the sports gambling markets, the modeling objective is to forecast the gambling shocks effectively.

Definition 2.5.1: *Gambling shock is a measure of the gambling public's level of irrationality/rationality.* Let $|GS|$ denote the absolute value of a gambling shock and suppose that GS can be modeled effectively. Then larger values of $|GS|$ are indicative of increasing levels of irrationality, whereas smaller values of $|GS|$ are indicative of increasing levels of rationality. If GS is not predictable, the crowd's I/R level is uncertain—in the sense that nonpredictable GS may be due to unforeseen events that accompany the course of a game and its outcome.

Definition 2.5.2: *Market efficiency and inefficiency.* Increasing (decreasing) levels of irrationality (rationality) are indicative of increasing (decreasing) levels of market inefficiency (efficiency): The greater (less) the level of market inefficiency, the greater (less) the likelihood of profitable modeling/forecasting rules and the more (less) predictable the value of GS.

2.6 STATISTICAL SHOCKS: ALIAS VARIABLES

Whereas a gambling shock is the difference between a game outcome and the public line, a *statistical shock* is the difference between the outcome and its expected value based on a statistical forecasting model. In the former case, the public line is known, and in the latter case the expected value is unknown and must be estimated. Thus, a forecasting model for, say, an NFL game may contain both lagged gambling and statistical shocks, especially when the lagged statistical shocks may reflect significant variables that are not reflected by the lagged gambling shocks.

Lagged statistical shocks, known as *moving average* (MA) *variables* in ARMA (autoregressive moving average) modeling and introduced in Section 4.3, are critical in modeling price changes in financial markets where there is no public line on an anticipated price change and hence no gambling shocks. And as with gambling shocks, the statistical shocks in financial markets are likely to reflect physiological–psychological–sociological variables.

In a study of active traders at a London bank, testosterone and cortisol levels were monitored through saliva tests in 17 male traders for eight consecutive days, during which time each trader recorded his profits and losses. Given that past studies indicated that testosterone plays a role in winning and losing and cortisol (a hormone with the opposite effect, of dampening exuberance) plays a role in responding to stress and uncertainty, the researchers hypothesized that these steroids would respond to financial risk taking. The specific hypothesis was that *testosterone would rise on days when traders made an above-average gain in the markets, and cortisol would rise on days when traders were stressed by an above-average loss.... The study indicated that men with higher testosterone levels in the morning made above-average profits during the day. Success raised testosterone levels further, leading to higher confidence and greater risk-taking.... In the short term, in a rising market, this feedback loop tends to have a positive effect.... But other studies show that eventually the "winner effect" goes too far and effective risk taking turns into dangerous behavior. Cortisol rose in the traders as market volatility rose and profits and losses became more variable. Chronic cortisol exposure ... promotes feelings of anxiety, a selective recall of disturbing memories and a tendency to find danger where none exists. Cortisol is likely to rise in a crash, make traders dramatically and perhaps irrationally risk-averse, and exaggerate the sell-off* (Coates, 4/15/08). Although this study showed that increasing testosterone levels increase a trader's appetite for risk, it did not answer the question of whether testosterone was having beneficial effects by increasing the trader's skills.[11]

The latter question was addressed in a follow-on study and published in PLos One on 11/25/09. Evaluating a trader's skills in terms of his profit and loss (P&L) may be misleading; that is, knowing that a trader made \$100 million says nothing about the skill involved unless we know the risk involved (e.g., what if the trader could just as easily lost \$500 million?). Instead, trader skills can be evaluated in terms of the *Sharpe ratio* (SR; Sharpe, 1994), which measures how well the return of an asset compensates the investor for the risk taken. *SR* is the ratio of P&L to risk where risk is measured by the standard deviation P&L [denoted sd(P&L)]. The higher (lower) the risk, the lower (higher) the *SR* value. A trader making \$100 million but having sd(P&L) = \$500 million will have a low $SR = 0.20$, whereas a trader making \$100 million and having sd(P&L) = \$100 million will have an $SR = 1.0$.

The high-frequency male traders in the study group had an average $SR = 1.02$ between 2005 and 2007. This average was significantly higher than their benchmark index, Germany's Dax, which averaged 0.53. Although testosterone levels did not have a significant effect on *SR*, it was found that the *a trader's Sharpe ratios increased markedly with the number of years they had traded.... Moreover, traders increased their Sharpe ratios significantly during the two years of the study—indicating that they were*

[11]The trader–testosterone phenomenon in finance has drawn analogies with the use of illicit performance-enhancing drugs in sports. Should the sports officialdom enforce a ban on such drugs? Or *should sport open the door to enhancements that appear to be relatively innocuous? Prohibiting them would be justified if they were not widely available, on the ground that there is no point in adding to the unfairness in sport already created by differences in natural talent, luck or wealth. A ban also can be supported simply on the basis that enhancements violate the rules. But sports organizations should weigh the costs of enforcing the rules against the impact of changing them to permit the use of safe enhancements. A ban against the use of nitrogen tents may be futile, for example, since it may be next to impossible to develop tests to determine if an athlete had done so.... Even in cases where accurate tests could be developed, sport should consider letting the market decide whether relatively safe enhancements should be permitted, especially if they are widely available and the costs of prohibiting them would be substantial*. For example, *enough people enjoy watching or participating in power-lifting, an offshoot of weightlifting that does not test competitors for enhancement drugs. The sport has survived since the early 1960s. On the other hand, XFL football died when people refused to watch it* (Mehlman, 2005).

learning to make more money per unit of risk. Learning was encouraged by the compensation scheme at the trading company where they worked (Coates, 11/25/09). It would be of considerable interest to continue the study with these traders over the long term to determine the extent to which decreased testosterone levels with increasing age are compensated by increased skills.

Media coverage of the earlier study recalled that in October 2007 *a New York hedge fund was sued by a trader who had allegedly been made to take female hormones for behaving too aggressively.... Financial institutions could do something to reduce the excesses of testosterone-fueled trading without taking such extreme action. Financial trading is dominated by young men. You could stabilize the markets by including more women and older men who would be less susceptible to the testosterone feedback loop* (Cookson, 4/15/08). Longitudinal studies are obviously in order to compare the Sharpe ratios of younger male traders with those of women and older men.

Gambling shocks (Section 1.2) and related statistical shocks are probably reflections (or aliases) of the physiological–psychological–sociological variables that come into play in studies that examine the effects of testosterone and cortisol on trader performance.

3

Opacity and Present-Day Dynamics

3.1 DILEMMAS BETWEEN SOCIAL AND ECONOMIC EFFICIENCY

It is said that we live in a world shaped by capitalism. For three centuries it has been argued that communal well-being—defined in terms of economic and social efficiency—is best advanced through the enlightened self-interest of capitalists.[1]

[1]In 2001, a conference was held in Bayonne, in southwestern France, celebrating the 200th birthday of Frederic Bastiat (1801–1850), a pioneer of free market capitalism. *Through his writings and speeches, and as a member of the French Chamber of Deputies, Bastiat fought valiantly against the protectionism and socialism of his time. His weapons were wit and satire; his weapons were reducto and absurdum. ... Despite the publication of Adam Smith's "The Wealth of Nations," decades earlier, Bastiat was still fighting the mercantilist view of exports as good and imports as bad. He pointed out that under this view, the ideal situation would be for a ship loaded with exports to sink at sea. One nation gets the benefit of exporting and no nation has to bear the burden of importing. ... The most famous example of Bastiat's satire was his petition to the French parliament on behalf of candle makers and related industries. He was seeking relief from "ruinous competition of a foreign rival who works under conditions so far superior to our own for the protection of light that he is flooding the domestic market with it at an incredibly low price." The foreign rival was the sun. The relief sought was to a law requiring the closing of all blinds to shut out sunlight and stimulate the domestic candle industry. ... Bastiat stressed that because we have limited resources and*

It is not from the benevolence of the butcher, the brewer, or the baker, that we expect our dinner, but from their regard to their own interest. We address ourselves, not to their humanity but to their self-love, and never talk to them of our own necessities but of their advantages. Nobody but a beggar chuses to depend chiefly upon the benevolence of his fellow-citizens. (Adam Smith, *Wealth of Nations*)

Smith's commentary was intended to distinguish between mutually rewarding commercial exchanges and the pursuit of self-interest outside a framework of cooperation and regulation. Smith wrote further: *Monkeys when they rob a garden throw the fruit from one to another till they deposit it in the hoard, but there is always a scramble about the division of the booty, and usually some of them are killed*. Smith might have been talking about bonus time at an investment bank (Kay, 2/25/09).

Effective cultural pluralism (defined as effective social efficiency) is said to be the embodiment of capitalism (defined as effective economic efficiency).

Come to the London Exchange, a place more respectable than many a court. You will see assembled representatives of every nation for the betterment of mankind. Here the Jew, the Mohametan and the Christian deal with one another as if they were of the same religion, and reserve the name "infidel" for those who go bankrupt. Here the Presbyterian puts his trust in the Anabaptist, and the Anglican accepts the Quaker's promissory note. Upon leaving these peaceful and free assemblies, one goes to the synagogue, the other for a drink; yet another goes to have himself baptized in a large tub in the name of the Father through the Son to the Holy Ghost; another has his son's foreskin cut off, and over the infant he has muttered some Hebrew words that he doesn't understand at all. Some others go to their church to await divine inspiration with their hat on their head. And all are content. (Voltaire: On his visit to the London Exchange in 1651)

In light of recent corporate scandals (e.g., Enron, WorldCom), the subprime mortgage crisis, and Ponzi schemes, obvious dilemmas exist between economic and social efficiency. Prowse (6/16/02) notes that *economists typically assume that individuals, rational or irrational, are self interested but honest*. They regard people as *personal utility maximizers* who take whatever steps they can to promote their own welfare and *accept limits to*

[1](Continued) *unlimited wants, it's foolish to contrive inefficiencies just to create jobs. Progress comes from reducing the work need to produce, not increasing it. Yet, a day doesn't pass that we don't hear of some proposal to "create jobs," as if there's no work to be done otherwise. If it's jobs we want, let's just replace all bulldozers with shovels. If we want even more work, replace shovels with spoons. Bastiat suggested working only with our left hand* (McTeer, 7/5/01).

their maximizing behavior; i.e., *they try to get rich while obeying the law and without cheating or deceiving others*. The dilemma is that *true utility maximization is not consistent with behavior that respects social norms and rules. If taken to its logical limits, it naturally turns into wholly opportunistic behavior. After all, if I regard my goal in life as to promote my own personal interests, and if I am logical, I should lie, steal and cheat whenever I calculate that the likely gains will outweigh the likely losses. In other words, I should deceive others and break the law whenever I expect to get away with it. The true utility maximizer is wholly immoral. In extending the role of markets and in promoting market-style incentives even in the public sector, policy makers are thus inadvertently helping to destroy moral fabric built up over generations. In domain after domain, they are instructing people to act "economically"—in other words to put self interest ahead of everything else. The sad reality is that an economic orientation in which morality has no place has become, for many, a total philosophy of life. And it has become so with the blessing of our political leaders. We must accept, in short, that economic efficiency is not equivalent to social efficiency, and that one can rely too much, as well as too little, on markets*.

There is an adage that 10% of the people are completely honest, 10% are completely dishonest, and 80% are somewhere in between. Toward promoting greater social efficiency, it has been suggested that remedial actions should distinguish between changing the mindset of the quasi-utility maximizers (the potentially dangerous 80%) and chasing after the true utility maximizers (the bad minority) with a bigger stick.

3.2 TOWARD A MORE VISIBLE HIDDEN HAND

The efficient market hypothesis has roots in Adam Smith's metaphor of the *invisible hand* of the market, which reconciled supply and demand and brought about economic well being without the visible hand of government. Interest in Smith's work increased when Fredrich von Hayek claimed Smith as one of the precursors of *spontaneous order,* the notion that social institutions come into being without deliberate human intention and function without comprehensive planning and design.

> *From the time of Hume and Adam Smith, the effect of every attempt to understand economic phenomena—that is to say, of every theoretical analysis—has been to show that, in large part, the coordination of individual efforts in society is not the product of deliberate planning, but has been brought about... by means which nobody wanted or understood, and which in isolation might be regarded as some of the most objectionable features of the system. It showed*

> *that changes implied, and made necessary, by changes in our wishes, or in the available means, were brought about without anyone realizing their necessity. In short, it showed that an immensely complicated mechanism existed, worked and solved problems, frequently by means which proved to be the only possible means by which the result could be accomplished but which could not possibly be the result of deliberate regulation, because nobody understood them.* (Frederick von Hayek; see C. Smith, 2006)

Given the aftermath of financial deregulation in the late 1990s and innovations under the guise of efficient markets, indications are that Keynes' *demand management*, a visible government fist in the world financial system, is replacing Smith's *invisible hand* and that Hayek's *spontaneous order* is facing an unprecedented regulatory backlash (Tett et al., 10/10/08).

In light of the recent financial crisis, Wadhwani (12/17/08) provides the following commentary on policy mistakes by central bankers due to implementations of the EMH and its hidden hand. *In recent years, many countries made their central banks independent and these typically run* by economists whose policy mistakes were influenced by their belief in the EMH. *While bubbles were formed over the last decade, it was frequently argued that central bankers had neither more information nor greater expertise in valuing an asset than private market participants. This was often one of the primary explanations for central banks not attempting to "lean against the wind" with respect to emerging bubbles.... It is likely that had central banks raised interest rate by more than was justified by a fixed-horizon inflation target while house prices were rising above most conventional valuation measures, the size of the bubble would have been smaller.... Once the bubbles burst in 2007, some central banks were surprisingly slow to cut interest rates... a policy mistake that may well lead the current recession to be longer and deeper than it might have been.... One reason for their reluctance to cut interest rates was the significant rise in commodity prices. In relying on the EMH yet again, policymakers used longer-dated futures prices for these commodities in preparing their inflation projections. Their failure to allow for the possibility that a "bubble" had developed in the commodity markets... led them to significantly overestimate prospective inflationary pressures.*

Without attempting to analyze the complex financial instruments that unraveled so catastrophically, Atwood (2008) provides a summary of the subprime mortgage crisis. *Some large financial institutions peddled mortgages to people who could not possible pay the monthly rates and then put this snake-oil debt into cardboard boxes with impressive labels on them and sold them to institutions and hedge funds that thought they were worth something.* Atwood makes reference to Samuel Johnson's remarks on the efficacy of debtors' prisons: *We have now imprisoned one generation of debtors after another but we do not find that their numbers lessen. We have now learned that rashness and imprudence will not be deterred from taking credit: let us try whether fraud and avarice may be more easily restrained from giving it*.

Regarding the subprime mortgage crisis, another Samuel Johnson commentary is particularly appropriate.

> *Those who made the laws have apparently supposed, that every deficiency of payment is the crime of the debtor. But the truth is, that the creditor always shares the act, and often more than shares the guilt, of improper trust. It seldom happens that any man imprisons another but for debts which he suffered to be contracted in hope of advantage to himself, and for bargains in which proportioned his own profit to his own opinion of the hazard; and there is no reason, why one should punish the other for a contract in which both concurred.* (Samuel Johnson, Idler 22, September 16, 1758)

3.3 HEDGE FUNDS, GALAPÁGOS, AND EVOLUTION

Recounting the adaptive market hypothesis, periods of relative market stability and incremental change are followed by periods of intense change as participants evolve to take account of new realities. Applications of the evolutionary principles to the markets have led to analogies between the markets and Galápagos.

During his visit to the Galápagos Islands in 1835, Darwin recorded observations on finch varieties with different-shaped beaks scattered across the archipelago's some 100 islands—islands with differing ecosystems. The observations provided him with a cross-sectional view of evolution that enabled him to develop his theory of natural selection.

By analogy, it has been said that regulatory body environs correspond to the islands, while market participants—hedge funds, mutual funds, private equity funds, sovereign wealth funds, investment banks, etc.—correspond to the islands' incumbent species. Hedge funds, hitherto subject to very limited government regulation, are analogous to occupants of islands with few predators and a friendly ecosystem. Such islands became experimental laboratories for testing moneymaking ideas. In contrast, the highly regulated mutual funds are analogous to occupants of islands with many predators.

Unlike mutual funds that are limited to long-only positions and that sustained severe losses following the bursting of the Internet bubble, hedge funds can short stocks or commodities to profit from price declines. Hedge funds can also leverage. *A trade that makes not even 1% is worth doing if you can borrow enough money to make the same trade 10 times. And they can limit withdrawals by investors, allowing them more flexibility than funds that must be prepared for redemptions every day.*

The 2007–2008 market disequilibrium was an example of intense change, especially for the hedge funds, a change that reflects Gould's punctuated equilibrium (see Section 4.3). *In July 2008 hedge funds started to lose money and then lost it in a big way the following September. The end of September gave hedge fund investors one of their periodic opportunities to remove money.* (Estimates of withdrawals ranged from $31 billion to $43 billion, while investment losses reduced hedge fund assets by $210 billion.) *Meanwhile, the Lehman Brothers bankruptcy in mid-September prompted a sudden increase in the price of leverage as investments on whom hedge funds rely for their short term funding applied much tighter restrictions. Then came the ban on shorting financial stocks and all the hedge funds' critical evolutionary advantages had been removed. Leverage, it appears, was vital to the eco-system of the Galápagos.... The removal of leverage may not administer the same shock to hedge funds as the asteroid that eliminated the dinosaurs, but they now face a new phase of evolution—in a far more hostile environment* (Authers, 10/18/08).

3.4 LOTTERIES: MARKET FOR LOSERS

> *The art of taxation consists in so plucking the goose as to obtain the largest number of feathers with the least possible amount of hissing.* (Jean-Baptiste Colbert, Chief Minister of Louis XIV[2])

Following Colbert's wisdom, governments established lottery markets—a form of stealth taxation (Brittan, 11/25/99)—as a tax on stupidity (Brooks, 6/10/08). Lottery markets are a means of establishing the process of *creative destructive benevolence*. Lotteries generate needed government revenues largely at the expense of those who can least afford the tax and thus become a destructive force on segments of society. Then, in its benevolence, the government encourages rehabilitation programs for those succumbing to the destructive force of the lotteries. Colbert would argue that the lottery process

[2]Colbert, described as a *sustained polemic*, espoused protection in external economic policy and detailed regulation at home. *But Colbert was also a successful conventional finance minister. The combination of dirigisme with the search for sound finance is familiar to students of present day French policy. Colbert was distinguished from contemporaries not by any originality in his opinions but because he had success in enforcing them* (The New Palgrave Dictionary Online).

has greater benefit than no lottery process. After all, vast government revenues are generated for the benefit of its citizens, while jobs are created by the rehabilitation programs to ameliorate the effects of the destructive force of lotteries.

Whether through naivety or self-serving interests, sports gambling markets have been linked erroneously to lotteries. As discussed earlier, player success in sports gambling markets requires skills, whereas lottery outcomes are independent of player skills. The independence of outcomes and player skills and the minute probabilities of winning lotteries assure large profits for agencies sponsoring the lotteries. For government lotteries, the take is roughly 67 to 75% of the total amount bet. Those who play the lottery have been described as irrational, hapless fools.[3]

> *When gasoline prices shot up in 2008, Peggy S. thought about saving the $10 she spends weekly on lottery tickets.... But the prospect that the $10 could become $100 million or more was too appealing. So rather than stop buying Mega Millions tickets... she saved money instead by packing her lunch a few days a week, keeping alive her dreams of hitting a jackpot and retiring as a multimillionaire.* (Zezima, 9/13/08)
>
> *In October 2005, the Ohio Lottery will attempt to combat falling tickets sales by introducing a new game that offers a lower jackpot but a higher probability of winning.... For lottery operators... the 18–25 year-old demographic has been a key target market.... The odds of winning the new Ohio game will be 6 million to 1, compared with 14 million to 1 odds of winning the current Super Lotto Plus.... The quest is to persuade youth to overlook the odds.* (As reported by G. Malkani in the *Financial Times,* 9/13/05)

In a 2009 voter initiative, California Governor Arnold Schwarzenegger proposed a major expansion in the existing state lottery. *The California Lottery is replacing its "Big Spin" television game... with "Make Me a Millionaire"* (Lawrence, 1/16/09). In his State of the State address, Schwarzenegger stated that *our state is incapacitated until we address the budget crisis*. The "Big Spin" was intended to address the crisis. The initiative was rejected decisively by the voters.

[3]George Bernard Shaw's assessment of gambling, though misguided in general terms, is appropriate in reference to the lottery and slot machine players. *Gambling... is a vice which is essentially... ruinous. In extreme cases it is a madness which persons of the highest intelligence are unable to resist: they will stake all they possess though they know that the chances are against them. When they have beggared themselves in a half an hour or half a minute, they sit wondering at the folly of people who are doing the same thing, and at their own folly in having done it themselves* (George Bernard Shaw, *The Vice of Gambling and the Virtue of Insurance*).

A report by the Institute of American Values (IAV, 2008) and other contributing think tanks documents the deterioration of financial mores in the United States, a transformation that has led to a stark financial polarization. The report contrasts the investor class with the lottery class, the former with tax-deferred savings and financial advisors and the latter with little access to 401k's or financial planning, but plenty of access to payday lenders, credit cards, and lotteries. The agents of destruction are said to include many government agencies, including the Congress, the White House, Wall Street, and state governments. It is estimated that 20% of Americans are frequent lottery players, spending $60 billion a year. The spending is starkly regressive. A household with an income under $13,000 spends, on average, $645 a year on lottery tickets, about 9% of all income (Brooks, 6/10/08).

The IAV report includes the following recommendations:

1. Raise public consciousness about debt the way that antismoking activists did with their campaign.
2. Create institutions that encourage thrift.
3. Have foundations and churches issue short-term loans to cut into payday lenders' business.
4. Establish programs that give the poor and middle class access to financial planners.

In marked contrast to the IAV recommendations, Groz (2006) proposes increasing lottery tickets sales through *no-loss lotteries*. Under this scheme, 30 cents out of every dollar in ticket sales is set aside into a special account for the benefit of the lottery player and never put at risk in the game. It is assumed that this money is placed into a long-term investment account with an average rate of return of 10% yearly. *The remaining 70 cents is divided as usual among the prize pool, money raised for education or other socially useful purpose, and administrative costs.... Ticket sales could double or triple as players begin to realize that they're saving or investing every time the play. This would translate to greater revenues and bigger prize pools than existing lotteries*. Methods of tracking the amount wagered by player participants include smart cards, cell phones, and eventually by biometric sensors used to identify and authenticate individuals.

Aside from the questionable *average yearly rate of return 10%*, the Groz proposal would probably expand the lottery class and deepen its quagmire. On the other hand, recommendations in the IAV report are deficient in that they do not directly address the root causes of the financial polarization. By any measure, education is the ultimate poverty eradicator and financial literacy is one of the major educational challenges facing the United States today (see Chapter 12).

4

Adaptive Modeling Concepts in Dynamic Markets

4.1 QUANT FUNDS AND ALGORITHMIC TRADING

Algorithmic trading—the use of quantitative (*quant*) rules to finesse trade execution—is widely used by hedge funds, pension funds, mutual funds, and other institutional traders.[1] Large trades are divided into several smaller trades in order to manage market impact, opportunity cost, and risk. It may be used in any investment strategy, including market making, intermarket spreading, arbitrage, or pure speculation. The investment decision and implementation may be augmented at any stage with algorithmic support or may operate completely automatically. Dramatic price swings in the final hour of trading sessions are a sign of computer-driven quant activity (Gangahar, 3/30/09).

Hedge funds and similar traders use algorithmic trading to make decisions to initiate orders based on information that is received electronically, before human traders are even aware of the information. In *algorithmic* or *flash trading*, trades are executed in a few thousands of a second. *High-frequency traders often issue and then cancel orders almost simultaneously and get an early peek at how others are trading. On 9/17/09, the Securities and*

[1]For a pure quant fund, the decision to execute the order is made by the model. For non-pure quant funds, the fund manager incorporates both quant modeling and human judgment.

Forecasting in Financial and Sports Gambling Markets: Adaptive Drift Modeling, By William S. Mallios

Exchange Commission (SEC) proposed banning flash orders.[2] *Proponents of* [flash trading] *argue that such trading enhances the liquidity and efficiency of the markets* (Anderson, 9/18/09).

Of particular concern to SEC regulators is *the naked sponsored access, whereby a high-frequency trader's activities are not monitored in real time by the sponsoring broker*. (Estimates are that naked access comprises about half of U.S. daily share turnover as of early 2010.) *People familiar with the SEC's discussions say the regulator will likely propose eliminating naked access and require broker-dealers to take steps to prevent erroneous trading and check that investors do not breach credit and capital limits* (Mackenzie and Chung, 1/13/10).

During the 2007–2008 crises, many *quant* funds reported steep losses. [Very quickly, however, the loss cycle was subsequently reversed. By August 2009, hedge funds had made their best start to a year in a decade (Jones, 8/13/09).] Several explanations have been given for the *quant* losses in 2007–2008.

1. Parsimonious forecasting models are inevitably doomed. (Economics is not a natural science but, rather, a science of human behavior.)
2. Quant funds have failed to incorporate in their model development recent advances from allied sciences, particularly in terms of incorporating biological–psychological–sociological variables in models that adapt to dynamic market conditions. Quite apart from the biological–psychological–sociological variables, the weather is known to have a dominant effect on the price of produce such as sugar and cocoa and crops such as wheat and corn. Meteorologists are thus in great demand by hedge funds and investment banks. For agricultural commodities, weather research is now as important as research on consumer trends (Blas, 8/28/09).
3. With so few people qualified in effective quant modeling, competing firms lure these individuals, who then use the same strategies in the same markets and chase the same money once they depart.
4. Computer programs based on proprietary modeling used in daily decision-making are known as *black box investment*. Users of the system need not know or understand *black box* contents—contents

[2]In the protracted debate over how to reform our accident-prone financial system, little has been said on the cost of running the system. *For years, much of the best young talent in the western world has gone to private financial firms. Perversely, the largest individual returns seem to flow to those whose job is to ensure that microscopically small deviations from asset price relationships persist for only one millisecond instead of three. These talented and energetic young citizens could surely be doing something more useful* (Friedman, 8/27/09).

that have short shelf lives. Given such obsolescence, especially in the absence of adaptive-type drift modeling, black box output may quickly become worthless.

5. Forecasts based on publicly available information cannot compete with the certainties of insider trading. Analyses of the UK Financial Service Authority in 2008 suggested *that large scale cheating takes place between and within big institutions: 28.7% of all takeovers were preceded by suspect trading*.

In Section 5.3, recent cases of insider trading are summarized in terms of candlestick charts that depict price movements during the trading periods in question. But perhaps the most telling commentary on insider trading was provided by Bernard Madoff, who brazenly asserted that it was *impossible for* [an insider trading] *violation to go undetected, certainly not for a considerable period of time* because of regulatory safeguards. He added, however, that this was *something that the public doesn't understand. If you read things in the newspaper and you see someone violate a rule, you say "Well, they're always doing this"* (Chung, 12/15/08). This is the same Bernard Madoff who was later convicted of having perpetuated the greatest Ponzi scheme in history and sentenced to 150 years in prison. The size of Madoff's fraud, estimated to be $65 billion, dwarfs all other frauds.

The cancer continued to metastasize. One week following the Madoff revelation, charges of alleged fraud continued but on a far less grandiose level. Former Lehman Brothers executive Matthew Devlin was charged with *misappropriating* material nonpublic information concerning at least 13 acquisitions or attempted acquisitions based on information from his unsuspecting wife, a partner in a public relations firm involved in acquisitions. Trading on those tips, he and others involved in the scheme made more than a paltry $4.8 million in profits from at least March 2004 to July 2008. The lawyer for Devlin's wife said: *She was completely unaware that confidential information about her job was being used as the basis for security trading* (Chung, 12/15/08).

Faulty model forecasts are largely due to erroneous modeling assumptions that should have been revealed through model validation procedures. The Black–Scholes model, developed in 1973 to price options and still used extensively, assumes that present prices are independent of past prices and that the probability of extreme price changes is negligible. *Twenty years ago, unwarranted use of the* [Black–Scholes] *model spiraled into the worldwide October 1987 crash; the Dow Jones index dropped 23% in a single day, dwarfing recent market hiccups. Ironically, it was the very use of a crash-free model that helped to trigger a crash* (Bouchaud, 2008).

Under the modeling mantra of parsimony, *it is assumed that* [economic behavior] *exists just as electricity or gravitation exists and that it is capable of analysis in much the same way* (Spengler, 1932). In fact, it is foolish to suppose that simplistic modeling of complex economic/human behavior would lead to anything but nonviable forecasting. Aside from the natural sciences, nothing good can come out of the principle of Occam's razor[3] as applied to market forecasts: *Hypotheses and models: Cut'em thin*. (commentary by W.G. Cochran in the late 1950s in praise of R.A. Fisher).

4.2 MARKET VOLATILITY AND FAT-TAILED DISTRIBUTIONS

Regarding the impact of market volatility on price changes, irrational behavior has been exacerbated by the defining trend of our time—increasing volatility. When markets become volatile, the last price quoted for a security is indicative of next to nothing (Kaufman, 2001). It has long been recognized that asset returns exhibit positive sample excess kurtosis or follow fat-tailed distributions, such as power law distributions (Clauset et al., 2009). This means that extreme market volatility—usually described as market anomalies—occurs more frequently than would be expected under the assumption of bell-shaped curves such as a normal distribution.[4]

Applications resulting in fat-tailed distributions motivated the derivation of extreme value distributions (Coles, 2001) that date back to R.A. Fisher. At a later date, Mandelbrot and Hudson (2004) took note of fat-tailed anomalies in the financial markets. In Figure 4.2.1 Mandlebrot and Taleb (3/24/06) demonstrate the effects of extreme price movements on earnings. [The extreme price movements in this figure are indicative of Gould's punctuated equilibrium that is used to characterize the abrupt model drift (see Section 4.3).] And with reference to Figure 4.2.1, Mandlebrot and Taleb make the following statement. *Any attempts to refine the tools of modern portfolio theory by relaxing the bell curve assumptions, or by "fudging" and adding occasional "jumps" will not be sufficient. We live in a world primarily driven by random jumps, and tools designed for random walks address the*

[3]This principle states that the explanation of any phenomenon should make as few assumptions as possible and eliminate those that make no difference in the observable predictions of the explanatory hypothesis or model. The principle is often expressed in Latin as the *lex parsimoniae* (law of parsimony).

[4]Regarding descriptions of price changes in terms of the normal-type distributions, perhaps the discussion should start with the assumption that the variance of a price change is proportional to the price change; that is, the greater the price change, the greater its variance and hence the greater the risk. This assumption is addressed in terms of GARCH-type modeling of volatility introduced in Section 4.4.

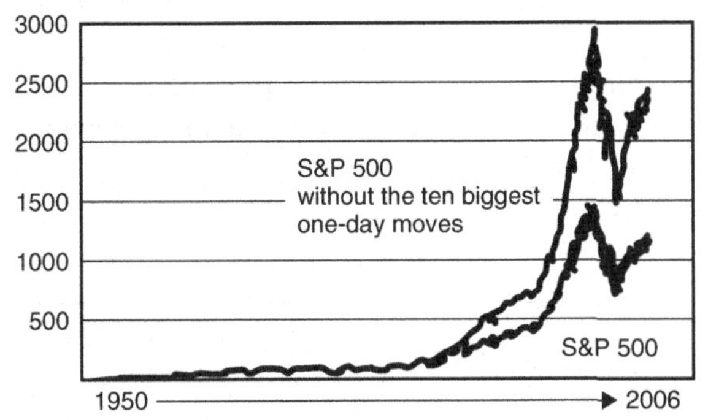

Figure 4.2.1 *A focus on the exceptions that prove the rule. (*Source: *Mandlebrot and Taleb, 3/24/06)*

wrong problem. It would be like tinkering with models of gases in an attempt to characterize them as solids and call them "a good approximation."

Mandelbrot–Taleb reasoning is in conflict with the basic tenets of statistical time series modeling. Specifically, when a particular model is identified and then estimated for purposes of forecasting future price changes, what matters is not the observed distribution of prices changes but, rather, (1) the distributional properties of the residuals associated with the estimated model and (2) the volatility associated with the residuals. Objectives of adaptive drift modeling are to build effective forecasting models for both price changes and volatility. For the markets under study, conventional modeling in terms of nondynamic, first-order ARMA-type processes typically lead to ineffective forecasting models and/or to conclusions of random walk.

The term *volatility* is typically defined as *the rate and magnitude* of price changes. The market *VIX Index*, also known as the *fear index*, is a popular measure of implied volatility. The VIX Index is a forward-looking measure of S&P 500 volatility. Implied volatility, a forward-looking measure, is the market price of an option based on an option pricing model. In contrast to implied volatility, historical volatility is based on the past prices of a security.

When the market is calm and trading is in a moderate trading range, volatility is typically described as low. During these periods—which reflect complacency or of lack of fear—call option buying is generally greater than put option buying. However, when prices drop sharply, investor anxiety increases and traders rush to buy puts, which increases the price of put options and consequently increases the value of the VIX Index. Since 2003, the VIX value has been based on S&P 500 option series.

Figure 4.2.2 demonstrates the divergence in percent changes between the VIX and the S&P 500 (GSPC) Indexes from April through September 2009, a period following the *March 2009 nadir* (see Section 9.10). Figure 4.2.3 demonstrates the divergence between the VIX Index and the Dow Jones Index following the collapse of Lehman Brothers on September 15, 2008. Figure 4.2.4, an alternative presentation of Figure 4.2.3, is given in terms of a daily candlestick chart for price changes with moving averages.

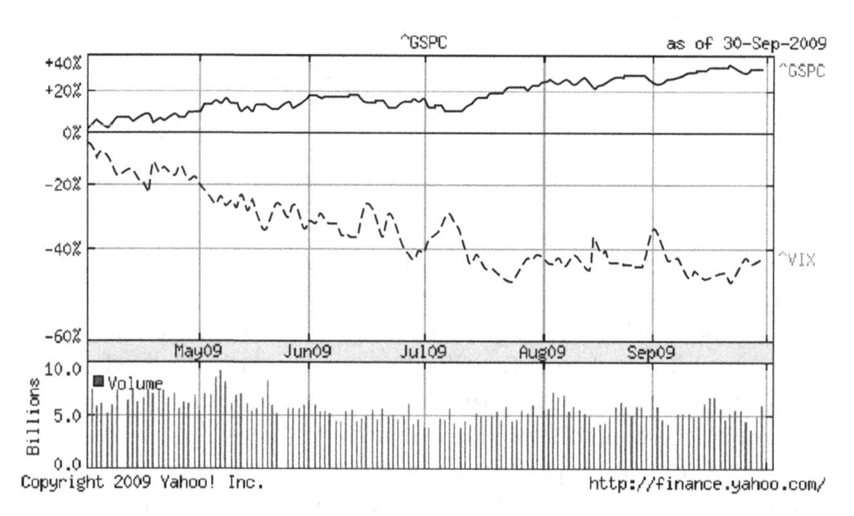

Figure 4.2.2 *Relative changes in the S&P 500 and VIX Indexes following the March 2009 nadir.*

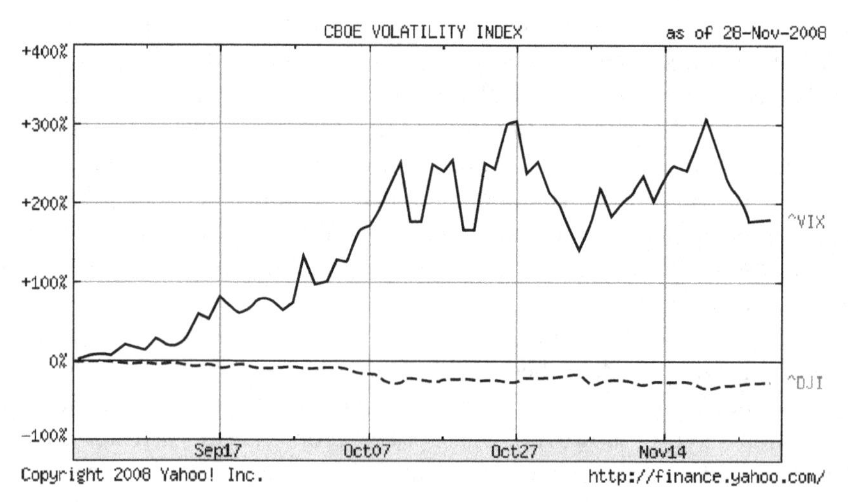

Figure 4.2.3 *Daily candlestick chart for VIX, the* fear index, *including the relative change in the DJI Index following the collapse of Lehman Brothers.*

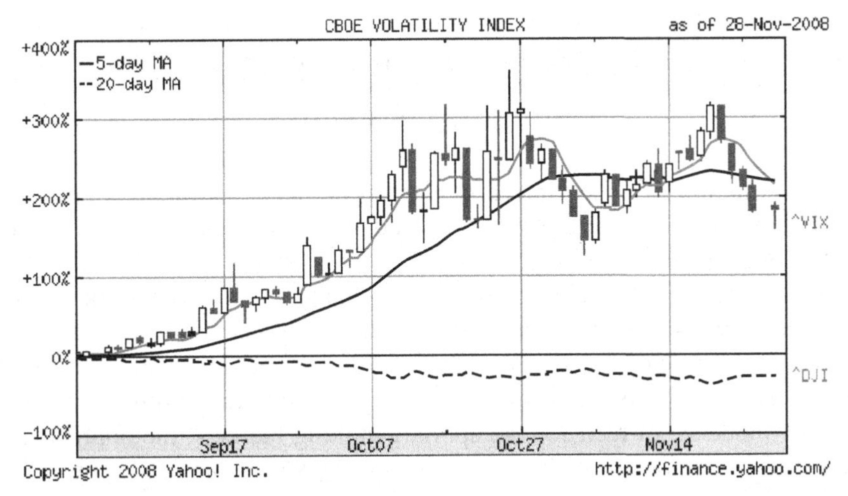

Figure 4.2.4 *Daily candlestick chart for VIX, including the relative change in the DJI Index following the collapse of Lehman Brothers.*

4.3 ADAPTIVE ARMA(1, 1) DRIFT PROCESSES

Let $D(t)$ denote a price change over the fixed time interval $[t-1, t]$, where t can be defined in terms of, say, hours, days weeks, or months. In the context of sports, $D(t)$ may denote the winning/losing margin in game t for a particular team. Disregarding effects of covariates (such as the line on a game in the sports gambling markets), suppose that that $D(t)$ is generated by the time-varying ARMA(1, 1) process:

$$D(t) = \alpha(t)D(t-1) + \gamma(t)\varepsilon(t-1) + \varepsilon(t), \tag{4.3.1}$$

where $\alpha(t)$ and $\gamma(t)$ denote, respectively, time-varying coefficients of AR(1) and MA(1) variables, and $\varepsilon(t)$ denotes the model error (or contemporaneous statistical shock). When the interaction between $D(t-1)$ and $\varepsilon(t-1)$ affects $D(t)$ directly, the following bilinear process (Granger and Anderson, 1978) may be conjectured to generate $D(t)$:

$$D(t) = \alpha(t)D(t-1) + \gamma(t)\varepsilon(t-1) + \alpha_1(t)D(t-1)\varepsilon(t-1) + \varepsilon(t). \tag{4.3.2}$$

One approach to dynamic modeling assumes that $\alpha(t)$ and $\gamma(t)$ in (4.3.1) are generated by random walk (West and Harrison, 1997):

$$\begin{aligned} \alpha(t) &= \alpha(t-1) + \delta^*_\alpha(t), \\ \gamma(t) &= \gamma(t-1) + \delta^*_\gamma(t). \end{aligned} \tag{4.3.3}$$

The vector of model errors $\boldsymbol{\delta}^*(\mathbf{t}) = (\delta^*_\alpha(t), \delta^*_\gamma(t))'$ is assumed normally distributed with $E(\boldsymbol{\delta}^*(\mathbf{t})) = \mathbf{0}$ and $E(\boldsymbol{\delta}^*(\mathbf{t})\boldsymbol{\delta}^*(\mathbf{t})') = V_{\delta^*}(t)$; that is,

$$\boldsymbol{\delta}^*(\mathbf{t}) : \text{normal}(\mathbf{0}, V_{\delta^*}(t)). \tag{4.3.4}$$

The assumption of a possible time-varying variance–covariance structure allows for the time-varying volatility and the estimation of GARCH processes (see Section 4.4).

An alternative dynamic modeling approach is to assume that each coefficient in (4.3.1) is generated by an MA process; that is, for the markets under study, changes in the coefficients at time t are the result of unforeseen shocks occurring at time $t-1$ (and earlier times for more general models). In fact, for the sports gambling markets, the applications will clearly demonstrate that lagged gambling shocks are instrumental in affecting coefficient changes.

Accordingly, our working hypothesis is that coefficients are generated by lagged shocks, the magnitude of which may alter the structure of model (4.3.1): for example,

$$\begin{aligned} \alpha(t) &= \alpha + \alpha_1 \varepsilon^*(t-1) + \delta_\alpha(t), \\ \gamma(t) &= \gamma + \gamma_1 \varepsilon^*(t-1) + \delta_\gamma(t). \end{aligned} \tag{4.3.5}$$

For *gradual adaptive mean drift*, coefficient changes tend to be incremental such that

$$\varepsilon^*(t-1) = \varepsilon(t-1). \tag{4.3.6}$$

For *abrupt adaptive mean drift*,

$$\varepsilon^*(t-1) = \varepsilon(t-1) + \xi v(t-1) \qquad \text{if } |\varepsilon(t-1)| \text{ is sufficiently large,} \tag{4.3.7}$$

where ξ denotes the effect of $v(t-1)$, a covariate associated with the process whose effect becomes significant in the particular abrupt drift scenario; $v(t-1)$ may or may not retain its significance for an indefinite period of time. The vector of model errors, $\boldsymbol{\delta}(\mathbf{t}) = (\delta_\alpha(t), \delta_\gamma(t))'$ is assumed to have a time-varying variance–covariance structure as in (4.3.4).

Substituting expressions for $\alpha(t)$ and $\gamma(t)$ in (4.3.5) into (4.3.1), we have a reduced, second-order ARMA(1, 1) model:

$$\begin{aligned} D(t) = {} & \alpha D(t-1) + \gamma\varepsilon(t-1) + \alpha_1 D(t-1)\varepsilon^*(t-1) \\ & + \gamma_1 \varepsilon^*(t-1)\varepsilon(t-1) + \varepsilon_R(t), \end{aligned} \tag{4.3.8}$$

where $\varepsilon_R(t)$ denotes the reduced model error. Notice that for abrupt drift scenarios, the predictor variables in the reduced model may change from one time to the next (in the sense that the significance or nonsignificance of predictors may change as the model is updated from one time to the next). Note also that lagged first and second moments affect subsequent first moments, indicating the existence of an interactive feedback between the first and second moments.

For $\gamma_1 = 0$, (4.3.8) takes the form of a bilinear process. Model (4.3.8) is not invertible when $\gamma_1 \neq 0$ (Granger and Anderson, 1978). However, estimation biases resulting from the noninvertibility condition may be inconsequential if the fitted reduced model provides effective forecasts. Note that (4.3.8) could have resulted from a variety of time-varying coefficient models and not necessarily (4.3.5). When model structure for time-varying coefficients is of specific interest, the validity of equations in (4.3.5) can be evaluated by fitting each equation to the data for successive values of t (where each fitting is within prespecified time windows). The fitting of (4.3.5) is also necessary for purposes of estimating prior distributions of reduced model coefficients when applying empirical Bayesian estimation.

A complication associated with abrupt drift scenarios is that *large* values of $|\varepsilon(t-1)|$ may disrupt model structure to the extent that forecasts are temporarily unreliable. In such cases, risk management considerations suggest that (1) forecasting decisions should be delayed or altered until the model stabilizes in terms of gradual or more moderate abrupt drift[5] and/or that (2) greater emphasis should be placed on a volatility modeling forecast, as discussed in Section 4.4.

Abrupt shocks disrupted the model structure in a study that reported a significant negative relation between the performances of commodities and equities. Gorton and Rouwenhorst (2004)[6] constructed an equally weighted index of commodity futures monthly returns from July 1949 to March 2004. Their purpose was to study simple properties of commodity futures as an asset class. Commodity returns were found to be negatively correlated

[5]At an earlier date, Forrester (1971) provided a prescient connection between model destabilization and evolution. He argued that *evolutionary processes have not given us the mental skill needed to interpret properly the dynamic behavior of the systems of which we have now become a part* and that the behavior of complex systems is counterintuitive. In many instances it emerges that the known policies describe a system that actually causes the troubles. In fact, a downward spiral develops in which the presumed solution makes the difficulty worse and thereby causes redoubling of the presumed solution.

[6]The paper by Gorton and Rouwenhorst, "Facts and Fantasies About Commodity Markets," was published while the authors *were consultants to AIG Financial Products, the insurance company subsidiary better known for its fateful push into credit default swaps* (Meyer, 2/10/10).

with equity and bond returns, a relation *due in significant part to different behavior over the business cycle*. Performances of commodities were said to *work well when they are needed most*: when stock market returns are disappointing.

The authors' view *was embraced en mass by institutional investors and helped transform commodities from a niche investment into a proper standalone asset class.... But the diversification benefits of commodities* [became] *increasingly tenuous as prices in the years following* [publication of the finding] *moved in tandem with other major asset classes, including equities and bonds. As stock markets plummeted worldwide in 2008, commodities fared just as badly*. The theory that commodities and equities respond to different phases of the business cycle was obviously challenged when both bottomed during the March 2009 nadir. *These negative correlations have proved nasty for investor, such as pension funds, that piled into commodities as a way to broaden portfolios and spread risk* (Meyer, 2/10/10). A reanalysis of the data used by Gorton and Rouwenhorst is necessary to determine whether coefficients were subject to mostly gradual drift during the 45 years of study with abrupt drift setting in thereafter. An added question is whether volatility modeling would have predicted the abrupt drift.

Adaptive model drift (AMD), whether gradual or abrupt, has parallels in other fields, such as evolutionary and molecular biology. For the latter, gradual drift reflects *antigenic drift*—minor changes in antigens due to gene mutations in influenza virus—while abrupt drift reflects *antigenic shift*—major changes in antigens due to gene reassessment in influenza virus.

In evolution, gradual drift is analogous to Darwinian evolution, while abrupt drift follows the Gould–Eldredge (1977) thesis of *punctuated equilibrium*. (*Punctuated* is used in the sense of a large meteorite hitting Chicago.) *At some point in time a small isolate of that larger species moves away anatomically so that not very much time elapsed but the structure has shifted dramatically.... In this model there are periods of stasis interrupted by sudden surges of change—usually cladogenic*. Whereas the adaptive market hypothesis attempts to explain the markets in terms of evolutionary biology, adaptive model drift allows modeling procedures to accommodate dynamic market changes whatever the evolutionary mechanism.

The foregoing parallels between adaptive drift modeling and evolution bring to mind disparate views on evolution. Lay opposition to Darwin's theory has been and continues to be in terms of intelligent design, whereas Gould was linked to Marxism. Apparently motivated by Engle's prophesy that Marxism will do for society what Darwin did for biology, an English social scientist named Halsted—who presumably was appalled by the abrupt shock of *Ten Days that Shook the World* (Reed, 1919)—stated: *The concept of punctuated equilibrium is undoubtedly linked to Marxism. Gould* [replaces] *gradualism with the flip-like style of change which has been appreciated within Marxist philosophy for a long time*. Gould (Gould and Eldredge, 1977) replied: *I did not develop the theory of punctuated equilibrium as part of a sinister plot to ferment world revolution but rather as an attempt to resolve the oldest empirical dilemma impeding an integration of paleontology into modern evolutionary thought*.

Commentary: *Who cares if Gould's theory is Marxist? In the late 19th century, social anthropologists embraced Darwin's theories and extended them to explain cultural change. Similarly, some modern anthropologists and sociologists have embraced this cladogenic theory and predicted immediate culture change and immediate overthrow of governments.... People are not independent of culture and vice versa. If scholars use physical evolution as a springboard for social advocacy, then the metaphor of biological change becomes a tool for politicians, revolutionaries, and many others* (Seth Mallios, private correspondence).

4.4 TIME-VARYING VOLATILITY

Let $D(t) = E[D(t)] + \varepsilon_R(t) = D(t)^\wedge + e_R(t)$, where, for example, $E[D(t)] + \varepsilon_R(t)$ is given by the right-hand side of (4.3.8). $D(t)^\wedge$ denotes a sample (fitted) form of the model, and $e_R(t)$ denotes the residual corresponding to the model error $\varepsilon_R(t)$. Unconditionally, the variance of $\varepsilon_R(t)$ is typically assumed to be homogeneous: variance$[\varepsilon_R(t)] = \sigma^2_{\varepsilon R}$. Conditionally, however, the variance may vary with time so that variance$[\varepsilon(t)] = \sigma^2_{\varepsilon R}(t)$. One approach to modeling volatility is to assume that $\varepsilon^2_R(t)$, as an estimate of $\sigma^2_{\varepsilon R}(t)$, is generated in terms of an ARMA(p, q) process, which is known as a GARCH(p, q) process. *GARCH modeling* (Bollerslev, 1986), an acronym for "generalized autoregressive conditional heteroskedasticity," is a generalization of ARCH(p, q) modeling (Engle, 1982). For example, in a GARCH(1,1) process, $\varepsilon^2_R(t)$ is generated in terms of an ARMA(1, 1) model:

$$e_R(t)^2 = \mu + \phi e_R(t-1)^2 + \psi\Delta(t-1) + \Delta(t), \qquad (4.4.1)$$

where ϕ and ψ represent the respective effects of the autoregressive and moving average variables, μ the mean, and $\Delta(t)$ the contemporaneous model error.

Similar to the approach of structured stochastic volatility modeling, GARCH processes may be generalized by assuming that coefficients are time varying; for example, (4.4.1) is replaced by

$$e_R(t)^2 = \mu(t) + \phi(t)e_R(t-1)^2 + \psi(t)\Delta(t-1) + \Delta(t). \quad (4.4.2)$$

AR, MA, ARMA, or bilinear processes are likely candidates for generating the time-varying coefficients; for example, for MA(1) processes,

$$\begin{aligned} \mu(t) &= \mu + \mu_1\Delta(t-1) + \delta_\mu(t), \\ \phi(t) &= \phi + \phi_1\Delta(t-1) + \delta_\phi(t), \\ \psi(t) &= \psi + \psi_1\Delta(t-1) + \delta_\psi(t). \end{aligned} \quad (4.4.3)$$

Substituting expressions for $\mu(t)$, $\phi(t)$, and $\psi(t)$ in (4.4.3) into (4.4.2), we have a reduced, second-order GARCH(1, 1) model:

$$\begin{aligned} e_R(t)^2 = \mu &+ \phi e_R(t-1)^2 + \psi\Delta(t-1) + \mu_1\Delta(t-1) \\ &+ \phi_1 e_R(t-1)^2\Delta(t-1) + \psi_1\Delta(t-1)^2 + \Delta(t). \end{aligned} \quad (4.4.4)$$

An alternative model for $e_R(t)^2$ is to replace the first-order ARMA model in (4.4.2) with a bilinear model and explore whether the time-varying coefficients are generated by AR, MA, ARMA or even bilinear processes.

For the reduced variance model in (4.4.4), lagged second, third, and fourth moments affect subsequent second moments, while for the reduced mean model in (4.3.8), lagged first and second moments affect subsequent first moments (means). In terms of a path diagram (Wright, 1921), an implication is that lags of kurtosis, skewness, and volatility affect subsequent volatility, and in turn, lags of volatility and the mean affect subsequent means.

There is a major shortcoming in these volatility modeling approaches. Since $e_R(t)^2$ in (4.4.4) is based on the fitted model $E[D((t)^\wedge]$, the volatility model may be misleading when the model for $E[D(t)]$ is specified incorrectly. For example, if $D(t)$ is modeled as a first-order ARMA process when, in fact, a higher-order model is appropriate, the resulting volatility model may erroneously reflect effects of higher-order terms that should have been included in the $D(t)$ model. This situation is illustrated in Chapter 9, where alternative volatility models are explored.

5

Studies in Japanese Candlestick Charts

5.1 BULLISH AND BEARISH CONFIGURATIONS FROM CHARTIST PERSPECTIVES

Japanese candlestick charting was introduced in feudal Japan in the mid-18th century to forecast price movements of forward contracts on the Osaka Rice Exchange. It is said that a legendary rice trader named Munehisa Homma devised the charts to develop insights into market psychology, which, in turn, allowed him to amass a considerable fortune in the rice market and eventually to rise to the rank of Samurai. This was nearly 100 years before the origin of traditional methods of chart analysis (Nisson, 1991).

Candlestick charts depict the opening $[O(t))$, high $(H(t))$, low $(L(t)]$ and closing $[C(t)]$ prices defined over fixed time (t) intervals (e.g., minutes, hours, days, weeks). The candlestick body is determined by $O(t)$ and $C(t)$, where $C(t) > O(t)$ and $O(t) > C(t)$ are distinguished by white and dark bodies, respectively. The upper extreme of the candlestick wick represents $H(t)$ and the lower extreme, $L(t)$. Three hypothetical candlesticks were illustrated in Figure 1.1.2.

Figures 5.1.1 and 5.1.2 present daily candlestick charts for Quality Systems (stock symbol: QSII), the BLDRS Emerging Markets 50 ADR Index (ADRE) The charts depict successive trading days from 9/21/05 to

Forecasting in Financial and Sports Gambling Markets: Adaptive Drift Modeling, By William S. Mallios

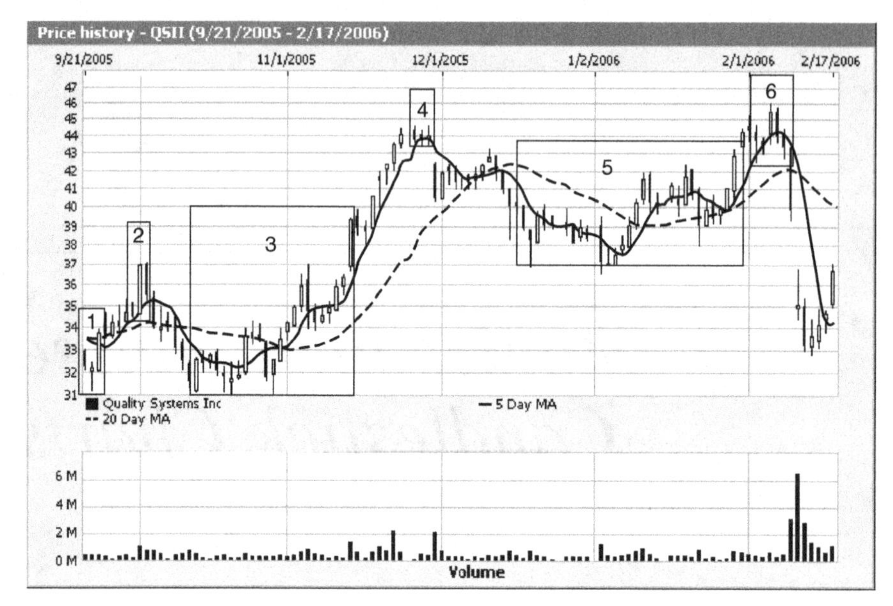

Figure 5.1.1 *Daily candlestick chart for Quality Systems (QSII) from 9/21/05 to 2/17/06.* (Source: *MSN Money*)

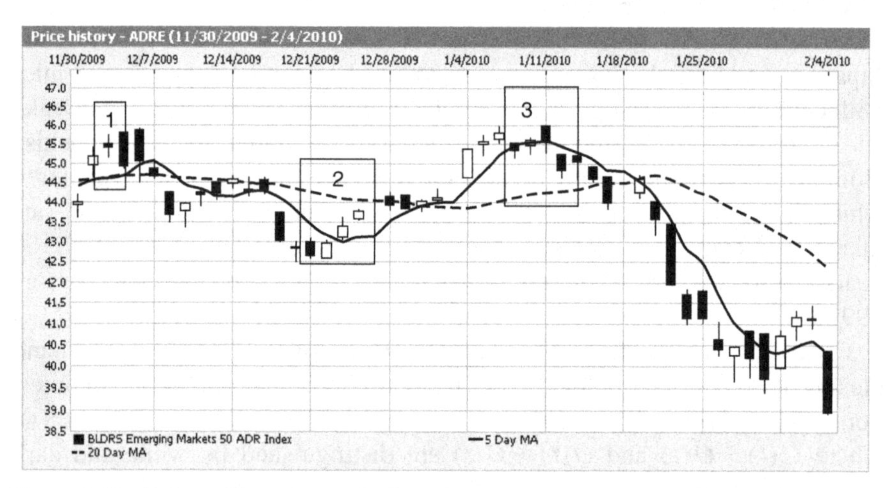

Figure 5.1.2 *Daily candlestick chart for BLDRS Emerging Markets 50 ADR Index (ADRE) from 11/30/2009 to 2/4/2010.* (Source: *MSN Money*)

2/17/06 for QSII and from 11/30/09 to 2/17/06 for ADRE. Also included are the 5- and 20-day moving averages of $C(t)$, denoted by $Cb5(t-1)$ and $Cb20(t-1)$, respectively [e.g., $Cb5(t-1)$ is the average of $C(t-1)$, $C(t-2), \ldots, C(t-5)$]. Daily trading volumes are depicted at the bottom of the chart and denoted by $V(t)$.

Candlestick charts provide a convenient method of graphing simultaneous (usually, cointegrated) time series (see Chapter 8) as well as price volatility. They may also aid in identifying periods of likely market efficiency and inefficiency as well as market irregularities. Moreover, in view of the ongoing pandemic of financial and mathematical illiteracy, the charts are effective means of teaching mathematics and finance, especially when augmented by mixed-market (financial and sports gambling markets) speculation and investment games for high school and undergraduate students (see Chapter 12).

Candlestick chartists use the charts to detect likely turning points (or relative maxima and minima) described by a series of peaks and troughs obscured by erratic disturbances. Such peaks and troughs are said to be identified by a variety of candlestick reversal patterns. Table 5.1.1 presents a condensed summary of selected patterns that are said to reflect likely turning points. Complete listings of such patterns may be found in the numerous Web sites on *candlestick forecasting*.

The bearish dark cloud cover patterns are said to be more prevalent than the bullish piercing line patterns. The reason may be related to a reknown Wall Street saying: *In with greed, out with fear. Although both are strong emotions, fear is more likely related to market volatility than greed. During market bottoms, traders or investors usually have the opportunity to wait for an opportunity to enter the market. They may bide their time and wait for the pullback or for the market to build a base, or to see how the market reacts to news. Fear is more prevalent at tops. Fear is saying "I want out—now"* (Nisson, 1991).

The daily candlestick chart in Figure 5.1.3 illustrates gaps in prices that may be due to earnings reports or dividend announcements. Box 1 contains a bullish engulfing pattern. Box 2 depicts the first gap and is a continuation of the bullish trend. Following a flat period, box 3 depicts a bullish breakaway pattern followed by the second gap.

Since the predictive validity of these candlestick patterns is certainly open to question, a simple method of assessing their predictive validity is through the tabulation of historical frequencies of subsequent, near-term price changes following each specific pattern. Unfortunately, selected empirical tabulations have not been convincing regarding forecasting claims. However, on a more positive note, selected patterns may be of value in identifying favorable near-term trading situations if such patterns are conditional on the following factors:

1. Locations of the specific pattern with respect to the 5-and 20-day moving average bands (particularly if specific configuration lies completely above or below the band)

TABLE 5.1.1 Overview of Selected Bullish and Bearish Candlestick Patterns

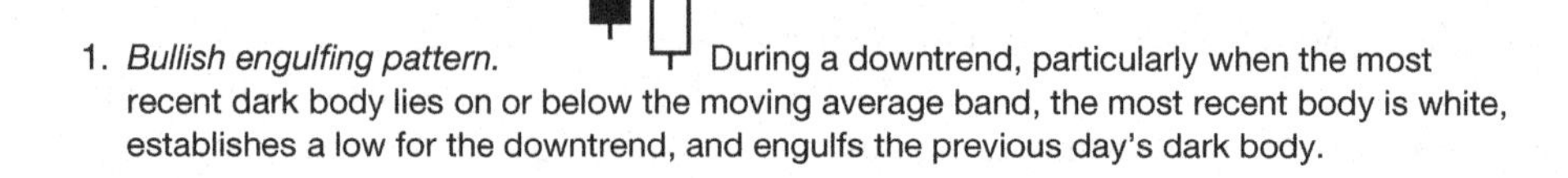

1. *Bullish engulfing pattern.* During a downtrend, particularly when the most recent dark body lies on or below the moving average band, the most recent body is white, establishes a low for the downtrend, and engulfs the previous day's dark body.

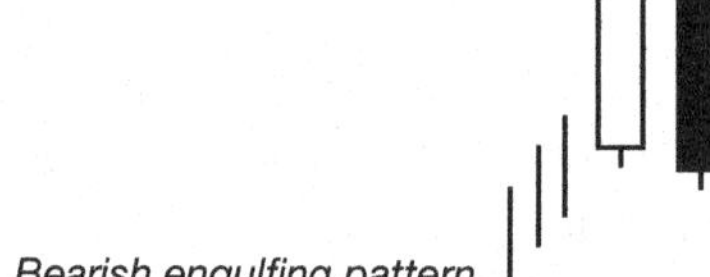

2. *Bearish engulfing pattern.* As the opposite of the bullish engulfing pattern, the most recent body is dark, establishes a high for the uptrend, and engulfs the previous day's white body.

3. *Bullish piercing pattern.* Following a downtrend, the most recent body is white, forms a new low, and closes above the midpoint of the previous body, which is dark.

4. *Bearish dark cloud cover.* As the opposite of the bullish piercing pattern, the most recent body is dark, forms a new high, and closes below the midpoint of the previous body, which is white.

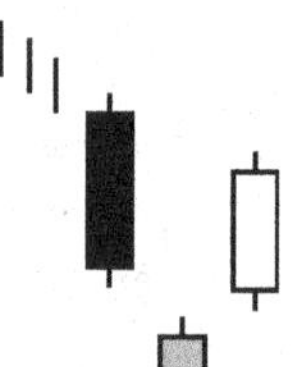

5. *Bullish morning star.* This reversal is identified by the three most recent bodies. There is a gap between the first two bodies, which continue the downtrend. The second body may be either white or dark. The most recent body is white and closes above the midpoint of the first body.

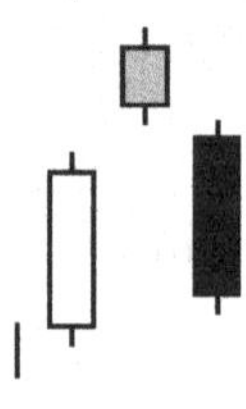

6. *Bearish evening star.* As the opposite of the bullish morning star, this reversal is again identified by the three most recent bodies. There is a gap between the first two bodies, which continue the uptrend. The second body may be either white or dark. The most recent body is dark and closes below the midpoint of the first body.

TABLE 5.1.1 (*Continued*)

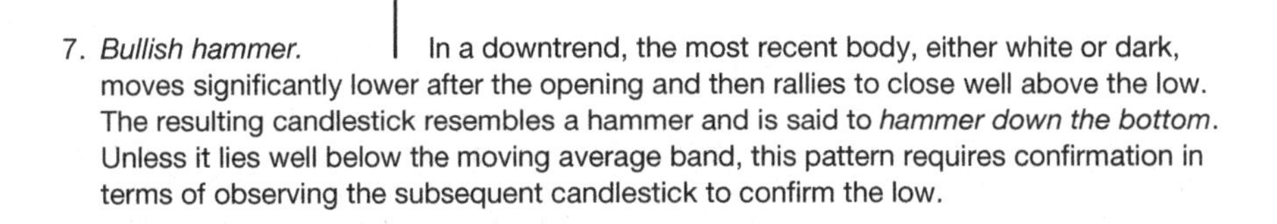

7. *Bullish hammer.* In a downtrend, the most recent body, either white or dark, moves significantly lower after the opening and then rallies to close well above the low. The resulting candlestick resembles a hammer and is said to *hammer down the bottom*. Unless it lies well below the moving average band, this pattern requires confirmation in terms of observing the subsequent candlestick to confirm the low.

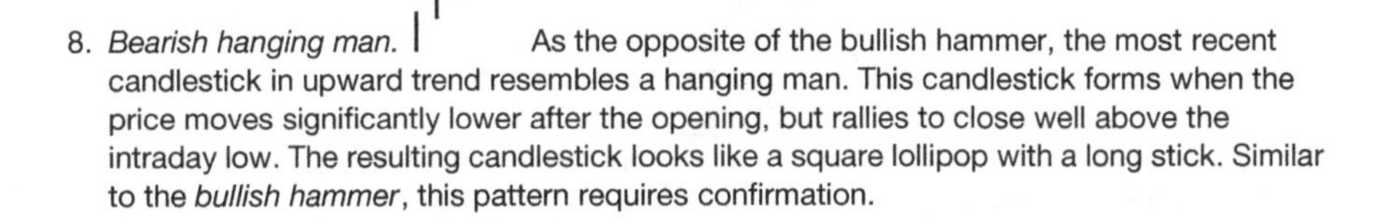

8. *Bearish hanging man.* As the opposite of the bullish hammer, the most recent candlestick in upward trend resembles a hanging man. This candlestick forms when the price moves significantly lower after the opening, but rallies to close well above the intraday low. The resulting candlestick looks like a square lollipop with a long stick. Similar to the *bullish hammer*, this pattern requires confirmation.

9. *Bullish harami pattern.* Harami translates as *pregnant*, where one candlestick is completely contained in the subsequent candlestick body. The body of the most recent candlestick may be either white or dark, but preferably white.

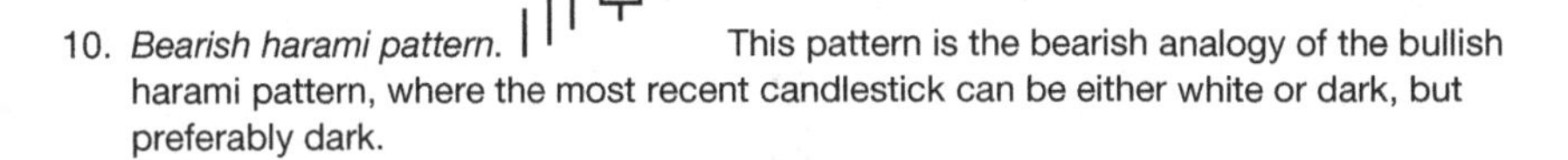

10. *Bearish harami pattern.* This pattern is the bearish analogy of the bullish harami pattern, where the most recent candlestick can be either white or dark, but preferably dark.

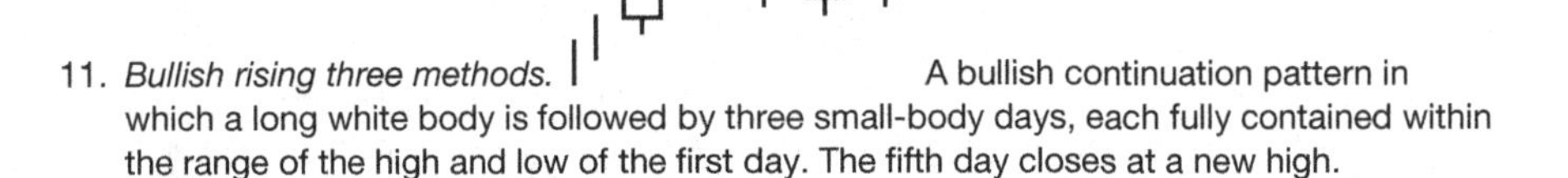

11. *Bullish rising three methods.* A bullish continuation pattern in which a long white body is followed by three small-body days, each fully contained within the range of the high and low of the first day. The fifth day closes at a new high.

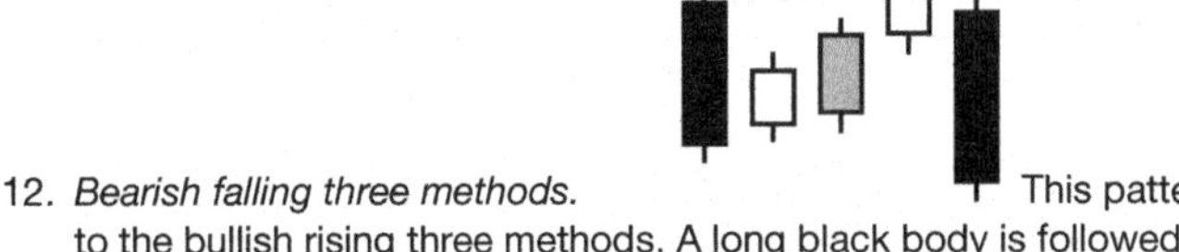

12. *Bearish falling three methods.* This pattern is the bearish analogy to the bullish rising three methods. A long black body is followed by three small-body days, each fully contained within the range of the high and low of the first day. The fifth day closes at a new low.

(*continued overleaf*)

TABLE 5.1.1 (*Continued*)

13. *Bullish three white soldiers.*  The second white body defines a bullish piercing pattern. The third and most recent candlestick adds to the likelihood of a bullish uptrend. This bullish reversal pattern consists of three consecutive long white bodies. Each should open within the previous body and the close should be near the high of the day.

14. *Bearish three black crows.* 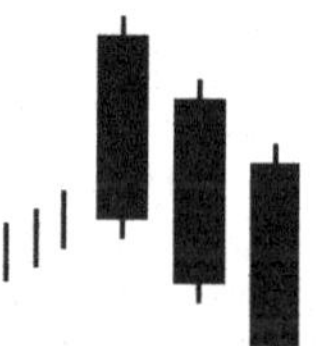 This pattern is the bearish analogy of the bullish three white soldiers. This reversal pattern consists of three consecutive long black bodies, where each day closes at or near its low and opens within the body of the previous day.

15. *Bullish stick sandwich.* 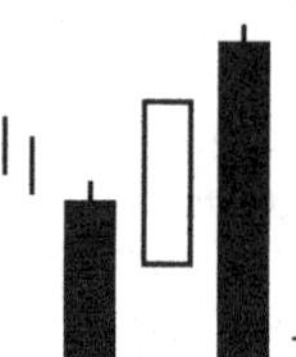 This bullish reversal pattern has two black bodies surrounding a white body. The closing prices of the two black bodies must be equal. A support price is apparent, and the opportunity for prices to reverse is said to be good.

16. *Bearish upside gap two crows.* 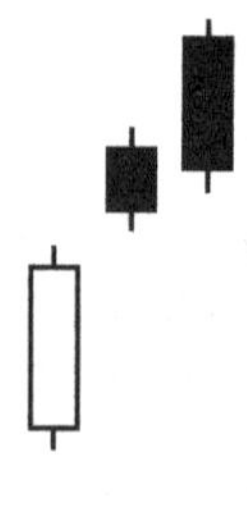 This three-day bearish pattern is said to occur only in an uptrend. The first day is a long white body followed by a gapped opening with the small black body remaining gapped above the first day. The third day is also a black day, whose body is larger than that on the second day and engulfs it. The close of the last day is still above the first long white day.

TABLE 5.1.1 (*Continued*)

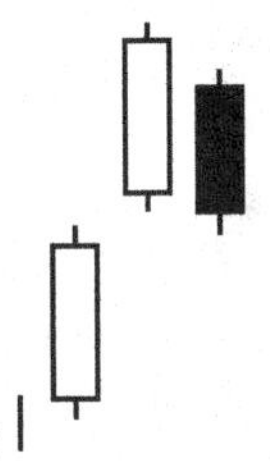

17. *Continued bullish upside tasuki gap.* A bullish continuation pattern with a long white body followed by another white body that has gapped above the first one. The third day is black and opens within the body of the second day, then closes in the gap between the first two days, but does not close the gap.

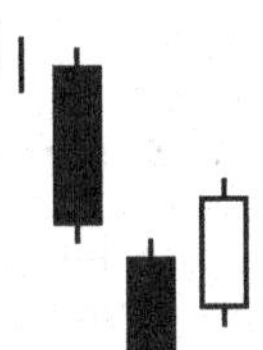

18. *Continued bearish downside tasuki gap.* A bearish continuation pattern with a long, black body, followed by another black body that has gapped below the first one. The third day is white and opens within the body of the second day, then closes in the gap between the first two days, but does not close the gap.

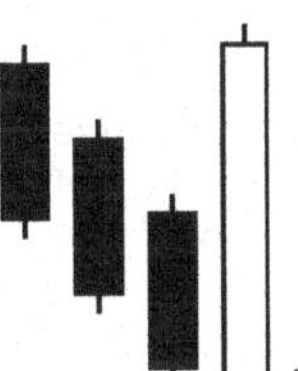

19. *Bearish three-line strike.* The most recent white body encompasses the three previous black bodies. An implication is that the strong rally during the most recent time period will lead to profit taking.

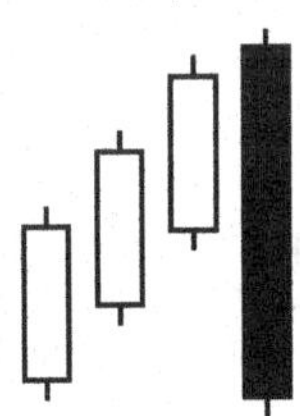

20. *Bullish three-line strike.* The most recent black body encompasses the three previous white bodies. An implication is that the sell-off provides a buying opportunity.

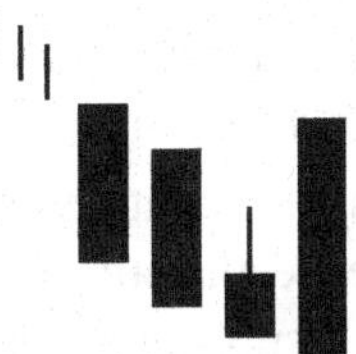

21. *Bullish concealing baby swallow.* The four most recent time periods are comprised of bearish candlesticks. The pattern is said to indicate a trend reversal because of the extreme bearishness of the trend.

(*continued overleaf*)

TABLE 5.1.1 (*Continued*)

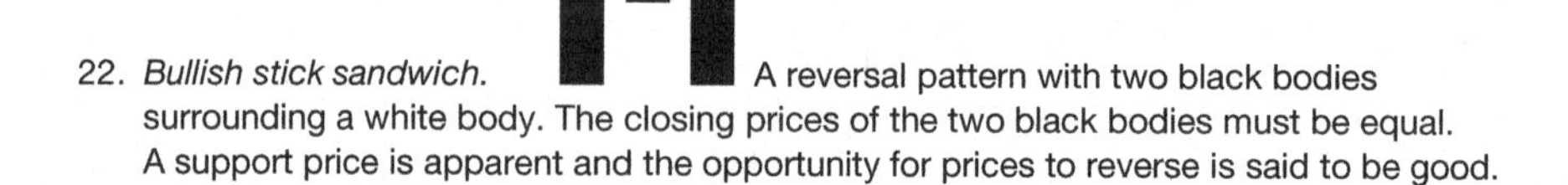

22. *Bullish stick sandwich.* A reversal pattern with two black bodies surrounding a white body. The closing prices of the two black bodies must be equal. A support price is apparent and the opportunity for prices to reverse is said to be good.

23. *Doji.* Doji form when a security's opening and closing are virtually equal. The length of the upper and lower shadows can vary, and the resulting candlestick looks like either a cross, an inverted cross, or a plus sign. Doji convey a sense of indecision or a tug-of-war between buyers and sellers. Prices move above and below the opening level during the session, but close at or near the opening level.

24. *Bullish inverted hammer pattern.* In a downtrend, the most recent candlestick is either white or dark, lies beneath the close of the previous dark body, and has a wick that is at least twice the length of the body.

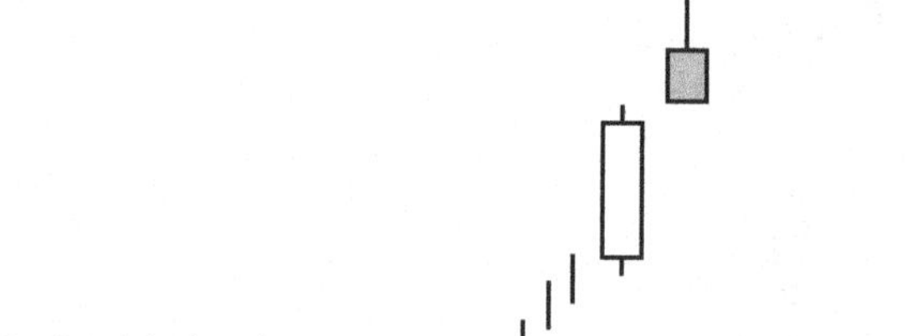

25. *Bearish shooting-star pattern.* This pattern is the bearish counterpart to the bullish inverted hammer pattern; that is, in an uptrend, the most recent candlestick is either white or dark, lies above the close of the previous white body, and has a wick that is at least twice the length of the body.

2. Distances between the two moving price averages (particularly if such distances become relatively large)
3. Elapsed time since the most recent crossover between the two moving price averages (in the sense that the longer the time lapse, the more likely the occurrence of a subsequent near-term crossover)
4. The number of relative maxima or minima that occur between successive crossovers of the moving average bands

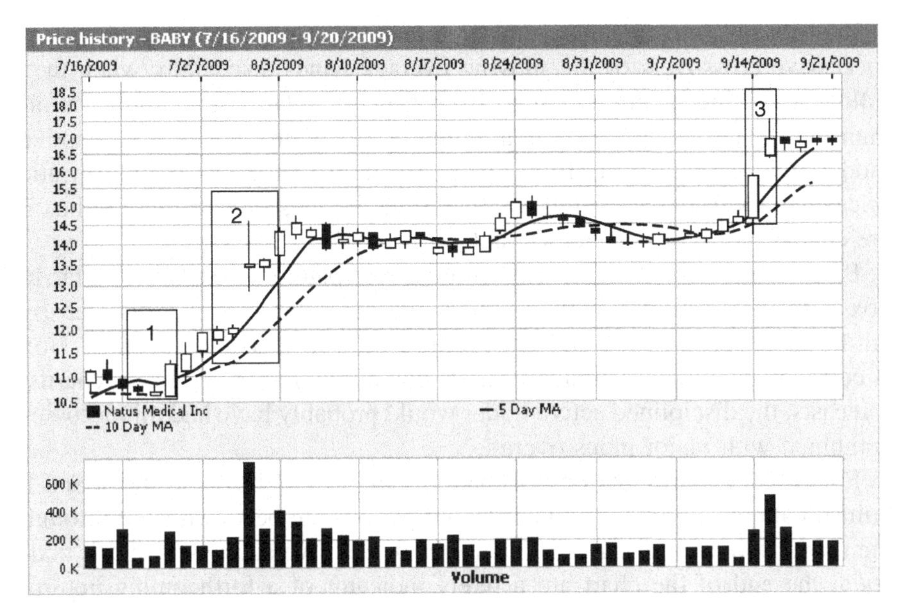

Figure 5.1.3 Daily candlestick chart for Natus Medical Inc. (BABY) from 7/16/09 to 9/21/09. (Source: MSN Money)

5. Volume trends that accompany the specific patterns
6. Relationships between near-term straight-line, quadratic, and cubic trends for each of the seven individual time series presented in the candlestick charts

These factors will be shown to be critical to effective forecasting during periods of market inefficiency.

Figures 5.1.1 and 5.1.2 illustrate several of the patterns in Table 5.1.1. For Figure 5.1.1, note that a long-term purchase of QSII stock in mid-September 2005 (the beginning of the chart) at approximately $32/share would have resulted in a gain of just over $4/share by mid-February 2006 (the end of the chart). In contrast, effective active trading (alternating between near-term long and short positions) may have led to a gain of 75% or more in the initial investment. Box 1 in Figure 5.1.1 contains a bullish hammer pattern that is confirmed by the following day's white body. A bearish dark cloud pattern appears in box 2. Box 3 begins with a bullish engulfing pattern and ends with a bullish breakaway pattern. Although the overall trend is bullish, several bearish patterns occur between the latter bullish patterns. From the beginning of box 3 through 11/1/05, three relative minima are observed before the 5-day moving average band crosses above the 20-day moving average band.

It is not uncommon for two to three relative extremes to occur between successive crossovers of the moving average bands, especially when they follow a significant price change from a major relative maximum (minimum) to the next major relative minimum (maximum). For box 3, adaptive modeling is particularly useful for purposes of discerning, if possible, erratic price changes from the overall positive trend. Periods such as those in box 3 are characteristic of both losses and gains for the active trader.

Box 4 contains a bearish dark cloud cover pattern. Box 5 is similar to box 3 in that it begins with a bullish morning star pattern, followed by erratic price movements, and ends with a bullish breakaway pattern. Box 6 correctly identifies a major bearish downturn. Based on these chartist forecasts, the disciplined active trader would probably have had minor losses combined with major gains overall.

Figure 5.1.2 illustrates bearish patterns in each of boxes 1 and 3 and a bullish pattern in box 2. The intervening periods are best evaluated through the model forecasts discussed in Chapter 9. The five successive black bodies at the end of the chart are a likely indicator of a forthcoming upturn, especially if the volumes do not support a continued downturn and if the distance between the moving averages continues to increase.

5.2 BLACK MONDAY

From August 1982 to the peak in August 1987, the Dow Jones Industrial Average (DJIA) increased from 776 to 2711. The increase in market indices for the 19 largest markets in the world averaged 296% during this period. The average number of shares traded on the New York Stock Exchange had risen from 65 million to 181 million.

October 19, 1987, a date known as Black Monday, marked the culmination of a market decline that started five days earlier. On Black Monday, the DJIA plummeted 508 points, losing 22.6% of its value in one day. The crash was the greatest single-day loss that Wall Street had ever suffered in continuous trading up to that point. Between the start of trading on October 14 and the close on October 19, the DJIA lost 760 points, a decline of over 31%. The crash was a worldwide phenomenon, with all major world markets declined substantially. Aside from herd behavior and worries of stock market overvaluation, program trading, derivatives, a worsening U.S. trade deficit, and a falling U.S. dollar were said to be factors contributing to the crash.

The daily candlestick chart for the period in question is presented in Figure 5.2.1. Prominent bearish configurations appeared prior to the crash, beginning with a bearish engulfing body on the day following 10/5, followed

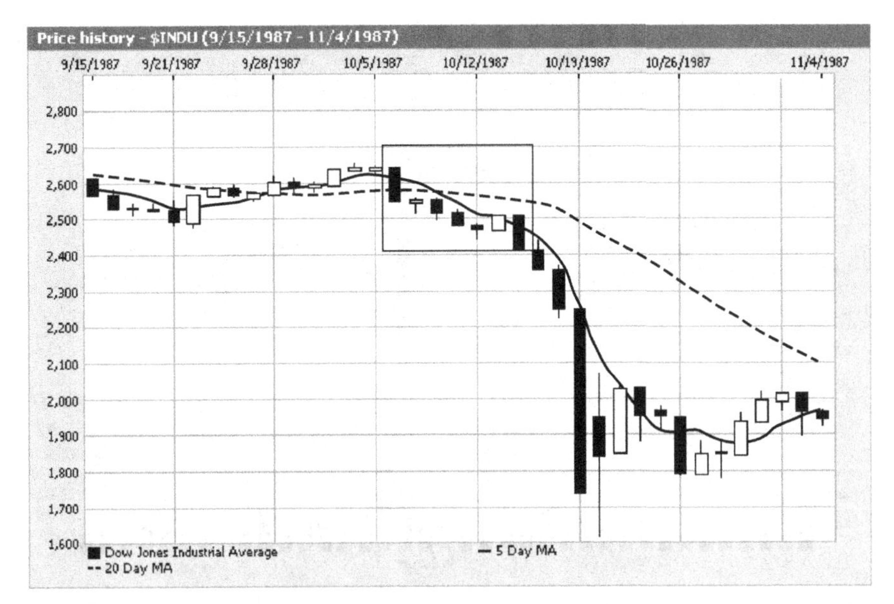

Figure 5.2.1 *Daily candlestick chart of Dow Jones Industrial Average during the period of the October 1987 crash.* (Source: *MSN Money*)

by a second bearish engulfing body two days following 10/12; see the box. The crash occurred three days later.

In the two-year period following Black Monday, the markets had fully recovered. However, nearly two years later, on 10/16/89, there was a similar crash; see Figure 5.2.2. Similar to Figure 5.2.1, bearish patterns were formed the day following 10/9 (see the box). These bearish patterns followed six successive white bodies (all above or nearly above the 5-day moving average) and a 10-day stretch in which nine of the 10 candlestick bodies were white and led to a gain in the Dow from approximately 2670 to 2790. Such gains are conducive to profit taking and shorting.

The fourth worse drop in the Dow occurred on December 1, 2008; see Figure 5.2.3. A bullish engulfing pattern occurred one day prior to November 24, 2008. However, the crash that occurred six days later came without the forewarnings that preceded the crashes depicted in Figures 5.2.1 and 5.2.2. The crash on December 1 was preceded by five white bodies that led to a 600-point short-term gain in the Dow; see the box. However, there was no obvious pattern that foretold the downturn on 12/1/08.

The Crash in Figure 5.2.3 is examined in terms the intraday, 5-minute candlestick chart for 12/1/08, as presented in Figure 5.2.4. It is seen that the Dow dropped over 200 points in the first 5 minutes of trading (see box 1) and then stabilized around 8400 until shortly after 3 p.m., at which time the

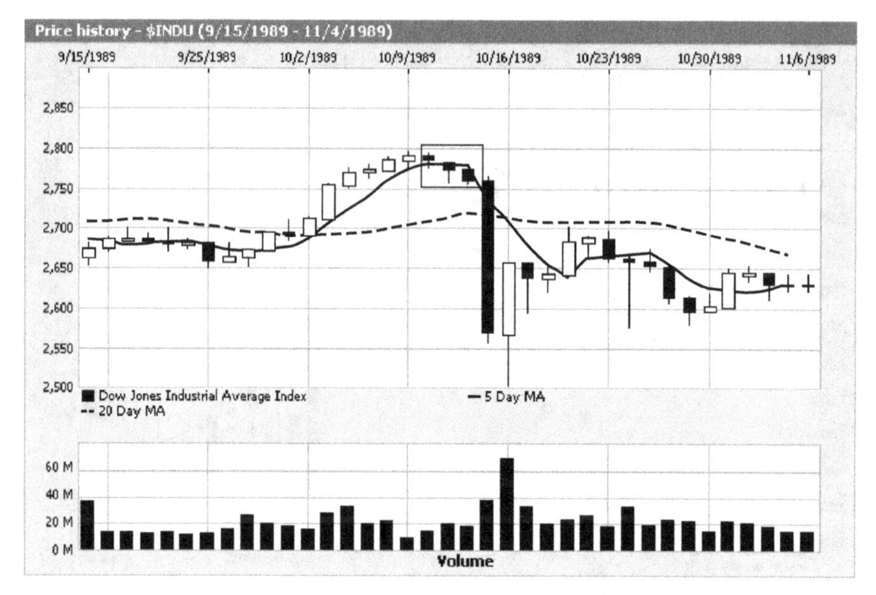

Figure 5.2.2 *Daily candlestick chart of Dow Jones Industrial Average during the period of the October 1989 crash.*

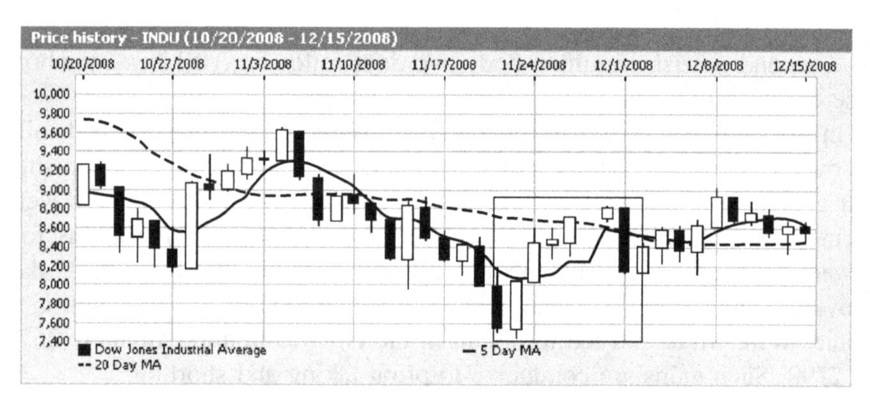

Figure 5.2.3 *Daily candlestick chart of Dow Jones Industrial Average during the fourth worse drop that occurred on 12/1/08.* (Source: *MSN Money*)

Dow dropped another 200 points (see box 2). The latter drop was preceded by a bearish pattern.

The 1929 stock market crash is said to have occurred on Thursday, October 24 (*Black Thursday*) and the following Tuesday, October 29 (*Black Tuesday*). After reaching a record high of 381.2 on 9/3/29, the DJIA fell to 199 by 11/13/29. Today's observer might find the drop in DJIA rather trivial during the 1929 crash. However, an adjustment of the record high of 381.2

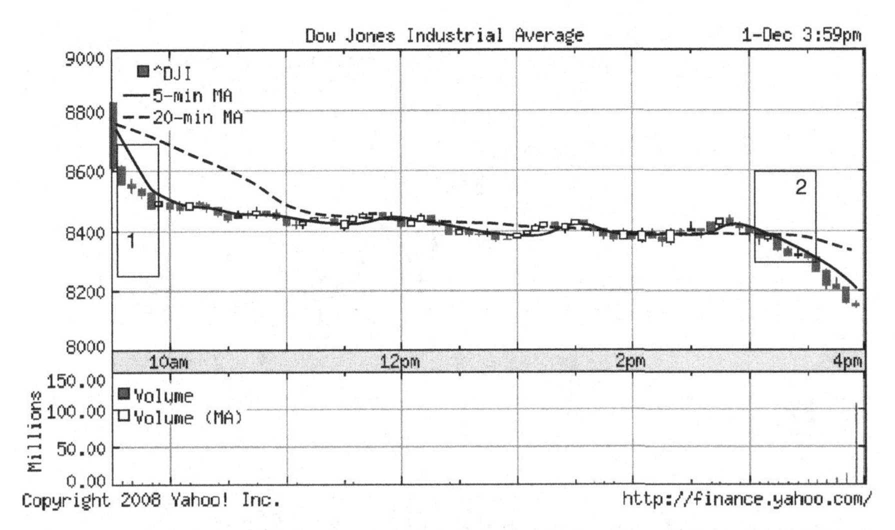

Figure 5.2.4 Intraday, five-minute candlestick chart for the Dow Jones Index on 12/1/08, the day of the fourth worse drop.

for nominal gross domestic product per head (a proxy for average wages) results in an astounding high of 21,000. It has become almost conventional wisdom today that the 2000 peak of 11,700 on the Dow was a record level of overvaluation, based on the twin measures of the 10-year rolling inflation-adjusted price–earnings ratio and the q ratio of price to book value (Jackson, 1/11/10). By the time the crash was completed in 1932, stocks had lost nearly 90% of their value.

The accompanying daily candlestick chart in Figure 5.2.5 shows two bearish breakaway patterns, the first of which occurred on 10/16/29; see the box. Following the low on 11/13/29, a bullish piercing configuration was followed by a brief rally.

In arguing that the volume of reserves in the banking system would fuel inflation, the Federal Reserve Board—with the concurrence of Treasury Secretary Henry Morgenthau, Jr., many financial experts, and leading academics—doubled commercial bank holding requirements in three stages from 8/36 to 5/37. The banks responded by calling in loans to build a liquidity cushion above legal requirements, thereby sharply contracting money, credit, and economic activity. The FED action, acknowledged as one of the gravest policy errors of the Depression era, reversed the earlier recovery and led to the 1937 recession. Even if the banks had used their excess reserves to expand credit and the money supply, unchecked inflation was not possible due to the high unemployment rates in 1936 and 1937. There was complete economic recovery only after the start of World War II.

Figure 5.2.5 *Daily candlestick chart of DJIA during the period of the market crash of October–November 1929.* (Source: *MSN Money*)

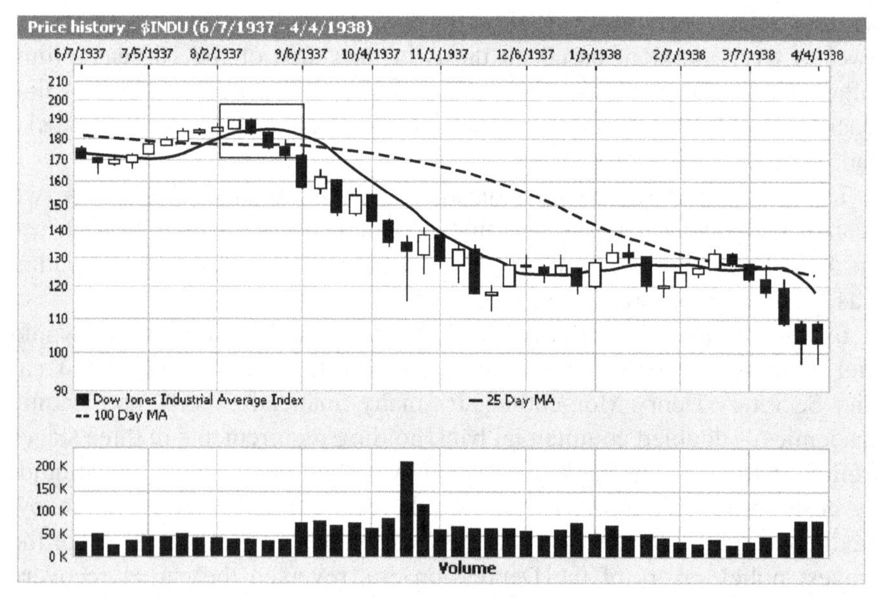

Figure 5.2.6 *Weekly candlestick chart for the weekly Dow Jones Industrial Average Index from 6/7/37 to 4/4/38.* (Source: *MSN Money*)

As indicated in Figure 5.2.6, the policy actions by the 1936 Fed had a lagged effect that led to a major drop in INDU. The downturn was flagged by the negative engulfing pattern the second week following 8/2/37; see the box. For those traders astute in shorting, this period provided major profit-making opportunities.

The apparent end of the 2008–2009 recession has pitted interest-rate doves—those who advocate keeping interest rates low for an extended period of time—against interest-rate hawks—those who advocate meaningful rate increases to prevent inflation from gaining a foothold. *The US New Deal recovery was turned into a mini-depression in 1937–1938 because of premature action to tighten money and balance the US budget. There is now more danger of economic stimuli across the world being reversed too soon than of their being continued too long* (Brittan, 7/24/09). The doves also point to Japan's experience in the 1990s regarding raising rates too early in the recovery. The hawks, motivated by fears of inflationary periods that date back to the Weimar Republic, maintain that the government's stimulus spending has made the likelihood of inflation all the more certain.

5.3 A MATTER OF ALLEGED INSIDER TRADING

For several publicized cases of alleged or proven insider trading, we present accompanying candlestick charts to explore whether such trading coincided with bullish or bearish patterns. If so, the unanswered question is whether the specific pattern may have been a partial reflection of insider trading.

Case 5.3.1: The Martha Stewart case ImClone Systems Incorporated, a biopharmaceutical company dedicated to developing biological medicines in the area of oncology, accepted a $6.5 billion acquisition offer from Eli Lilly and became a fully owned subsidiary in 2008. ImClone's stock price dropped sharply at the end of 2001 when its drug Erbitux, an experimental monoclonal antibody, failed to get the expected U.S. Food and Drug Administration (FDA) approval. It was later revealed by the Securities and Exchange Commission (SEC) that prior to the announcement (after the close of trading on December 28, 2001) of the FDA's decision, numerous executives sold their stock.

ImClone's founder, S. Waksal, was arrested in 2002 on insider trading charges for informing friends and family to sell their stock and attempting to sell his own. His daughter sold $2.5 million in shares on December 27. His father sold $8.1 million in shares over December 27 and 28. Company executives followed suit. ImClone's general counsel sold $2.5 million in shares on December 6, and the vice president for marketing and sales sold

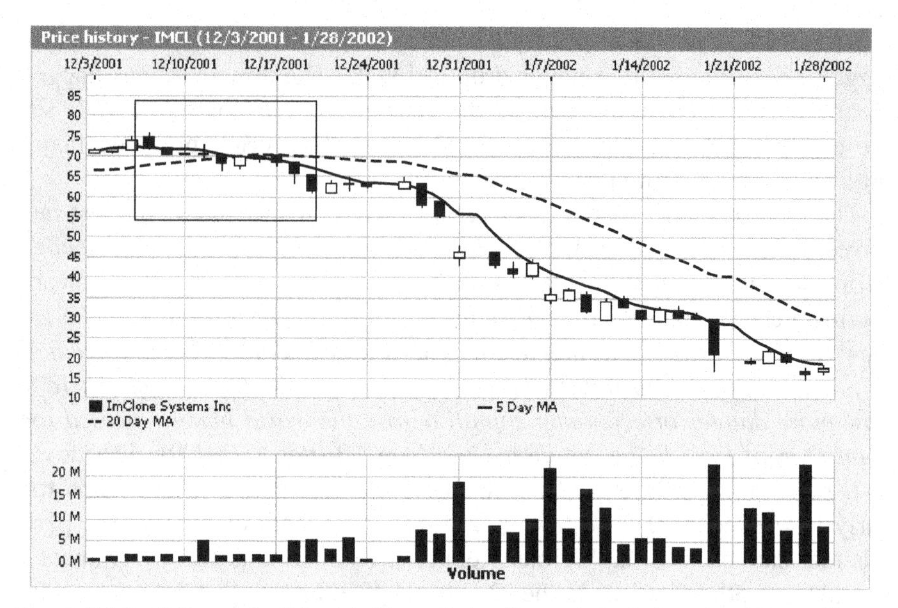

Figure 5.3.1 *Daily candlestick of ImClone during the period of alleged insider trading on 12/27/01.* (Source: *MSN Money)*

$2.1 million in shares on December 11. Four other executives sold shares in the following weeks as well. Later, founder Waksal pleaded guilty to various charges, including securities fraud, and on June 10, 2003 was sentenced to seven years and three months in prison.

Martha Stewart became embroiled in the scandal after it emerged that she sold about $230,000 in ImClone shares on December 27, just a day before the announcement of the FDA decision. Although Stewart maintained her innocence, she was found guilty and sentenced on July 16, 2004 to five months in prison, five months of home confinement, and two years of probation for lying about a stock sale, conspiracy, and obstruction of justice.

Figure 5.3.1 presents a daily candlestick chart during the period in question. Throughout December, bearish configurations dominate; see the box. A bearish dark cloud cover appears on December 6 (the day the ImClone general counsel sold his stock) followed by a bearish engulfing pattern on December 17 and a bearish breakaway from the norm on December 27 (the day that Stewart and others sold their stock).

Case 5.3.2: The Mark Cuban case On 11/17/08, the SEC accussed Mark Cuban, owner of the Dallas Mavericks, of insider trading. It is alleged that on June 28, 2004, Cuban sold his entire holdings in the Internet company Mamma.com (now Copernic, Inc.) within four hours of being told

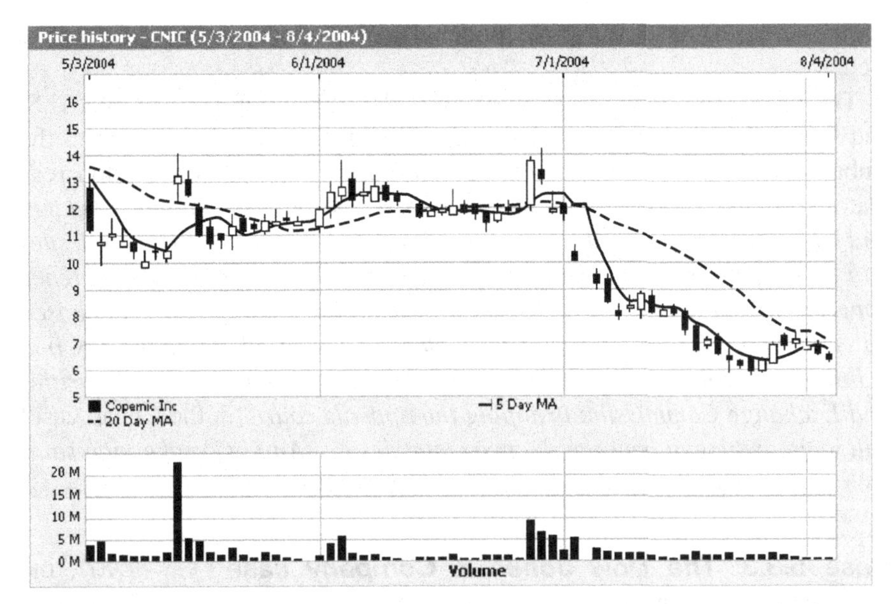

Figure 5.3.2 Daily candlestick chart for Copernic, Inc. (CNIC) during the period of alleged insider trading on June 28, 2004. (Source: MSN Money)

confidential information about an impending stock offering. According to the SEC, Cuban knew that the shares would be sold below the current market price and avoided losses in excess of $750,000 by selling 600,000 shares prior to the public offering announcement.

Figure 5.3.2 present the daily candlestick chart for Copernic, Inc. (stock symbol: CNIC) during the period of alleged insider trading. On the day in question, June 28, 2004, the candlestick displays a bullish, breakaway white body which is followed by a tombstone pattern; see the box. If Cuban sold on June 28, he, in fact, sold at an opportune time. If he were to have sold the stock on the following days through 7/1/24, it would have been under the sign of bearish patterns.

On 7/17/09, U.S. District Judge S. Fitzwater ruled that the SEC could not hold Cuban liable for insider trading because the agency did not allege that Cuban had agreed not to trade based on confidential information about CNIC. (*Note:* An *insider* is someone who has a position in a business or stock brokerage which makes him or her privy to confidential information—such as future changes in management, upcoming profit and loss reports, secret sales figures, and merger negotiations—which will affect the value of stocks or bonds. Although there is nothing wrong with being an insider, use of the confidential information unavailable to the investing public in order to profit through sale or purchase of stocks or bonds is unethical and a crime under the 1934 Securities and Exchange Act.[1]

The failed case against Cuban exposes the division in the way that U.S. and UK authorities enforce rules against insider trading. In contrast to the Cuban ruling, the Financial Services Authority, the City of London regulator, *brought several cases successfully in the past year against traders and their brokers for selling shares after learning of coming new issues that could depress the price of their holdings. . . . The essential difference between the two countries is that the US only bars trading when it is based on confidential information that is . . . stolen or misused, while the UK ban is broader, involving market-moving private information. . . . The Securities and Exchange Commission is hoping the appeals court* [in the Cuban case] *will narrow the gap between the two countries. . . . An SEC spokesman said: "We believe the district court erred"* (Masters, 10/12/09).

Case 5.3.3: The Dow Jones & Company case On 5/7/07, the SEC filed a lawsuit against a Hong Kong couple, Kan King Wong and Charlotte Ka Leung, for allegedly using insider information to make $8.2 million profit from trading 415,000 shares of Dow Jones & Company (DJ) through a joint Merrill Lynch (ML) account during the two weeks prior to Rupert Murdock's News Corp.'s takeover bid becoming public on May 1. A surge in volume in DJ June and September call options just to the public announcement suggested to some ML analysts that some investors knew that an offer was in the works.

Contacted by ML about the trading activity, the SEC said in its complaint that the *highly profitable and highly suspicious* trading was unlike any previous trading in which the couple had been involved. A federal judge granted the SEC request to freeze the account's assets. The SEC also asked that the couple forfeit all profits *realized from the unlawful trading alleged*, but did not say how the Leungs obtained advance notice of the News Corp. bid. However, David Li, chairman of Bank of East Asia and a former director of Dow Jones & Co., reached a tentative agreement with the SEC to pay $8 million to settle the suit. It was disclosed that Leung's

[1]In the 1920s, insider trading was rampant. Such trading was, at the time, legal, and Joseph Kennedy, father of President John F. Kennedy, was a master at the game. *It's easy to make money in this market*, said Kennedy to an associate. *We'd better get in before they pass a law against it*. By the mid-1920s, he had accumulated a net worth of $2 million—which, for those times, was considerable. Kennedy then became a key financial contributor to Franklin Roosevelt's 1932 presidential campaign. As a reward, Kennedy was appointed the inaugural chairman of the SEC. When Roosevelt was asked why he had tapped such a crook, he replied, *Takes one to catch one* (Legal dictionary.thefreedictionary.com, 2009).

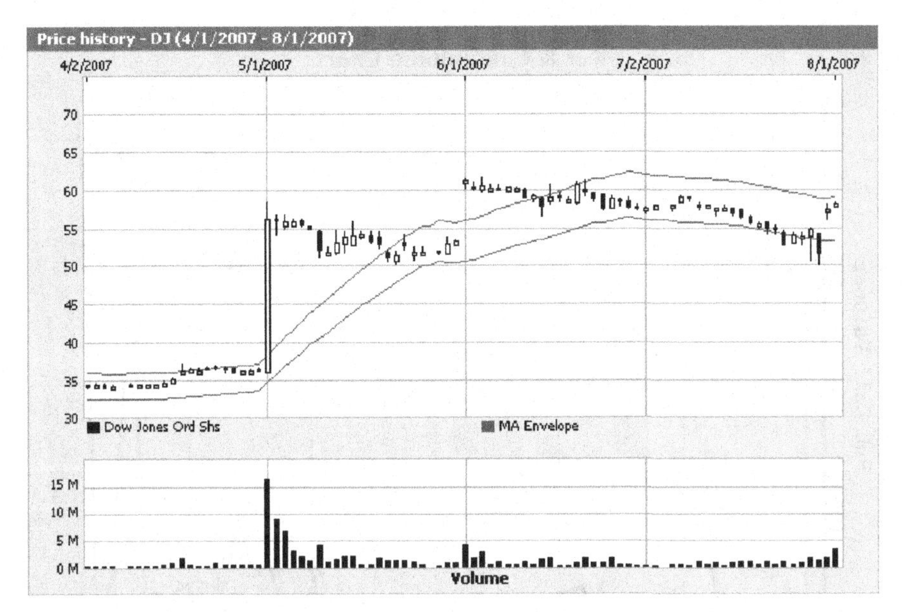

Figure 5.3.3 Daily candlestick chart for Dow Jones (DJ) during its takeover by Rupert Murdock's News Corp. The takeover bid became public on May 1, 2007, at which time the share price of DJ rose 58%. (Source: *MSN Money*)

father, Michael Leung, had close business and social ties with Li over the course of two investigations.

The daily candlestick chart in Figure 5.3.3, depicting DJ price and volume movements during the period of the 5/1 public announcement, shows no evidence of bullish indicators just prior to the announcement; see the box. (Dow Jones share prices surged by 58% when the bid was announced.) However, analyses by Tyler (5/3/07) on the DJ surge in September call options (see Figure 5.3.4) provide evidence that some investors knew that the Murdock offer was forthcoming.

The two suspicious option spikes depicted in Figure 5.3.4 occurred on 4/25 and 4/30. The first involved 3000 September 40 calls, which closed the day at approximately $0.80, and the second involved more that 3400 September 45 calls, which closed the day at approximately $0.30. More than 600 June 40 calls also traded on 4//30 and also closed at approximately $0.30. *Related to the activity just provided, one might say that that's pretty suspect and obviously somebody or some persons knew something big was about to happen. However, the observation here is that while the buying will likely turn up as the illegal variety, in all honesty, it wasn't enough to set off technical and optionable alerts worthy of providing a "sure thing" investment for other strategists* (Tyler, 5/3/07).

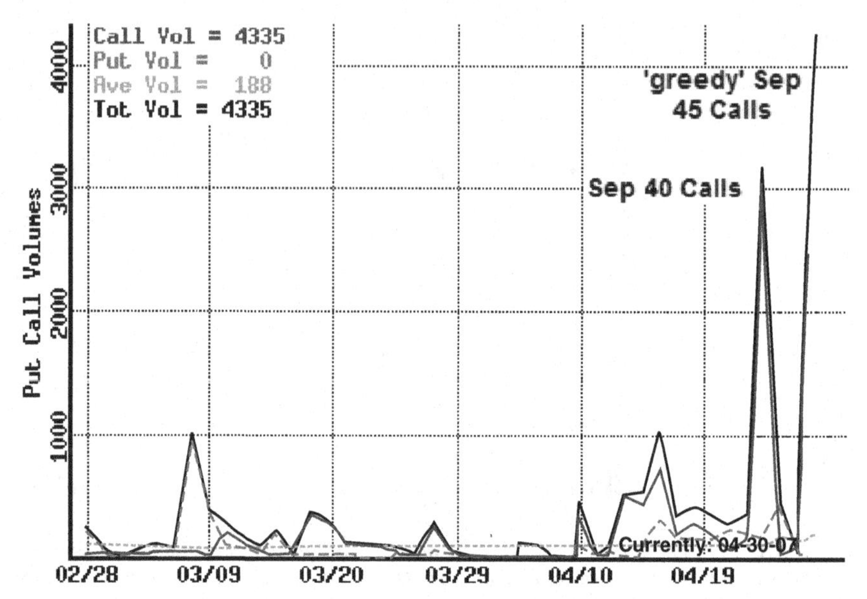

Figure 5.3.4 *Dow Jones options activity just prior to the Dow Jones takeover by Rupert Murdock's News Corp. (*Source: *Tyler, 5/3/07)*

Case 5.3.4: The EADS case On October 8, 2008, *France's stock market regulator* [known by its French acronym, AMF] *found itself yet again at the center of controversy after findings from its two-year investigation into controversial share dealings at EADS* [the parent company of Airbus] *were posted on the Internet.... AMF is alleging that 17 EADS executives, and their two core shareholders, had access to privileged information when they sold shares in a period stretching from November 2005 to April 2006* (Hollinger, 4/9/08). The AMF investigation was triggered when top management and shareholders carried out suspicious share trades in the time before EADS revealed severe delays in production of the Airbus A380 superjumbo jet in June 2006. Word of those delays caused shares in EADS to drop by a quarter at the time.

Figure 5.3.5 presents monthly candlestick chart for EADS from 8/24/04 to 11/24/08. During November 2005 (see box 1), several EADS and Airbus directors sold their EADS shares. During March 2006 (see box 2), EADS share prices reached a maximum as more executives either sold shares or exercised put options. Box 2 displays a pronounced bearish piercing pattern, which would have led active chartist traders to cover long positions and/or initiate short positions. On June 13 (box 3), share prices dropped 26%

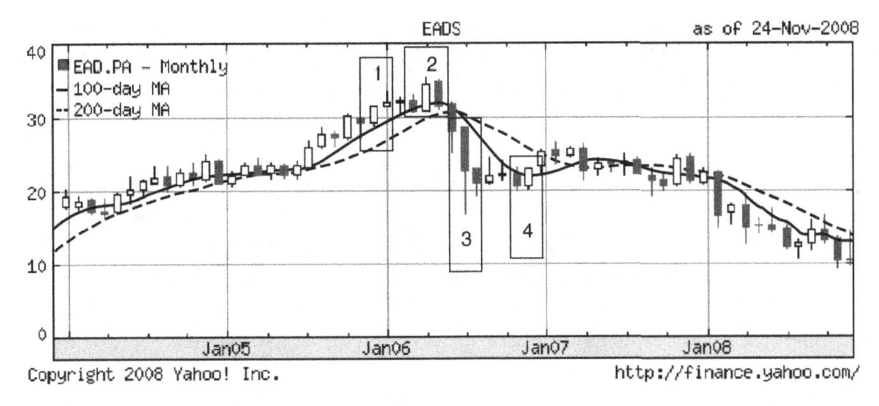

Figure 5.3.5 *Monthly candlestick chart for EADS from January 2004 through November 2008.*

when Airbus announced a second six-month delay in A380 deliveries. On November 24, 2006 (box 4), French authorities launched an investigation into possible insider trading.

In July 2009, a provisional report prepared for the French financial markets regulator by an independent examiner recommended $17.4 million in fines against seven current and former Airbus executives and $1 million in fines against EADS for alleged market information delays. The report was not binding, but the examiners' views were taken into account by the sanctions committee, which was to make its decision by the end of 2009 (Daneshku and Done, 7/29/07).

Case 5.3.5: The Galleon hedge fund case On 10/16/09, billionaire investor Raj Rajaratnam, founder of the Galleon hedge fund, and present and former executives of Bear Sterns, IBM, Intel, and McKinsey were charged with an insider trading scheme that prosecutors called *the biggest ever involving hedge funds.... The case marked the first time court-authorized wire taps—a traditional tool of investigators pursuing mob bosses and drug kingpins—had been used in a significant insider trading case*. Rajaratnam and others are alleged to have used insider information to trade ahead of earnings announcements, acquisitions, and joint venture deals involving companies such as Google, Hilton, and IBM.

Rajaratnam allegedly made a $4 million profit in July 2007 by buying 400,000 shares in Hilton after receiving nonpublic information about the impending purchase of the hotel chain by Blackstone, a private equity firm. Two weeks following the Hilton trade, Rajaratnam allegedly made $500,000 by trading Google stock based on inside information that Google would announce disappointing quarterly earnings (Chung, 10/17/09).

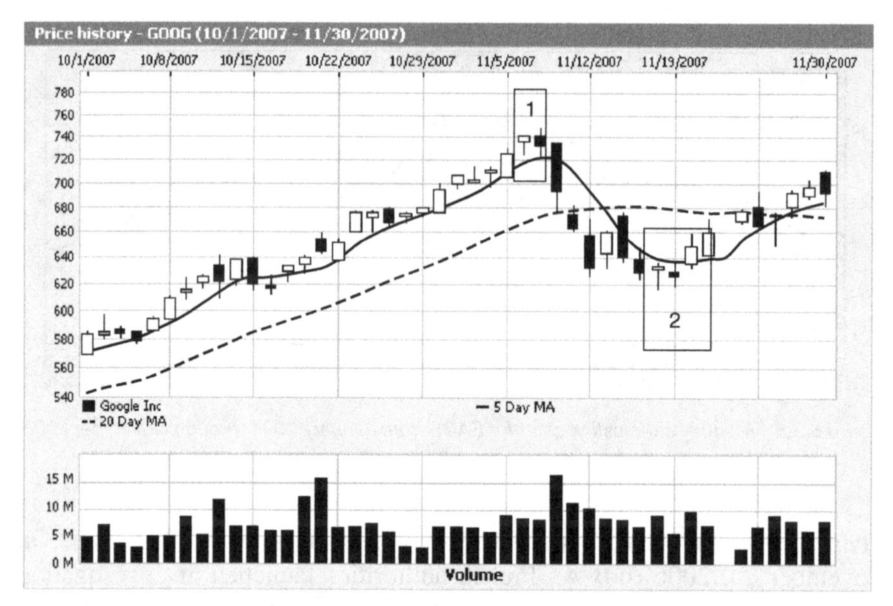

Figure 5.3.6 *Daily candlestick chart: insider trading and the decline of Google (GOOG) share prices following its disappointing earnings report on 11/7/07. (Source: MSN Money)*

Figure 5.3.6 presents a daily candlestick chart for Google (GOOG) during the alleged insider trading prior to the announcement of the disappointing third-quarter earnings reports. Note that in the five-week period following 10//1/07, Google share prices increased from $580 to over $740 with very little profit taking. The negative engulfing pattern on 11/7/07 is preceded by eight consecutive white bodies and an excessive period of time during which the width of the moving average band is large—all of which portend profit taking; see box 1. The announcement of disappointing earnings occurred following the close of business on 11/07/07. The share price then dropped to $630 before a cover short and buy pattern resulted. In this particular situation, the actions of an astute active trader would probably have been very similar to those of insider traders.

There were many other revelations in the Galleon case. Advanced Micro Devices (AMD) spun off its manufacturing arm in October 2008 into a new chipmaker company, Globalfoundries, with an investment of up to $8.4 billion from Abu Dhabi. Government legal filings state an unnamed ADM executive passed on nonpublic information about the transaction to Danielle Chiesi, a hedge fund manager who was one of the six people arrested in the Galleon case. Chiesi was told that it was 99% certain that the spin-off would take place before AMD's third-quarter earnings were announced on October 2008. *You know, we're gonna shock the hell out*

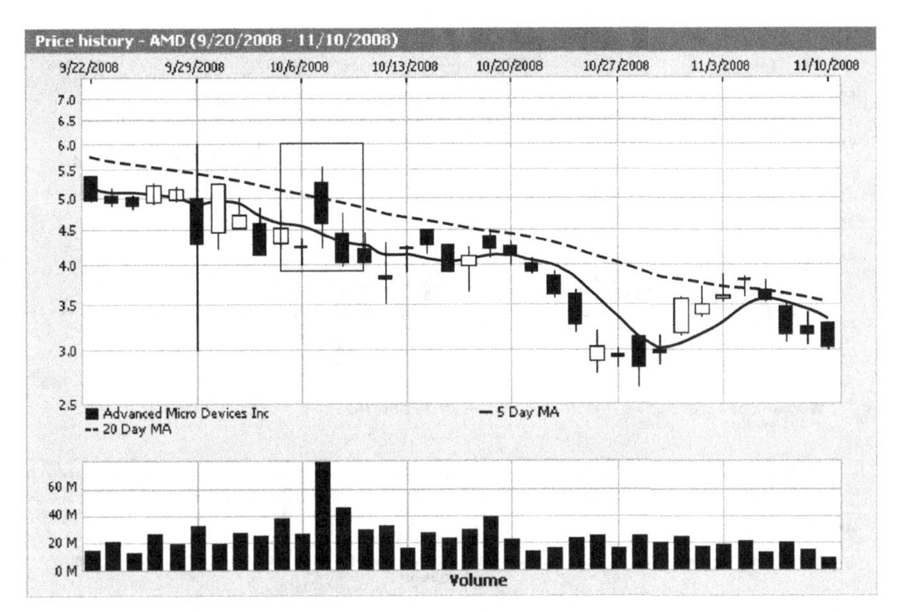

Figure 5.3.7 Daily candlestick chart for AMD during announcement of the spin-off of Globalfoundries on 10/7/09. (Source: MSN Money)

of everybody, the ADM executive said. The formation of Globalfoundries was announced on 10/8/08 and ADM's share prices rose more than 8.5%: from $4.23/share on 10/6/08 to $4.59/share the following day (Nuttall, 11/3/09).

Figure 5.3.7 presents a daily AMD candlestick chart for the late August–early November period. The candlestick pattern through 10/6 provides no forewarning of the price increase on 10/7, which was accompanied by a massive volume of nearly 80 million shares. A buy order during the late trading hours of 10/6 coupled with a sell order in the early trading hours of 10/7 would have resulted in a one-days gain of nearly 25%; see the box.

Shock waves from the Galleon case continued as criminal charges were brought against 14 people for alleged participation in insider trading schemes. An attorney with the law firm of Ropes & Gray, Arthur Citillo, was one of those charged. He allegedly misappropriated price-sensitive, nonpublic information from his firm, as they advised hedge funds in acquisitions, such as the acquisition of 3Com by Baines Capital Partners (which was announced on 9/28/07).

Figure 5.3.8 presents a daily candlestick chart for 3Com that encompasses the public announcement of the acquisition on 9/28/07. From 9/27 to 9/28, the closing price increased 37%, from $3.60 to $4.94. From a chartist's

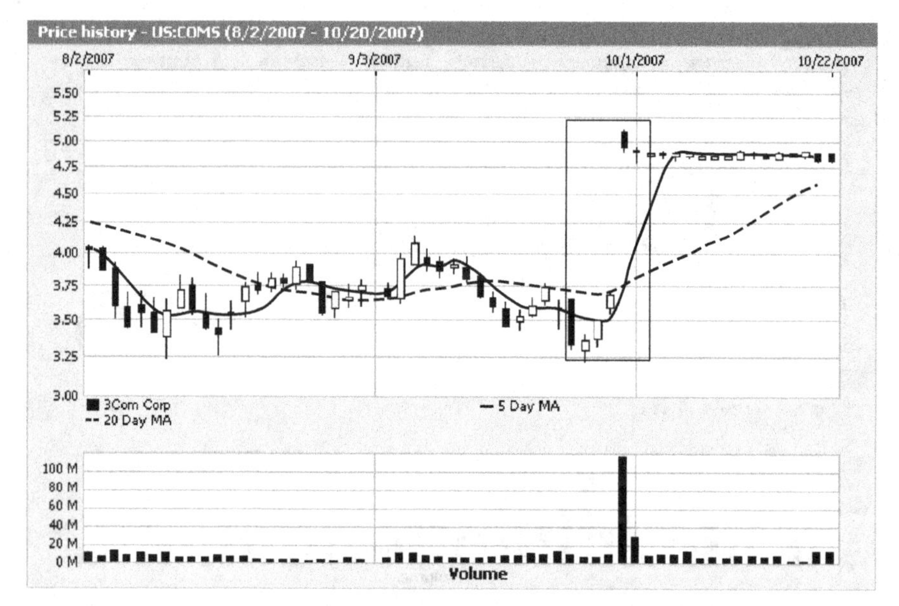

Figure 5.3.8 Daily candlestick chart for 3Com. The acquisition of 3Com by Bain Capital Partners was announced on 9/28/07. (Source: *MSN Money*)

perspective, the 9/28 announcement is preceded by the bullish candlestick pattern of three successive, nonoverlapping white bodies; see the box. Note also that the first of these white bodies forms a relative minimum that lies beneath the 20-day moving average.

The 3Com saga continued. On 11/11/09, Hewlett-Packard agreed to buy 3Com for $2.7 billion in cash. *Call options gained as much as 315% following the announced take-over. More than 8,000 3Com calls changed hands on 11/11/09, 17 times the four-week moving average. The most active were contracts conveying the right to purchase 3Com for $5/share through 11/20/09. "Somebody knew something was coming," said Stefen Choy, founder of Livevol Inc., a provider of options market data and analytics. "It looks like very unusual call buying. I see this very frequently when there's a take-over"* (Kearns, 11/12/09).

Figure 5.3.9 presents the daily 3Com candlestick chart for the eight-week period prior to the 11/12/09 takeover announcement. Note the bullish engulfing pattern the day following 11/2/09 and a bullish breakaway pattern (with an increased volume) the trading day prior to 11/11/09; see the box. These bullish patterns appear innocuous in the presence of insider trading—in contrast to the surge in the call volume on 11/11/09.

According to the court filings disclosed on 2/2/20, David Slaine, a former employee of the hedge fund Chelsey Capital, pleaded guilt to conspiracy and securities fraud charges in the continuing Galleon case. *The*

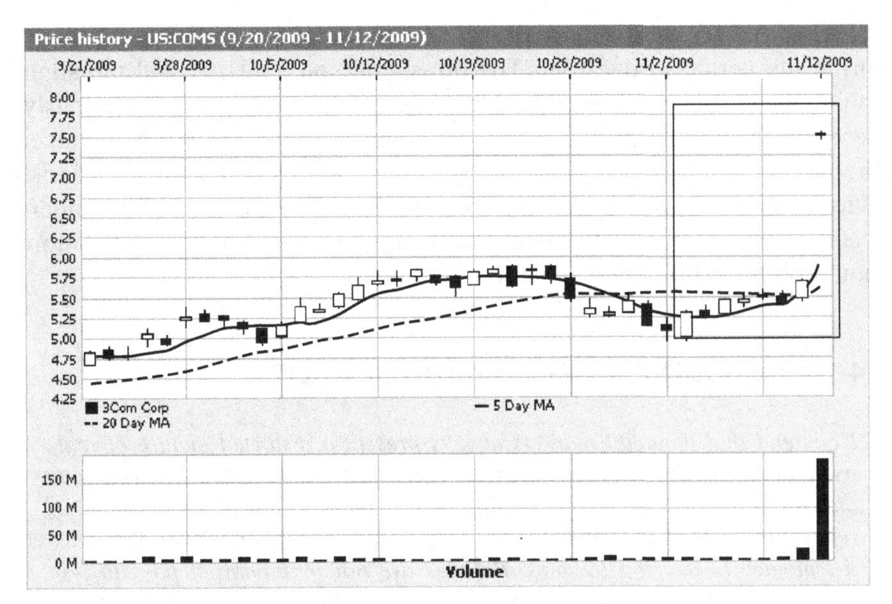

Figure 5.3.9 Daily candlestick chart for 3Com prior to the 11/12/09 takeover of 3Com by Hewlett-Packard. (Source: MSN Money)

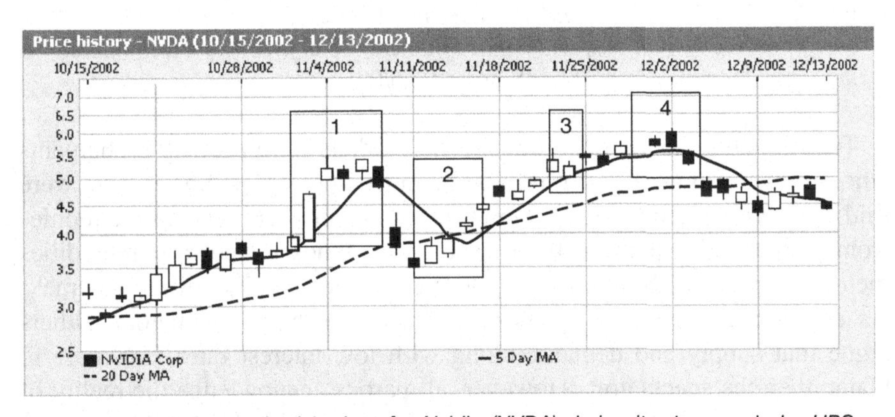

Figure 5.3.10 Daily candlestick chart for Nvidia (NVDA) during its downgrade by UBS on 11/8/02. (Source: MSN Money)

specific instances to which he pleaded guilty occurred in 2002, when Mr. Slaine allegedly received information from two insiders at UBC indicating that [on or about 11/21/02] *its research department would downgrade its ratings of Nvidia, a technology stock.... Slaine said he persuaded Chelsy Capital to sell short about 200,000 shares in the stock, while he shorted 75,000 shares. When the research was published, and Nvidia's shares fell, Chelsey and Mr. Slaine reaped profits of* $69,000 *and* $33,000 *respectively* (Farrell, 2/3/10).

Figure 5.3.10 presents a daily candlestick charts for Nvidia (NVDA) during the period in question. The downgrade on 11/21/02 and the slight drop in the closing price the following day are identified in box 3. Curiously, these insider trading profits are minor compared with the profits obtained through active chartist trading during this period. Box 1 shows a bearish pattern that is followed by a sharp drop in prices. Thereafter a bullish pattern occurs in box 2. Box 4 shows another bearish pattern where long positions should be traded for short positions.

5.4 COMMODITY BUBBLES AND VOLATILITY

I contend that financial markets always present a distorted picture of reality. The mispricing of financial assets can affect the so-called fundamentals that the price of those assets is supposed to reflect. That is the principle of reflexivity. Instead of a tendency towards equilibrium, financial markets have a tendency to develop bubbles. Bubbles are not irrational: it pays to join the crowd, at least for a while. So regulators cannot count on the market to correct its excesses.... The crash of 2008 was caused by the collapse of a super-bubble that was growing since 1980. This was composed of smaller bubbles. Each time a financial crisis occurred the authorities intervened, took care of failing institutions, and applied monetary and fiscal stimulus, inflating the super-bubble even further. (Soros, 10/26/09)

Traditionally, the commodities futures markets were used by participants who produce commodities or rely on them to do business. Speculators were said to provide price liquidity. However, in recent years speculative monies from hedge funds, pension funds, and index funds linked to commodities increased from $13 billion in 2003 to $260 billion in 2008. Many lawmakers now blame the price bubbling increases on these speculators. Others argue that supply and demand, along with low interest rates, are more to blame than the speculators. However, all parties acknowledge the reality of increased demand for commodities from emerging markets such as China and India.

Figures 5.4.1–5.4.5 present weekly candlesticks charts for gold (per ounce), sugar (per pound), the Australian dollar (U.S. dollar/Australian dollar) and monthly charts for crude oil (per barrel) and corn (per bushel). The charts (which include a single four-period moving average) are intended to illustrate the price volatility and bubbles rather than the predictive validity of reversal patterns—which, for Figures 5.4.1–5.4.5, is often hit and miss. The risks associated with price volatility led researchers to model the expected variability associated with price changes through GARCH-type modeling, introduced in Section 4.4.

Figure 5.4.1 *Weekly candlestick chart for Comex Gold (GC, Globex) from April 2007 to November 2009. Prices are in terms of U.S. dollars per ounce of gold. (Source: Tradingchart.com Inc.)*

Weekly prices for gold in Figure 5.4.1 ranged from under $700/ ounce in October 2008 to a high of over $1200/ ounce a year later. Assorted prognosticators have the price of gold extending above $2000/ ounce in the 2010–2011 period. Combined with periods of high volatility, bullish and bearish reversals appear to be indicated when prices deviate by more than $100/ ounce from the four-week moving average, such as for candlesticks within the four boxes.

For crude oil prices in Figure 5.4.2, the price went from under $20/barrel in late 2001 to a high of $147/barrel in mid-2008 and then crashed to a low of $32/barrel in late 2008. The bullish pattern in box 1 precedes the extreme price volatility in box 2. The latter contains a bearish pattern that precedes the collapse of the bubble. The question in box 2 is the extent to which the price volatility could be modeled effectively.

For the sugar prices in Figure 5.4.3, the peak of the bubble was yet to form. There is an initial bearish pattern in 2008 (see box 1) and at least three bullish patterns thereafter (see boxes 2 to 4). By late January 2010, *the crisis over a scarcity of sugar deepened after Indonesia, one of the world's leading importers, was unable to buy a single pound of the sweetener in its latest tender*. The setback sent the cost of raw sugar in New York to

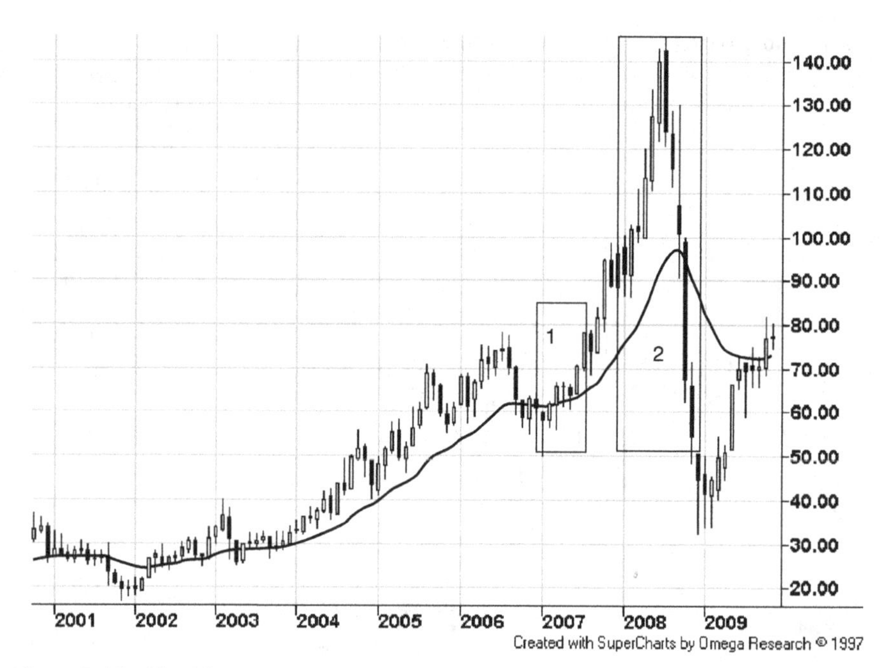

Figure 5.4.2 *Monthly candlestick chart for NYMEX Crude Oil (Light) (CL, Globex) from October 2000 through December 2009. Prices are in terms of U.S. dollars per barrel of oil; one barrel contains 40 gallons.* (Source: *Tradingchart.com Inc.)*

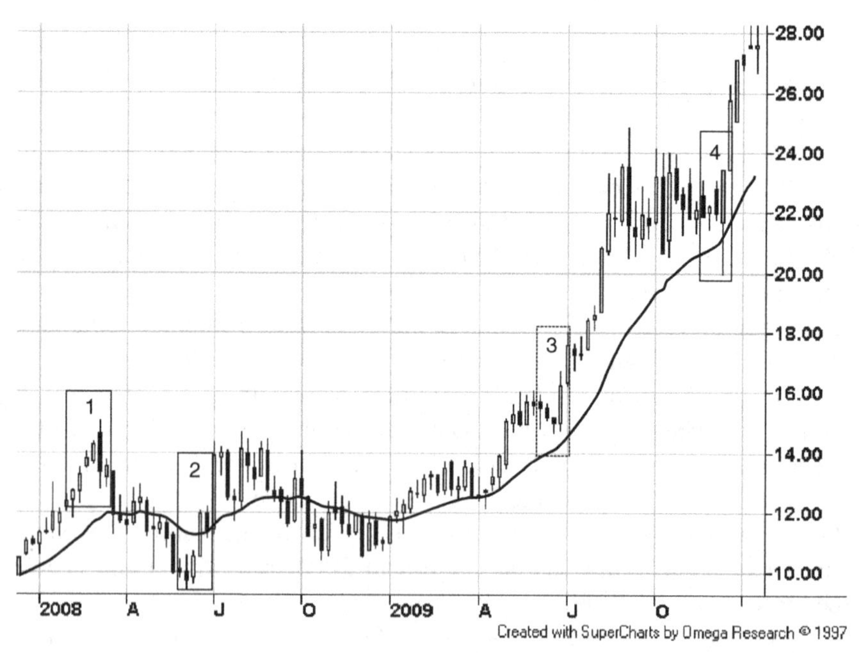

Figure 5.4.3 *Weekly candlestick chart for sugar (SB, ICE, NYBOT) from October 2008 through December 2009. Prices are in terms of cents per pound.* (Source: *Tradingchart.com Inc.)*

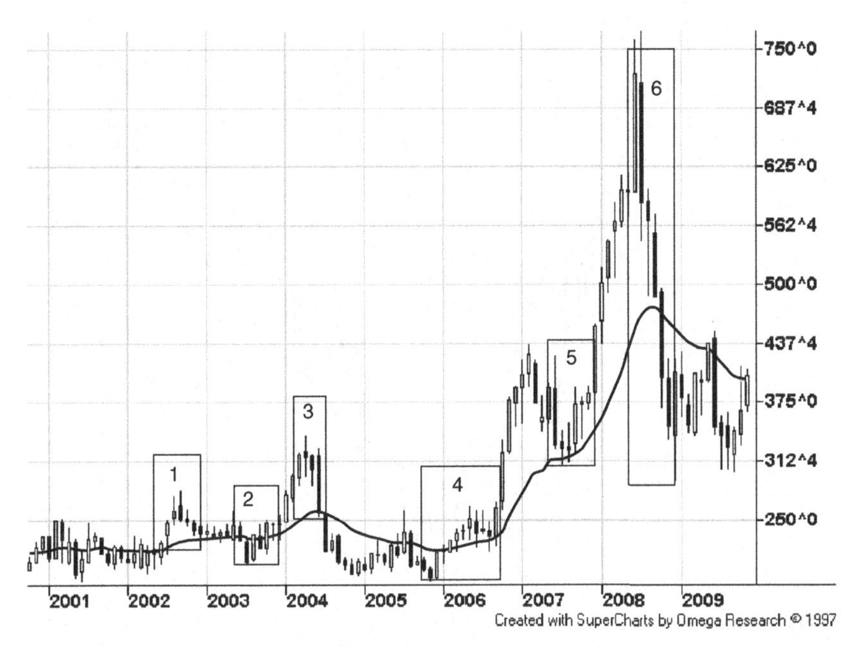

Figure 5.4.4 *Monthly candlestick chart for Corn Mini (YC, CBOT) from October 2008 through December 2009. Prices are in terms of price per bushel times 100.* (Source: *Tradingchart.com Inc.)*

a 29-year high of $0.29/pound. *Although sugar is no longer a key food commodity in developed countries, it is still a crucial source of calories in emerging countries, making its price a political issue. The sugar crisis has been caused by a large supply deficit due to disappointing crops in Brazil and India, the world's top producer, due to bad weather* (Blas, 1/22/10).

Regarding the price of a bushel of corn, Figure 5.4.4 depicts the most recent bubble and the excessive price volatility in the bubble's final formation and deflation; see box 6. Prior to the formation of the bubble, there are alternating bearish patterns (see boxes 1 and 3 and the beginning of box 5) followed by bullish patterns; see boxes 2 and 4 and the end of box 5. The bubble in corn prices has been attributed to a number of factors, including hedge fund speculation, bad weather conditions in corn-growing states such as Iowa, the demand for ethanol, and rising meat consumption in the developing world.

Figure 5.4.5 shows the severe decline in the value of the U.S. dollar ($US) relative to the Australian dollar ($A) during 2009. A bearish pattern occurred prior to the sharp decline in the U.S. dollar in August 2008; see box 1. This pattern was followed by excessive volatility in the final stages of the decline; see box 2. The second relative minimum reveals a bullish indicator; see box 3.

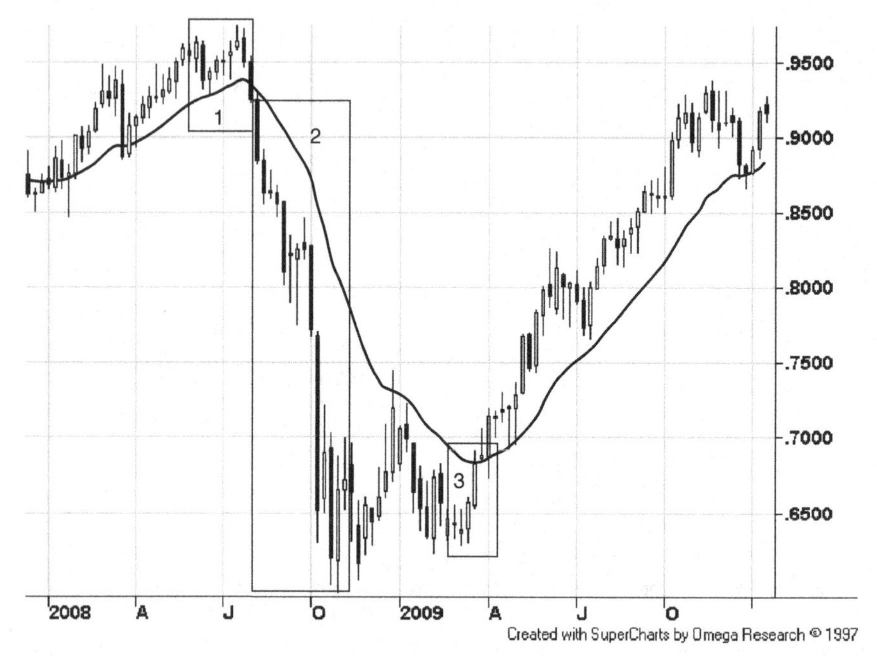

Figure 5.4.5 *Weekly candlestick chart for the CME Australian dollar (A6, Globex) from mid-December 2007 through mid-January 2010. Prices are in terms of U.S. dollar/Australian dollar. (Source: Tradingchart.com Inc.)*

Australian commentator Kenneth Davidson gave the following three reasons for the rise of the Australian dollar (as reported by Keen, 10/27/09):

1. *The government bailout funds in the U.S. and U.K. have cashed up financial institutions that don't want to lend any more to mortgages (and have long ago forgotten how to lend to fund productive enterprises), so they're looking for short term hot money gains.*
2. *The Reserve Bank of Australia's flagging that it intends raising rates from 2–3% above rates in the U.S. to possibly 4–5% above is a "sure thing" return on a currency that was already appreciating because of commodity sales to China.*
3. *This gives the hedge funds a sure fire double whammy gain: borrow in the U.S. and U.K. at 1% to buy $A and "invest" in floating rate bonds or shares (particularly in banks) and get a higher return (at least 2% better than the borrowing costs, and assured to rise); and drive up the $A in the process, so that when you sell out, you get both a higher return and an appreciated currency in which it is denominated.*

The most remarkable thing about this bubble is that the RBA's "we're raising rates now and we're going to keep on doing it for a few months" *messages are part of the cause, and yet they seem unaware of both the phenomenon and the dangers it poses. Davidson points out that Brazil, which is experiencing a similar commodity-driven currency appreciation bubble, is aware of the dangers and is doing something about it: "The rising value of the Brazilian real and the Australian dollar against the US dollar has had a disastrous impact on both countries' non-commodity export and import competing industries. Brazil's popular and largely economically successful left-wing government led by President Lula da Silva is meeting the problem head on. It has decided to impose a 2 per cent tax on all capital inflows to stop the real appreciating further." Of course, like any speculative bubble, this has an end-game—and that's when you think the rate rises have come to an end, sell out and watch the $A crash for those who are still holding it. Then the dollar (and Australian bank shares) will crash, and our economy will have acted as a dollar pump for the hedge funds.*

5.5 SHORT SELLING

To capitalize on the expectation that the price of a stock will decline, one borrows shares of the stock (usually from a broker) and then sells the stock—which is termed *short selling*. If the price declines, the borrower buys them back at a reduced price and returns them to the broker at a profit. If the price increases, the borrower loses money. The following cases pinpoint bearish configurations that are inductive to short selling.

Case 5.5.1: Citrix Systems Shares of Citrix Systems (CTXS), a company that develops virtualization software that allows a single computer to act like many "virtual" computers, soared 153% in 1999, after having gains of 92% and 95% the previous two years. CTXS reached a high of $122/share in early March 2000 and then collapsed as the high-tech bubble burst.

Figure 5.5.1 presents the CTXS daily candlestick chart for the May–June 2000 period. At the beginning of June, eight of 11 security analysts covering Citrix Systems had issued strong buy recommendations on the stock. In contrast, the Dallas-based hedge fund Maverick Capital Eight bet against CTXS and shorted 600,000 shares at $58/share the day following 6/5, a day defining a bearish engulfing body; see the box. Within three days CTXS fell to $41/share, at which time the Capital Eight short position jumped to 1.6 million shares. With the announcement of a bad second- quarter earnings report, CTXS shares collapsed to $22 on 6/12. At this point, Capital Eight covered their short positions, for a return of $38 million in less than a week.

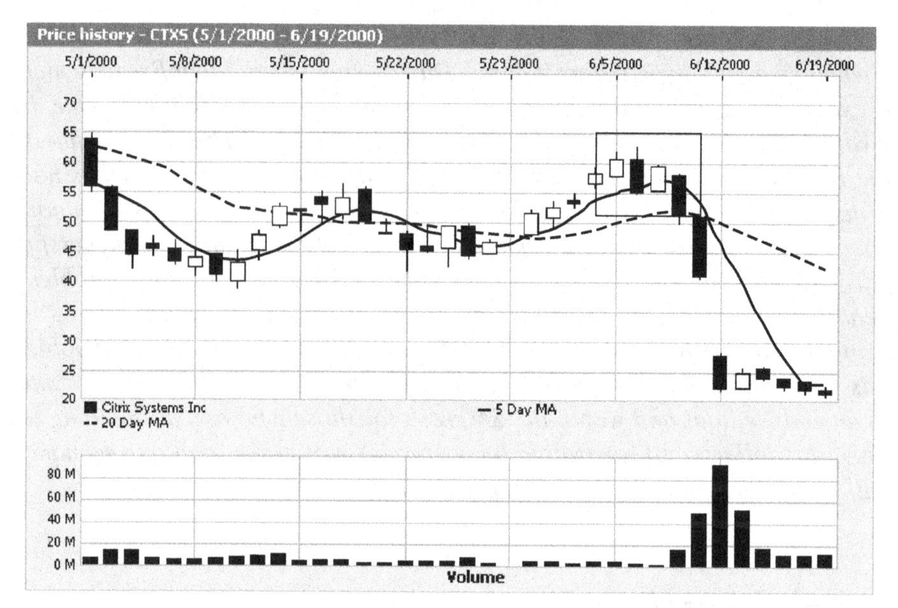

Figure 5.5.1 Daily candlestick chart for Citrix Systems (CTXS) during the May–June period of 2000. (Source: MSN Money)

Case 5.5.2: Online gambling businesses The background for this case is discussed in Section 2.4. In anticipation of the Frist amendment to the Safe Port Act (an amendment that banned online gambling), the Kynikos Associates hedge fund shorted shares of the online gambling businesses Sportingbet (SPBTF) and World Gaming (WGMGY) just prior to passage of the Act on 10/2/06. Daily candlestick charts for SPBTF and WGMGY are presented in Figures 5.5.2 and 5.5.3. There are no bearish patterns that precede the drop in share prices on 10/2/06. On that day, share prices of SPBTF and WGMGY plunged by 58% and 76%, respectively. The Kynikos Associates' short selling resulted in a financial killing.

Case 5.5.3: The pyrotechnics of Volkswagen share prices Figure 5.5.4 presents a daily candlestick chart for Volkswagen (DE:VOW) during the takeover of VW by Porsche. Volkswagen's shares rose 147% on October 27, 2008, after Porsche unexpectedly disclosed (on Sunday, October 26) that through the use of derivatives (settled in cash rather than shares) it had increased its stake in VW from 35% to 74.1%, sparking outcry among investors, analysts, and corporate governance experts. The ploy was made possible by Bafin, Germany's financial regulator, who had *recently ruled that companies were not obliged to disclose such positions where derivatives were settled in cash rather than shares.... The sudden disclosure meant there was a free float of only 5.8%* [of VW's total shares],

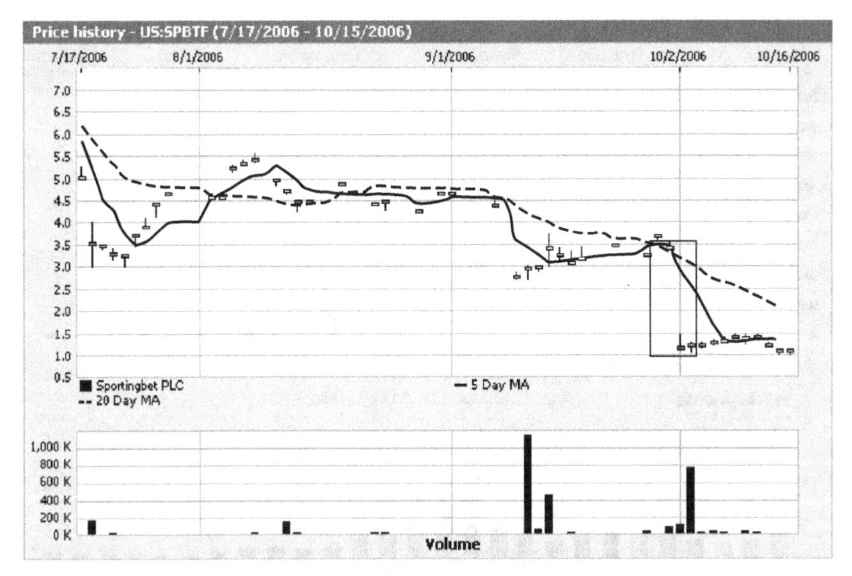

__Figure 5.5.2__ Daily candlestick chart of Sportingbet (SPBTF). Short positions were initiated by Chanos just prior to passage of the Frist amendment on 10/2/06. (Source: *MSN Money*)

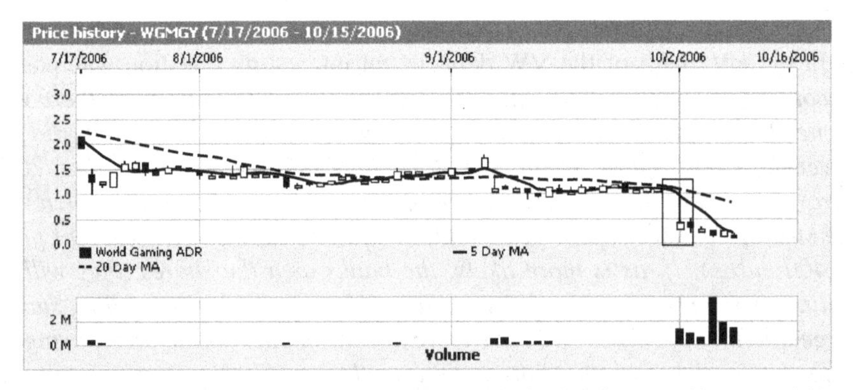

__Figure 5.5.3__ Daily candlestick chart of World Gaming (WGMGY). Short positions were initiated by Chanos just prior to passage of the Frist amendment on 10/2/06. (Source: *MSN Money*)

sparking panic among hedge funds (Milne, 10/28/08). After the close of business on 10/28/08, it was estimated that loses for some hedge funds and banks could be as high as $38 billion (Milne and Burgess, 10/29/08).

Many had shorted VW shares just prior to the price increases on 10/27 and 10/28. Notice the bearish engulfing patterns the day prior to 10/19 and 10/20. As in the previous case, the candlestick chart gave no warning of the October 27 price increase. For those with long positions on October 27, the question is whether they were astute enough to take profits on the following day.

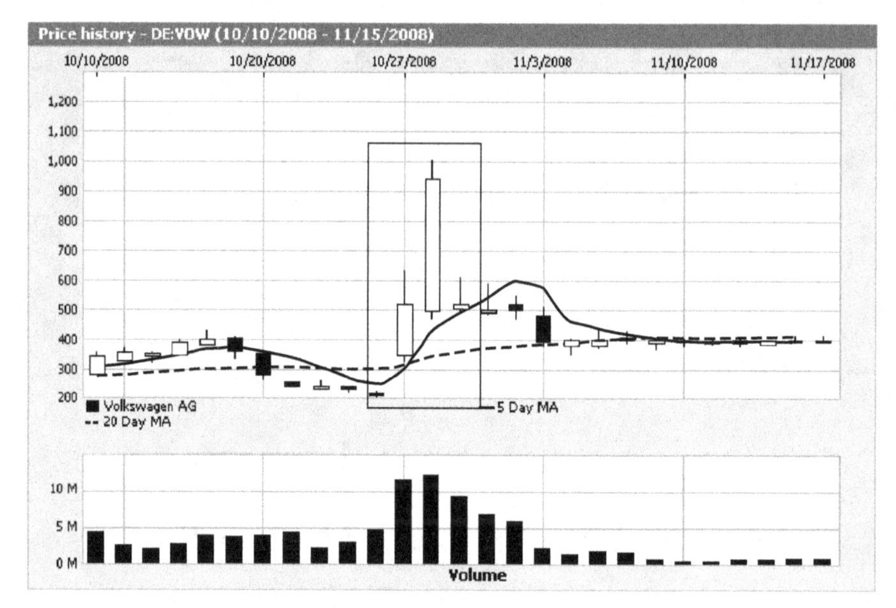

Figure 5.5.4 *Daily candlestick chart of Volkswagen (DE:VOW) during takeover of VW by Porsche. (Source: MSN Money)*

In the aftermath of the VW debacle, an interesting question was posed. *According to German law, since the options were settled in cash, there was no need to report the operation.... If the banks that sold the options did not cover their position by owning the stock, they stand to lose more than 20bn euros (assuming that the strike price of the options is 150 euros, which was the stock price most of the year, and the options expire at the current price of 400 euros). If, as is more likely, the banks own the shares, they will be selling them around the maturity of the options. Undoubtedly, selling such a large quantity of shares will have a significant impact on the price, bringing it to the level before the run-up or even lower. In this scenario too, the banks would have losses of tens of billions of euros.... The only scenario in which the banks do not lose is if they have a secret agreement with Porsche whereby the company promises to buy the shares right after the exercise of the options at the same price. Of course, such an agreement must be made public, even under lax German security laws* (Santa Clara, 11/5/08).

5.6 TERRORIST ATTACKS AND THE MARKETS

Figures 5.6.1 and 5.6.2 present daily candlestick charts for American Airlines and the Dow Jones Index during the 9/11 period. Both figures display bearish trends through 9/10. Chartist forecasts would have favored short

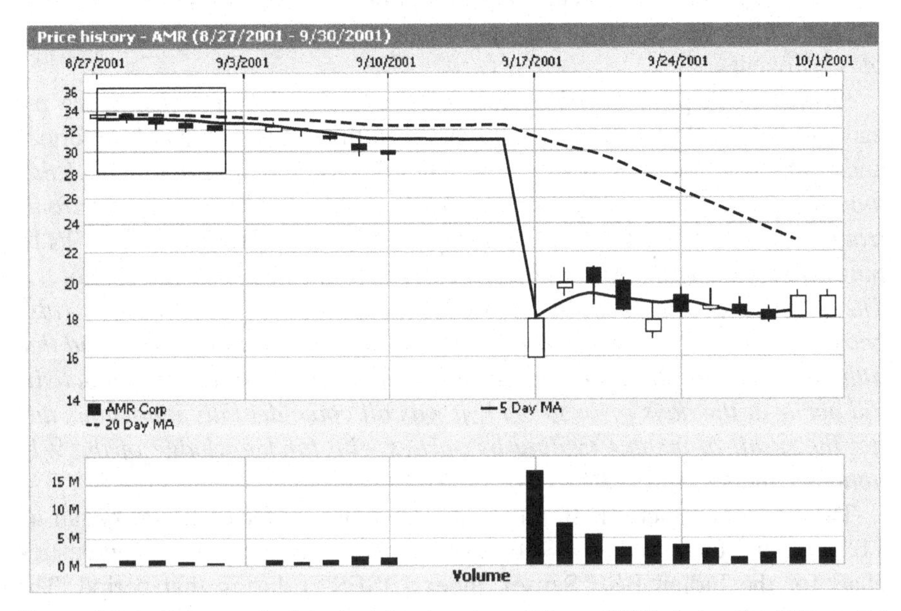

Figure 5.6.1 *Daily candlestick chart for American Airlines (AMR) during the 9/11 period. (Source: MSN Money)*

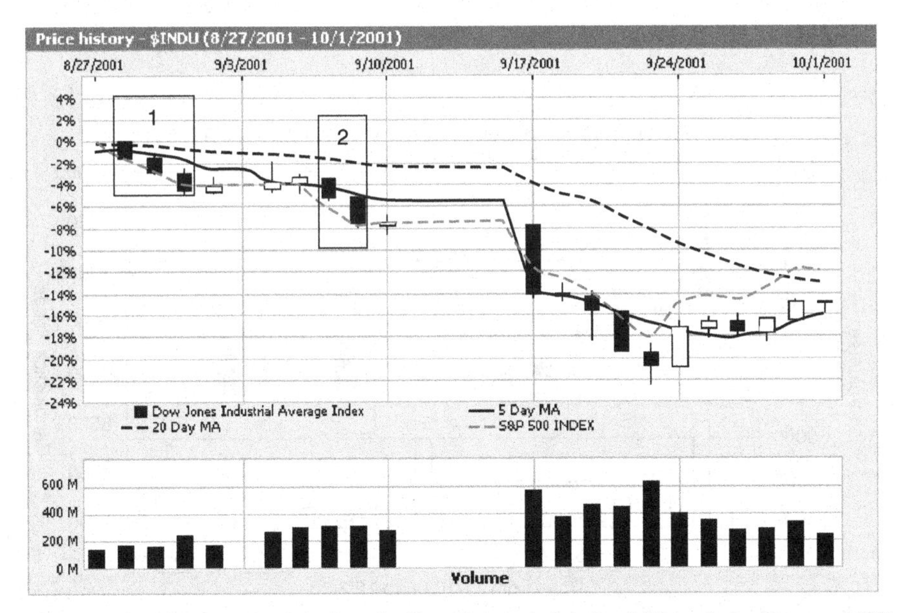

Figure 5.6.2 *Daily candlestick chart for Dow Jones during the 9/11 period. (Source: MSN Money)*

positions or purchases of near-term put options. The question under study by the authorities was whether profits were made by those with foreknowledge of 9/11 events.

In the month prior to 9/11, there were unusual discrepancies in the put and call ratio—25 to 100 times normal—in the stock options of American and United Airlines. *But it was in the final few trading days before 9/11 that unusual variances in activity occurred. Bloomberg's Trade Book electronic trading system showed that on the Thursday before the black Tuesday* [9/11], *put option volume in UAL stock was nearly 100 times higher than normal.... The National Commission on Terrorist Attacks Upon the United States (also known as the "9/11 Commission") investigated these rumors and found that although some unusual (and initially seemingly suspicious) trading activity did occur in the days prior to 9/11, it was all coincidentally innocuous and not the result of insider trading by parties with foreknowledge of the 9/11 attacks* (Snopes.com, 12/11/05).

The terrorist attack on Bombay, the financial center of India, began on 11/26/08 and ended on 11/28/08. Figure 5.6.3 presents a daily candlestick chart for the Indian BSE Sensex Index (BSESN) during that period. The bearish engulfing pattern on the first day of the attack disrupted the bullish market upturn two days earlier.

Figure 5.6.4 presents an intraday, 5-minute candlestick chart for BSESN corresponding to the day the crisis ended. The intraday chart indicates

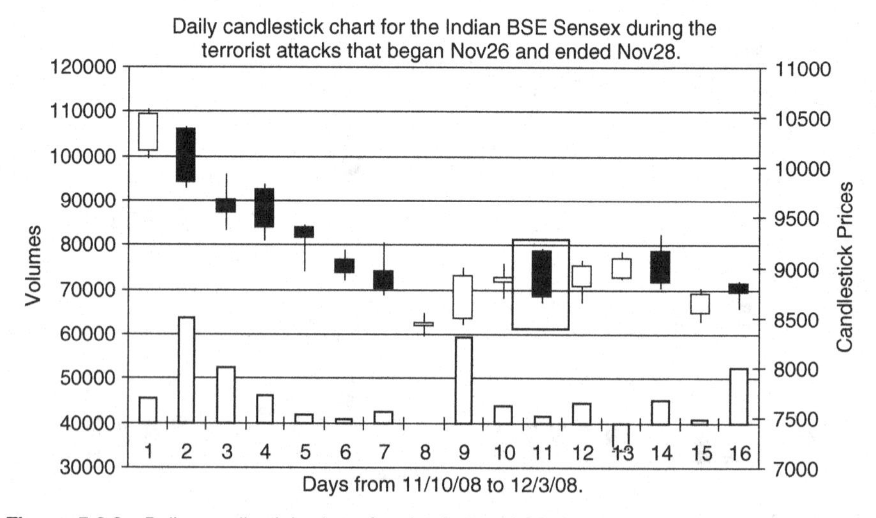

Figure 5.6.3 *Daily candlestick chart for the Indian BSE Sensex Index during the terrorist attacks in Bombay that began on November 26 and ended November 28. (Source: MSN Money)*

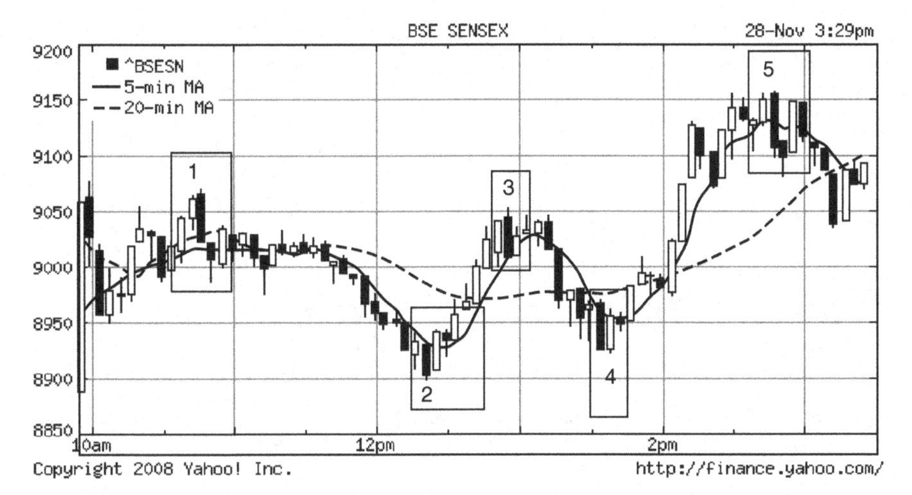

Figure 5.6.4 *Five-minute candlestick chart for the Indian BSE Sensex 30 Index on the day the Bombay hostage crisis ended.*

high-frequency trading opportunities, whereas the corresponding daily candlestick indicates a neutral position. The intraday chart shows bearish patterns in boxes 3 and 5 and bullish patterns in boxes 2 and 4.

From the perspective of Indian officials, *the Bombay attack was not the first terrorist atrocity and will not be the last. In fact, the more terrorist attacks there are, the less impact they have*. Given their impact on the markets, the terrorists would appear to have a convenient means of funding their causes.

5.7 A HOLLYWOOD ROMANCE: SPIDERMAN AND TINKERBELL

Shazam! On 8/28/09, Walt Disney Co. made a surprise $4 billion bid to acquire Marvel Entertainment (MVL) *and a range of characters that has resonated with boys for more than 40 years*. The purchase would be Disney's largest since the $7.6 billion acquisition of Pixar in 2006. Imagine *Spidy, Hulk*, and *Iron Man* joining ranks with *Mickey Mouse, Donald Duck, Snow White*, and *Hannah Montana. Think of the sexual appeal and sexual urges of the viewing audiences* (Garrahan and Edgecliffe-Johnson, 9/1/09).

Following the bid, Disney's shares closed down 2.98% while Marvel's shares jumped $10.35 to a record high of $49. Figures 5.7.1 and 5.7.2 present daily candlestick charts for MVL and Disney through August. Both charts display bullish patterns prior to the bid; see the first three candlesticks in the MVL chart and the first two candlesticks in box 1 of the DSN

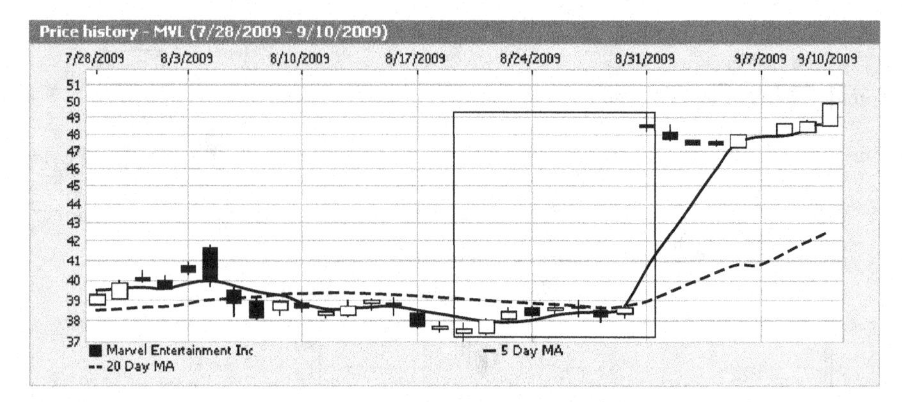

Figure 5.7.1 *Daily candlestick chart for MVL during August 2009, including the share price increase following the Disney bid on 8/31/09.* (Source: *MSN Money*)

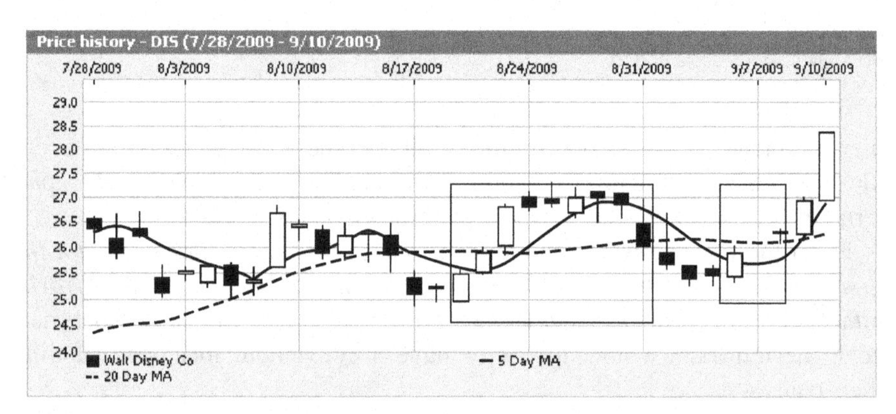

Figure 5.7.2 *Daily candlestick chart for Disney (DSN) during August 2009, including the share price decrease following the Disney bid for MVL on 8/31/09.* (Source: *MSN Money*)

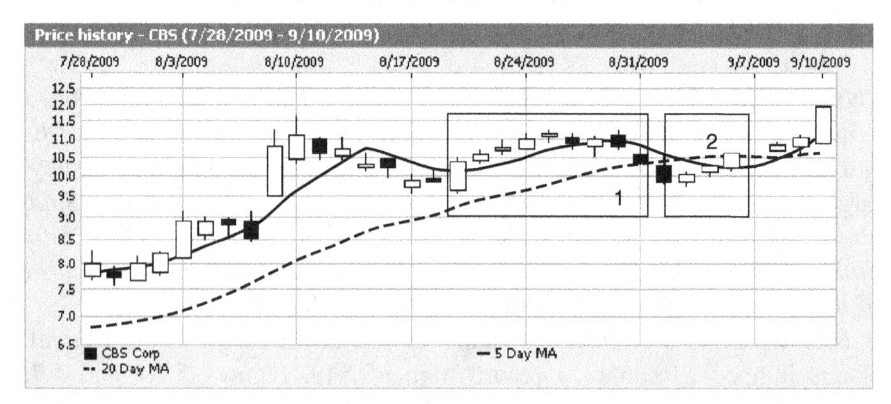

Figure 5.7.3 *Daily candlestick chart for CBS Corp (CBS), a Disney competitor during the Disney bid for MVL.* (Source: *MSN Money*)

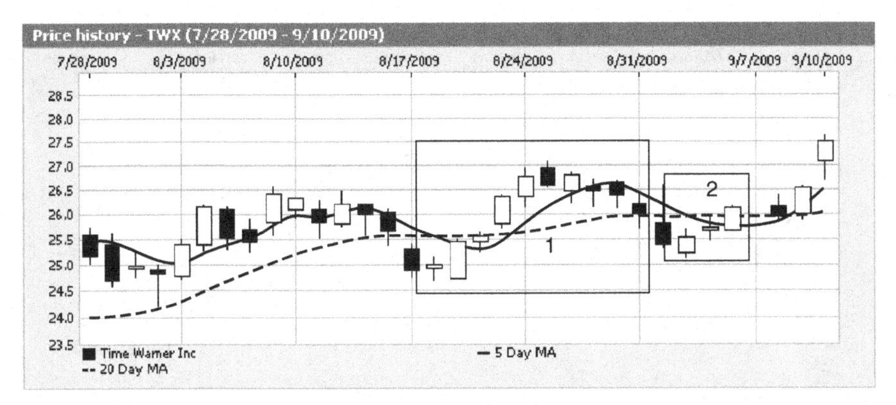

Figure 5.7.4 Daily candlestick chart for Time Warner (TWX), a Disney competitor, during the Disney bid for MVL. (Source: MSN Money)

chart. Note that both patterns are below or nearly below the moving average band, an indication of short-term buying opportunities that resulted in gains—albeit an unexpectedly large gain for MVL. The temporary drop in the acquiring company is followed by a bullish pattern; see box 2 in the DIS chart.

In detecting and reacting to bullish or bearish indicators, the active trader should also track the movements of like-kind stocks for purposes of confirmation. Like-kind stock issues tend to react in tandem. Daily candlestick charts for two Disney competitors, CBS and TWX, are presented in Figures 5.7.3 and 5.7.4. Both charts display the same bullish patterns (in boxes 1 and 2) as DIS.

5.8 COPENHAGEN AND CLIMATE CHANGE: EXXON MOBILE BUYS XTO ENERGY

There were many sequels to the Disney acquisition of Marvel Entertainment. Exxon Mobil Corporation, the world's largest publicly traded oil company, announced on 12/14/09 that it struck a $41 billion all-stock deal to buy Houston-based XTO Energy. The move marked a major bet on gas as a cleaner-burning fuel than oil, as governments around the world looked to Copenhagen climate-change talks the following week to stem the flow of greenhouse gases into the atmosphere to combat global warming. The acquisition, Exxon Mobil's largest in a decade, sparked widespread investor speculation that the energy sector could be poised for a new round of consolidation as its biggest players vie for independent producers pinched by low gas prices (Gelsi, 12/14/09).

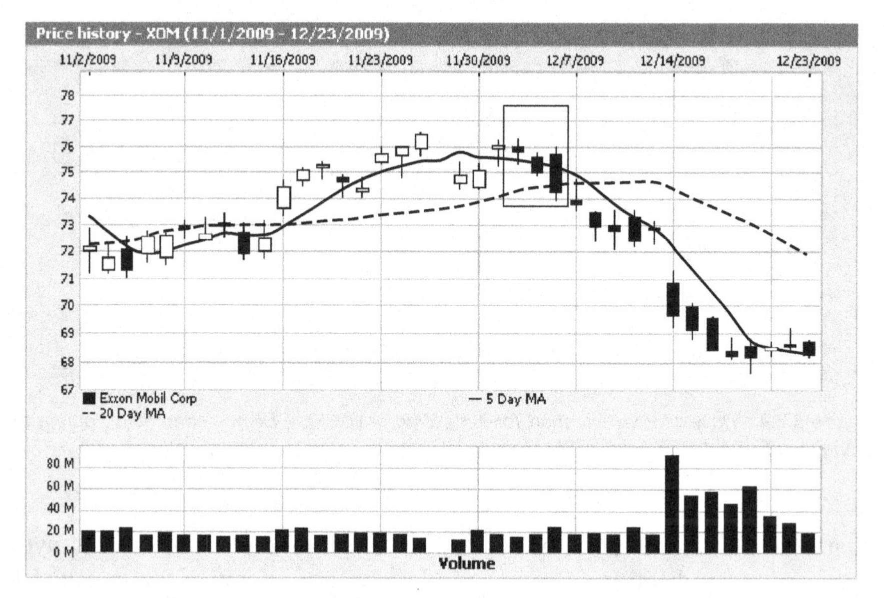

Figure 5.8.1 *Daily candlestick chart for Exxon Mobil (XOM) during their acquisition of XTO on 12/7/2009. (Source: MSN Money)*

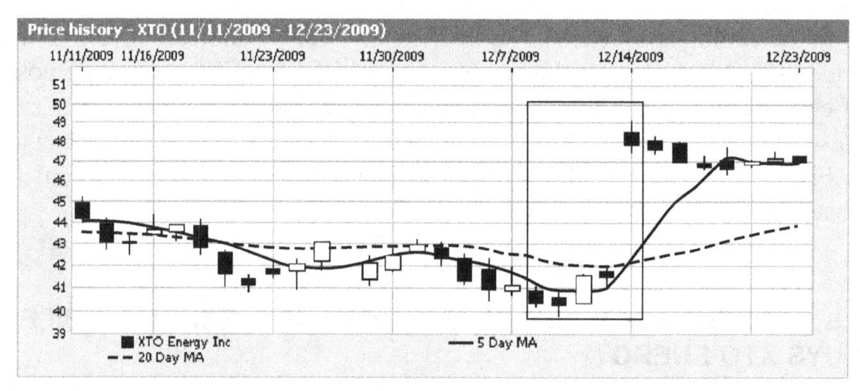

Figure 5.8.2 *Daily candlestick chart for XTO during their acquisition by XOM on 12/7/2009. (Source: MSN Money)*

Figures 5.8.1 and 5.8.2 present respective daily candlestick charts for XOM and XTO during the period of the acquisition. The effect of the acquisition announcement was to sharply increase the share price of XTO and to depress the XOM share price. For XOM a strong bearish pattern occurred earlier on the trading day before 12/7/09; see the box. For XOT a bullish pattern occurred two trading days prior to the acquisition announcement; see the box. Active traders who had been charting the two stocks would have been in enviable positions with the 12/14/09 announcement.

6

Pseudo-Candlesticks for Major League Baseball

6.1 THE 2008 WORLD SERIES: PHILADELPHIA VERSUS TAMPA BAY

For Major League Baseball games, the money line takes the place of point spreads and is given in terms of odds. The team wagered on has to win the game whatever the final score. For example, in the final game of the 2008 World Series, the winning Philadelphia Phillies were 0.64 to 1 favorites to beat Tampa Bay (i.e., every $1 bet on Philadelphia returned $0.64), which implies that the probability was 0.61 that Philadelphia would win the game (see Section 7.4). As with NBA and NFL games, a second line, denoted by $LTOT(i,t)$, is on the total runs scored by both teams. For Philadelphia games, $LTOT(i,t)$ usually ranged from 7 to 9.

In the present application of candlestick charts to MLB, the illustrations focus on $GST(i,t) = TOT(i,t) - LTOT(i,t)$, the gambling shock corresponding to the total line, and on the winning or losing margin on a per game basis. For Philadelphia (PHL), Figure 6.1.1 charts the last 32 games of the season, which concludes with their 14 postseason games. The latter includes the first-round series win over Milwaukee in four games, the National League championship series win over the Dodgers in five games, and the World Series win over Tampa Bay in five games.

Forecasting in Financial and Sports Gambling Markets: Adaptive Drift Modeling, By William S. Mallios

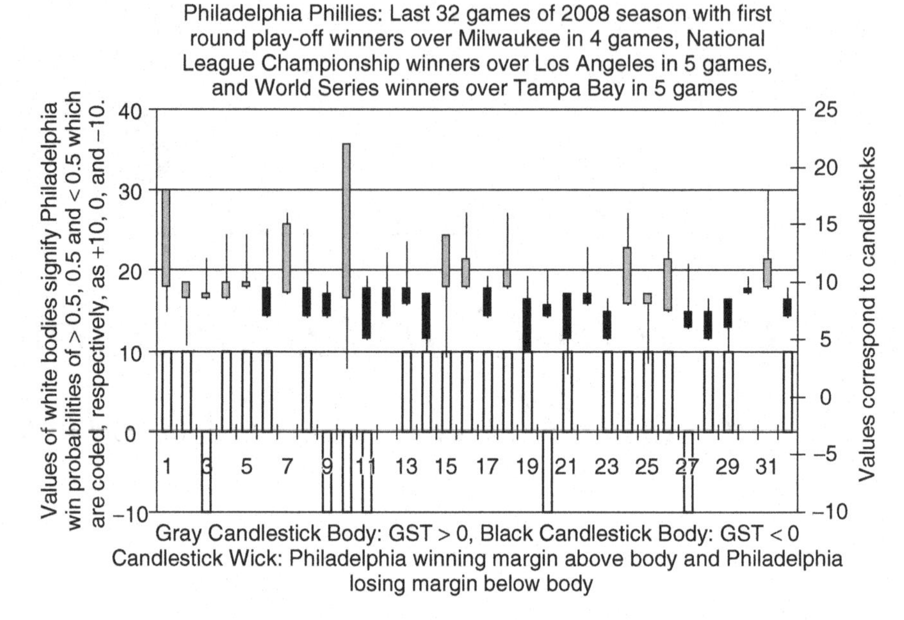

Figure 6.1.1 *Candlestick chart for the 2008 Philadelphia Phillies in their last 31 games.*

In Figure 6.1.1, $GST(i = \text{PHL}, t)$ is abbreviated as $GST(t)$. Of the 32 games depicted, Philadelphia lost the first two. A loss (win) in terms of the score differential is indicated by the candlestick wick lying below (above) the body. The bookmaker's probability that Philadelphia would win or lose is indicated by the white bodies. A probability of winning (losing) has a white body of +10 (−10). Gray and black bodies represent, respectively, the values of $GST(t) > 0$ and $GDT(t) < 0$.

The initial two losses were followed by a string of seven consecutive wins. During the time period under study, a loss by Philadelphia was always followed by a string of at least three successive wins—with one exception: Philadelphia losses in the 14th and 15th games were followed by three successive wins. The implication is that physiological–psychological–biological factors that accompanied a Philadelphia loss provided motivation for wins in subsequent games.

Note also that larger gray bodies $[GST(t) \gg 0]$ tend to be followed by smaller black bodies or at least smaller gray bodies. The implication is that offensive games (resulting in a larger than expected number of runs) tended to be followed by defensive games. In addition, Philadelphia won 78.5% of the 14 play-off games in which their pitching tended to prevail—in the sense that $GST < 0$ in 10 of the 14 games.

Philadelphia won 77.7% of the last 18 games of the regular season, compared to $15/27 = 55.6\%$ of the games for which the money line had an associated win probability for Philadelphia of greater than 0.5. Conversely, the money line was correct in predicting only one of six games for which the associated loss probability for Philadelphia was less than 0.5. Given that the money line reflects the gambling public's expectations, the money line appeared to be largely irrational and the market largely inefficient.

6.2 THE 2008 CHICAGO CUBS: VISIONS OF 1908 HEROICS

When the Bambino's curse was lifted with Boston's World Series win in 2003, the other reigning curse was in Chicago, where the Cubs hadn't won a World Series since 1908. (At the time, gas was selling for 20 cents a gallon.) Moreover, the Cubs hadn't won a National League Championship since World War II ended in 1945. In a letter to conservative columnist William F. Buckley, Jr., Ira Glasser, Executive Director of the American Civil Liberties Union, asked: *What explains 1945 as the dividing line in Cubs history? Can it have something to do with the end of World War II? Is the failure of the Cubs to win a National League pennant after 1945 somehow related to the failure of the U.S. Army to win a war after 1945? (We came close in the Persian Gulf, but then so did the Cubs in 1969, 1984, and 1989.) I know this connection seems fanciful but no more so than the relationship you once asserted between the drop in the S.A.T. scores and the Supreme Court's school prayer decision in* Engel *v*. Vitale (*Wall Street Journal*, April 24, 1994).

Now, in 2008, the Cubs and the Chicago White Sox were both in the play-offs in the same year for the first time in over 100 years. *Cubs fans, in their persistent and admirable hopefulness, continually and confidently proclaimed that this was the year that the team was going to take them where no Cubs teams had in 100 years, despite the fact that, uh, no Cubs team had taken them there in 100 years* (Isaacson, 10/7/08).

In the same format as Figure 6.1.1, Figure 6.2.1 presents the final 31 games of the Cubs' 2008 season. The pseudo-candlesticks portray the chart of an underachieving team—one not likely to be a World Series winner. The Cubs were favored to win in each of the first 20 games in the chart and managed to win half of those games. They won only four of the last 11 remaining regular-season games, including the final three games against the Dodgers in the first round of the play-offs.

The loss to the Dodgers could hardly have been surprising. Notice that larger gray bodies (describing games where the total runs scored is relatively

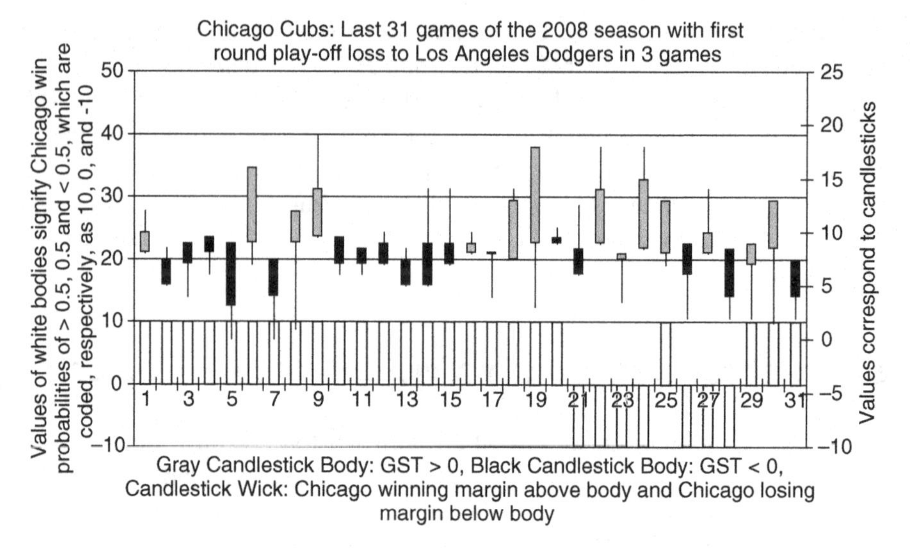

Figure 6.2.1 *Candlestick chart for the 2008 Chicago Cubs in their last 32 games.*

large relative to the line) tend to be followed by a Chicago loss (denoted by the candlestick wick extending below the body). The implication is that the Cubs' involvement in an offensive game (where more runs were scored than expected) portended a loss in their following game. Perhaps the overuse of Cub relief pitchers in offensive games is a reflection of relief pitcher nonavailability in the subsequent game. This conjecture can be evaluated by an analysis of pitcher performances, as described by the analysis of Boston and Oakland pitchers in Sections 8.7 and 8.8.

6.3 A STRANGE SET OF COINCIDENCES: A PLATE UMPIRE'S AFFINITY FOR A PITCHER

Some umpires can work their entire careers and never find themselves behind the plate for a no-hitter. Eric Cooper has been the plate umpire for three of them, including two by White Sox left-hander Mark Buehrle, who pitched a perfect game on July 22, 2009 against the Tampa Bay Rays. Cooper was also behind the plate for Buerhle's 2007 no-hitter against the Texas Rangers (and also Buerhle's 6–0 shutout victory against the Los Angeles Dodgers on June 17, 2005).

Mark Buehrle wears uniform number 56. Eric Cooper wears umpire uniform number 56. The time of the game in Buerhle's 2007 no-hitter was 2:03. The time of the game in Buehrle's 2009 perfect game was 2:03. *After the game, we certainly talked about this being the second one with Buehrle,*

Cooper said. *I thought about it more then than when I was actually working*. With the crowd buzzing on every pitch, was Cooper feeling the pressure? *I'm certainly aware of the situation, especially late in the game, just because of the way the fans are reacting*, he said. *But it doesn't differentiate how I call the game. I have the same approach whether the score is 0–0 or 10–0 or whether there is a perfect game or no-hitter going. I try to call strikes strikes and balls balls. Certainly late in the game, when the fans are on their feet and they're loud, they want every pitch to be a strike, and not every pitch is a strike. That's the time when you really need to kind of take a deep breath and really focus on what's going on and not get caught up in the moment* (De Luca, 7/24/07).

Cooper was also behind the plate for Boston Red Sox right-hander Hideo Nomo's no-hitter against the Baltimore Orioles on 4/6/01. The only other active major league umpire to work multiple no-hitters is Tim Tschida, who was behind the plate for Cubs' ace Carlos Zambrano's no-hitter against the Houston Astros on 9/14/08 and for Nolan Ryan's seventh career no-hitter, pitching the Rangers past the Toronto Blue Jays on 5/1/91.

7

Single-Equation Adaptive Drift Modeling

7.1 ADAPTIVE ARMA PROCESSES

The modeling procedures described in Section 4.3 are now generalized by assuming that $D(t)$ is generated by an ARMA(p, q) process with time-varying coefficients:

$$D(t) = \boldsymbol{\alpha}(\mathrm{t})'\mathbf{D}(\mathbf{t}-) + \boldsymbol{\gamma}(\mathbf{t})'\boldsymbol{\varepsilon}(\mathbf{t}-) + \boldsymbol{\beta}(\mathbf{t})'\mathbf{x}(\mathbf{t}) + \varepsilon(t). \qquad (7.1.1)$$

The p autoregressive (AR) and q moving average (MA) variables and r covariates are denoted, respectively, by the vectors $\mathbf{D}(\mathbf{t}-)$,$\boldsymbol{\varepsilon}(\mathbf{t}-)$, and $\mathbf{x}(\mathbf{t})$, with respective coefficients $\boldsymbol{\alpha}(\mathbf{t}) = (\alpha(i,t)), \boldsymbol{\gamma}(\mathbf{t}) = (\gamma(j,t))$, and $\boldsymbol{\beta}(\mathbf{t}) = (\beta(k,t))$. As in (4.3.5), the assumption is that the time-varying coefficients are generated by lagged shocks (or moving average terms):

$$\begin{aligned}
\boldsymbol{\alpha}(\mathbf{t}) &= \boldsymbol{\alpha} + A\boldsymbol{\varepsilon}^*_\alpha(\mathbf{t}-) + \boldsymbol{\delta}_\alpha(\mathbf{t}),\\
\boldsymbol{\gamma}(\mathbf{t}) &= \boldsymbol{\gamma} + \Gamma\boldsymbol{\varepsilon}^*_\gamma(\mathbf{t}-) + \boldsymbol{\delta}_\gamma(\mathbf{t}),\\
\boldsymbol{\beta}(\mathbf{t}) &= \boldsymbol{\beta} + B\boldsymbol{\varepsilon}^*_\beta(\mathbf{t}-) + \boldsymbol{\delta}_\beta(\mathbf{t}).
\end{aligned} \qquad (7.1.2)$$

$A(p \times q_\alpha), \Gamma(q \times q_\gamma)$, and $C(r \times q_\beta)$ are constant coefficient matrices for the vectors $\boldsymbol{\varepsilon}^*_\mathbf{v}(\mathbf{t}-) = [\varepsilon^*_v(t-j)]$ for $v = \alpha, \gamma, \beta$. Analogous to (4.3.3), the

Forecasting in Financial and Sports Gambling Markets: Adaptive Drift Modeling, By William S. Mallios

vector of model errors $\boldsymbol{\delta}(\mathbf{t}) = (\boldsymbol{\delta}_\alpha(\mathbf{t})', \boldsymbol{\delta}_\gamma(\mathbf{t})', \boldsymbol{\delta}_\beta(\mathbf{t})')'$ is assumed normally distributed with $E(\boldsymbol{\delta}(\mathbf{t})) = \mathbf{0}$ and $E(\boldsymbol{\delta}(\mathbf{t})\boldsymbol{\delta}(\mathbf{t})') = V_\delta(t)$; that is,

$$\boldsymbol{\delta}(\mathbf{t}) : \text{normal}(\mathbf{0}, V_\delta(t)), \tag{7.1.3}$$

where $V_\delta(t)$ is modeled in terms of GARCH-type processes in Section 7.5.

For *gradual mean drift*,

$$\varepsilon_v^*(t-j) = \varepsilon(t-j).$$

For *abrupt mean drift*,

$$\varepsilon_v^*(t-j) = \varepsilon(t-j) + \boldsymbol{\alpha}_v'\mathbf{v}(\mathbf{t-}) \text{ if } |\varepsilon(t-j)| \text{is sufficiently large}, \tag{7.1.4}$$

where $\boldsymbol{\alpha}_\mathbf{v}$ is the coefficient vector of $\mathbf{v}(\mathbf{t-})$, a vector of variables distinct from $\mathbf{D}(\mathbf{t-})$ and $\mathbf{x}(\mathbf{t})$, whose elements become significant in the particular abrupt drift scenario. Elements of $\mathbf{v}(\mathbf{t-})$ usually include interactions that tend to change the explanatory variables that appeared in the previous model update.

Substitutions of expressions of time-varying coefficients in (7.1.2) into (7.1.1) lead to a second-order reduced-form ARMA model:

$$\begin{aligned} D(t) = {} & \boldsymbol{\alpha}'\mathbf{D}(\mathbf{t-}) + \boldsymbol{\gamma}'\boldsymbol{\varepsilon}(\mathbf{t-}) + \boldsymbol{\beta}'\mathbf{x}(\mathbf{t}) + \mathbf{D}(\mathbf{t-})'\mathbf{A}\boldsymbol{\varepsilon}_\alpha^*(\mathbf{t-}) \\ & + \boldsymbol{\varepsilon}(\mathbf{t-})'\Gamma\boldsymbol{\varepsilon}_\gamma^*(\mathbf{t-}) + \mathbf{x}(\mathbf{t})'B\boldsymbol{\varepsilon}_\beta^*(\mathbf{t-}) + \varepsilon_R(t), \end{aligned} \tag{7.1.5}$$

where $\varepsilon_R(t)$ is the reduced model error. When $\Gamma = 0$, (7.1.5) takes the form of a bilinear model, while for $\Gamma \neq 0$, model (7.1.5) is not invertible. However, for fitted equations with $\Gamma \neq 0$, the resulting model bias may be inconsequential if the fitted model provides effective forecasts. (Perhaps George Box's remark that *essentially all models are wrong, but some are useful* applies to biased models that provide effective forecasts.)

Analogous to (4.3.8), (7.1.5) could have resulted from any number of time-varying coefficient models, not necessarily the one defined by (11.2.1) and (7.1.2). Model validity regarding time-varying coefficients can be evaluated by fitting the individual drift equations in (7.1.2).

Regarding model validity, selected applications have shown some evidence of significant three-factor interactions involving lagged shocks when included in a third-order reduced-form equation. Such interactions, reflecting possible direct effects of lagged distributional asymmetries, would result if a bilinear process were specified for $D(t)$ in (7.1.1):

$$\begin{aligned} D(t) = {} & \boldsymbol{\alpha}(\mathbf{t})'\mathbf{D}(\mathbf{t-}) + \boldsymbol{\gamma}(\mathbf{t})'\boldsymbol{\varepsilon}(\mathbf{t-}) + \boldsymbol{\beta}(\mathbf{t})'\mathbf{x}(\mathbf{t}) + \boldsymbol{\varepsilon}(\mathbf{t-})'\text{Д}_{\alpha\gamma}(\mathbf{t})\mathbf{D}(\mathbf{t-}) \\ & + \boldsymbol{\varepsilon}(\mathbf{t-})'\text{Д}_{\alpha\beta}(\mathbf{t})\mathbf{x}(\mathbf{t}) + \varepsilon(t). \end{aligned} \tag{7.1.6}$$

The matrices $Д_{\alpha\gamma}(\mathbf{t})$ and $Д_{\alpha\beta}(\mathbf{t})$ contain the time-varying coefficients of bilinear terms. Another modeling variation would be to include interactions between $\varepsilon(t-j)$ and elements of $\mathbf{v(t-)}$ in (7.1.4). These and other possibilities can be evaluated by fitting the drift equations in addition to fitting the reduced equation in (7.1.4).

7.2 VARIABLE SELECTION: IDENTIFYING THE REDUCED MODEL

For the markets under study, identifying effective forecasting models is 95% of the effort; estimating parameters of such models is 5% of the effort. Since the number of possible predictor variables in (7.1.5) is usually excessive, only a select few will be significant and retained in the process of identifying the reduced forecasting model. As such, suitable variable selection procedures are required.

Model (7.1.5) is rewritten in simplified form as follows:

$$D(t) = f[\mathbf{w(t-)}, \boldsymbol{\varepsilon}\mathbf{(t-)}; \boldsymbol{\Theta}] + \varepsilon_R(t). \tag{7.2.1}$$

Here $f[\cdot]$ denotes a second-order linear function of $\mathbf{w(t-)} = [\mathbf{D(t-)}', \mathbf{x(t)}', \mathbf{v(t-)}]'$, $\boldsymbol{\varepsilon}\mathbf{(t-)}$ is interpreted to contain all nonredundant elements of $\boldsymbol{\varepsilon}_{\alpha}{}^{*}\mathbf{(t-)}$,$\boldsymbol{\varepsilon}_{\gamma}^{*}\mathbf{(t-)}$, and $\boldsymbol{\varepsilon}_{\beta}^{*}\mathbf{(t-)}$, and $\boldsymbol{\Theta}$ is the vector of coefficients. A heuristic variable selection approach is described as follows:

Step 1. Perform a factor analysis on the correlation matrix associated with $\mathbf{w(t-)}$ (as illustrated in the analysis of a cointegrated time series in Section 9.1) and retain factors associated with sufficiently large eigenvalues.

Step 2. Apply the stepwise regression variable selection procedure [as available in SPSS (2008)] in the regression of $D(t)$ on the factor scores corresponding to the eigenvectors retained in step 1 and base the initial fitted model on the significant factor scores.

Step 3. Obtain an initial estimate of $\boldsymbol{\varepsilon}\mathbf{(t-)}$, say $\mathbf{e}_0\mathbf{(t-)}$, based on the residuals corresponding to the fitted model in step 2.

Step 4. Apply stepwise regression once more in the regression of $D(t)$ on those factor scores identified in step 2, all elements of $\mathbf{e}_0\mathbf{(t-)}$, and bilinear terms [or interactions between the factor scores and elements of $\mathbf{e}_0\mathbf{(t-)}$].

Step 5. If the significant factor scores identified in step 4 are not the same as those identified in step 2, regress $D(t)$ on the significant factor scores resulting in step 4, in order to obtain new, revised residuals, say $\mathbf{e}_1\mathbf{(t-)}$.

Step 6. This process is repeated until the significant factor scores obtained in the absence of $\boldsymbol{\varepsilon}(\mathbf{t}-)$ estimates are the same as the significant factor scores obtained in the presence of $\boldsymbol{\varepsilon}(\mathbf{t}-)$ estimates.

These six steps enable a tentative identification of significant variables from among the factor scores, elements of $\boldsymbol{\varepsilon}(\mathbf{t}-)$, and bilinear terms. This process is often tedious. Unfortunately, there appears to be no optimal stepwise procedure for selecting predictor variables among a large number of possible variables in bilinear and higher-order ARMA-type models. Nonlinear estimation of parameters follows the model identification stage when moving average and bilinear terms enter as predictors (see Section 7.3).

A second general approach to the selection of predictors is to forgo the factor analysis on the correlation matrix associated with $\mathbf{w}(\mathbf{t}-)$ and to scan all variables in $\mathbf{w}(\mathbf{t}-)$ directly through the stepwise procedure described by the six steps in the first approach. As with the first approach, this process of variable selection is completed when significant variables selected from $\mathbf{w}(\mathbf{t}-)$—obtained in the absence of $\boldsymbol{\varepsilon}(\mathbf{t}-)$ estimates and interactions involving $\boldsymbol{\varepsilon}(\mathbf{t}-)$ estimates—are the same as the significant variables in $\mathbf{w}(\mathbf{t}-)$ obtained in the presence of moving average and bilinear terms.

7.3 REDUCED MODEL ESTIMATION: SINGLE EQUATIONS

Given that significant predictors have been identified in (7.2.1), the model to be fitted is written in linear form as

$$\mathbf{D} = Z\boldsymbol{\Theta}^* + \boldsymbol{\varepsilon}_\mathbf{R}. \tag{7.3.1}$$

The vector $\mathbf{D} = (D(t))$ contains n successive observations on $D(t)$. The matrix Z contains corresponding values of all significant first- and second-order terms involving $\mathbf{w}(\mathbf{t}-)$ and $\boldsymbol{\varepsilon}(\mathbf{t}-)$; $\boldsymbol{\Theta}^*$ and $\boldsymbol{\varepsilon}_\mathbf{R}$ denote, respectively, the parameter vector and the reduced model error vector. Model (7.3.1) is termed pseudolinear when Z includes moving average and/or bilinear terms that must be estimated.

In minimizing $\boldsymbol{\varepsilon}_\mathbf{R}'\boldsymbol{\varepsilon}_\mathbf{R}$ with respect to $\boldsymbol{\Theta}^*$, the nonlinear least squares procedure is linearized as follows. Let $\boldsymbol{\Theta}_0^*$ denote an initial estimate of $\boldsymbol{\Theta}^*$ obtained, say, through the procedure described in Section 7.2. Substitution of $\boldsymbol{\Theta}_0^*$ for $\boldsymbol{\Theta}^*$ in (7.3.1) yields $\mathbf{D} = Z\boldsymbol{\Theta}^* + \boldsymbol{\varepsilon}_\mathbf{R} = Z_0\boldsymbol{\Theta}_0^* + \mathbf{e}_{\mathbf{R}0}$; Z_0 is an initial estimate of Z in the sense that elements of $\boldsymbol{\varepsilon}(\mathbf{t}-)$ in Z are replaced by $\mathbf{e}_0(\mathbf{t}-)$; $\mathbf{e}_{\mathbf{R}0}$ is the vector of residuals corresponding to $\boldsymbol{\varepsilon}_\mathbf{R}$. Since $\boldsymbol{\varepsilon}_\mathbf{R} = Z_0\boldsymbol{\Theta}_0^* - Z\boldsymbol{\Theta}^* + \mathbf{e}_{\mathbf{R}0}$, we then approximate Z by Z_0:

$$\mathbf{e}_{\mathbf{R}0} \approx Z_0(\boldsymbol{\Theta}^* - \boldsymbol{\Theta}_0^*) + \boldsymbol{\varepsilon}_\mathbf{R}. \tag{7.3.2}$$

Through (7.3.2), successive adjustments to $(\boldsymbol{\Theta}^* - \boldsymbol{\Theta}_0^*)$ may be obtained through linear least squares by regressing $\mathbf{e_{R0}}$ on Z_0. This process is continued until convergence occurs. If the initial estimate $\boldsymbol{\Theta}_0^*$ is not sufficiently close to the nonlinear least squares estimate, say $\boldsymbol{\Theta}^{*\wedge}$, convergence may be slow, or the process may not converge; see Box and Jenkins (1976) for discussions on linearized nonlinear estimation of nondynamic ARMA processes.

The fitted form of model (7.3.1) is written

$$\mathbf{D}^{\wedge} = Z^{\wedge}\boldsymbol{\Theta}^{*\wedge}. \tag{7.3.3}$$

Z is replaced by $Z^{\wedge}$ since elements of $Z(t)$ containing $\boldsymbol{\varepsilon}(\mathbf{t-})$ have been replaced by corresponding least squares estimates. If the model errors in (7.3.1) are assumed to be identically and independently distributed with zero expectation and constant variance $\sigma_{\varepsilon R}^2$, then, asymptotically,

$$\text{variance}(\boldsymbol{\Theta}^{*\wedge}) = [Z^{\wedge\prime}Z^{\wedge}]^{-1}\sigma_{\varepsilon R}^2. \tag{7.3.4}$$

The constant variance assumption may be relaxed to allow for volatility modeling.

7.4 REDUCED MODEL EMPIRICAL BAYESIAN ESTIMATION: SINGLE EQUATIONS

In contrast to $\boldsymbol{\Theta}^{*\wedge}$ in (7.3.3), we now define an empirical Bayesian estimator for $\boldsymbol{\Theta}*$ in (7.3.1), say $\boldsymbol{\Theta}_{\mathbf{B}}$, for purposes of comparing, confirming, and/or improving competing forecasts. If the forecasting inferences based $\boldsymbol{\Theta}^{*\wedge}$ and $\boldsymbol{\Theta}_{\mathbf{B}}$ agree, decision making is reinforced. When inferences differ, a reevaluation of the modeling assumptions and estimation procedures is in order.

The estimation of $\boldsymbol{\Theta}^*$ in (7.3.1) is based on the most recent window of n successive observations on $D(t)$ that end, say, at time T; that is, $\boldsymbol{\Theta}^{*\wedge}$ in (7.3.3) is based on the n observations $D(T), D(T-1), \ldots, D(T-n)$ with the objective of forecasting $D(T+1)$. A distribution of $\boldsymbol{\Theta}^*$ estimators is available from a systematic bootstrap sampling of N lagged windows of $D(t)$. Each lagged window contains n successive observations on $D(t)$ that end at times $T-1, T-2, \ldots, T-N$. Moreover, each lagged window provides an estimate of $\boldsymbol{\Theta}^*$. Let $\boldsymbol{\Theta}^{*\wedge}_{-1}, \boldsymbol{\Theta}^{*\wedge}_{-2}, \boldsymbol{\Theta}^{*\wedge}_{-3}, \ldots, \boldsymbol{\Theta}^{*\wedge}_{-\mathbf{N}}$; denote the N estimates of $\boldsymbol{\Theta}^*$. The unweighted average of these N estimates,

$$\boldsymbol{\Theta}_{\sim} = \Sigma_i \boldsymbol{\Theta}^{*\wedge}_{-\mathbf{i}}/\mathbf{N}, \tag{7.4.1}$$

is an empirical estimate of the mean of the prior distribution of $\mathbf{\Theta}$. The estimated variance–covariance estimate of $\mathbf{\Theta}_\sim$ is given by

$$\mathbf{S}_{\Theta\sim} = \Sigma_i(\mathbf{\Theta}^{*\wedge}_{-\mathbf{i}} - \mathbf{\Theta}_\sim)(\mathbf{\Theta}^{\wedge}_{-\mathbf{i}} - \mathbf{\Theta}_\sim)'/(\mathbf{N}-\mathbf{1}). \tag{7.4.2}$$

It is assumed that, approximately,

$$\mathbf{\Theta}_\sim : (\mathbf{\Theta}^*, \Sigma_{\Theta\sim}); \tag{7.4.3}$$

that is, $E(\mathbf{\Theta}_\sim) = \mathbf{\Theta}^*$, while variance$(\mathbf{\Theta}_\sim) = \Sigma_{\Theta\sim}$ is estimated by $\mathbf{S}_{\Theta\sim}$.

With I denoting the identity matrix, let $X' = [Z^{\wedge'}, I], \mathbf{y}' = [\mathbf{D}', \mathbf{\Theta}'_\sim]$ and assume that $\boldsymbol{\varepsilon}_\mathbf{R}$ in (7.3.1) and $\mathbf{\Theta}_\sim$ in (7.4.1) are distributed independently. Then, for

$$\mathbf{y} = X\mathbf{\Theta}^* + \boldsymbol{\delta},$$

where $\boldsymbol{\delta}$ denotes a vector of model errors,

$$y : [X\mathbf{\Theta}^*, \Sigma_\mathbf{B}] \tag{7.4.4}$$

and

$$\Sigma_\mathbf{B} = \begin{pmatrix} I\sigma^2_{\varepsilon R} & 0 \\ 0 & \Sigma_{\Theta\sim}) \end{pmatrix}. \tag{7.4.5}$$

Minimizing $\boldsymbol{\delta}'\Sigma_\mathbf{B}^{-1}\boldsymbol{\delta}$ yields the following empirical Bayesian estimator for $\mathbf{\Theta}^*$:

$$\begin{aligned} \mathbf{\Theta}_\mathbf{B} &= [\mathrm{X}'\Sigma_\mathbf{B}^{-1}X]^{-1}X'\Sigma_\mathbf{B}^{-1}\mathbf{y} \\ &= \{[\mathbf{\Sigma}_{\Theta\sim}]^{-1} + (\mathrm{Z}^{\wedge'}\mathrm{Z}^{\wedge})/\sigma^2_{\varepsilon R}\}^{-1}[\mathbf{\Sigma}_{\Theta\sim}]^{-1}\mathbf{\Theta}_\sim \\ &\quad + (\mathrm{Z}^{\wedge'}\mathrm{Z}^{\wedge})/\sigma^2_{\varepsilon R}\mathbf{\Theta}^{\wedge}], \end{aligned} \tag{7.4.6}$$

$$\text{variance}(\mathbf{\Theta}_\mathbf{B}) = \{[\mathbf{\Sigma}_{\Theta\sim}]^{-1} + (\mathrm{Z}^{\wedge'}\mathrm{Z}^{\wedge})/\sigma^2_{\varepsilon\mathrm{R}}\}^{-1}. \tag{7.4.7}$$

See Zellner (1971) for a derivation of (7.4.6)–(7.4.7) through maximum likelihood estimation.

In contrast to the forecast $D(t)^\wedge = Z(t)^{\wedge'}\mathbf{\Theta}^{*\wedge}$ in (7.3.3), a second forecast based on (7.4.6) for $D(t)$ is denoted by

$$D(t)^\wedge_B = Z(t)^{\wedge'}\mathbf{\Theta}_\mathbf{B}. \tag{7.4.8}$$

Decision making is reinforced if inferences based on (7.3.3) and (7.4.8) are the same: for example, both forecasts indicate that price shares will drop significantly in the forthcoming week. A red flag is signaled when forecasting inferences differ. Although many reasons can be given for such discrepancies, one should be emphasized regarding (7.4.8). Since model building is adaptive, changes in model structure may occur over time, particularly during periods of abrupt drift. The Bayesian estimator predictors used in the reduced model for $D(T)^\wedge$ are the same as those used in the models for $D(T-1)^\wedge, D(T-2)^\wedge$, and so on. However, allowances for changes in predictors variables over time indicate that whereas all coefficients of $\mathbf{\Theta}^{*\wedge}$ are significant, elements of $\mathbf{\Theta}_{-1}{}^{*\wedge}, \mathbf{\Theta}^{*\wedge}_{-2}, \mathbf{\Theta}^{*\wedge}_{-3}, \ldots, \mathbf{\Theta}^{*\wedge}_{-N}$ may be of greater or lesser significance and may deviate in value considerably from corresponding elements of $\mathbf{\Theta}^{*\wedge}$. Put simply, a model that is updated through time T might not be the same as models that are updated earlier. If so, the Bayesian estimator $\mathbf{\Theta_B}$ may be of only marginal value.

In the markets under study, another class of forecasts is probably of greater importance than the Bayesian forecasts. These include those defined by the updated, adaptive forecasting models for past forecasts, say for $D(T-1), D(T-2)$, and $D(T-3)$, if, for example, each of the forecasting models $D(T-1)^\wedge, D(T-2)^\wedge$, and $D(T-3)^\wedge$ can be used, along with $D(T)^\wedge$, to forecast $D(T+1)$ simply by updating the predictor variables, not the coefficients. If $D(T+1)^\wedge$ and updates of most recent forecasting models lead to the same inferences regarding $D(T+1)$, decision making is particularly enhanced. This situation is illustrated in forthcoming modeling exercises in both financial and sports gambling markets.

7.5 SINGLE-EQUATION VOLATILITY MODELING: ADAPTIVE GARCH PROCESSES

Volatility modeling concepts, introduced in Section 4.4, are now generalized with particular emphasis on scenarios of abrupt mean drift. As a first step, suppose that $e_R(t)^2$, the residual corresponding to $\varepsilon_R(t)$ in (7.1.5), is generated by a GARCH(p, q) process:

$$\mathrm{e}_R(t)^2 = \mu + \boldsymbol{\varphi}'\mathbf{e}^2_\mathbf{R}(\mathbf{t-}) + \boldsymbol{\psi}'\mathbf{\Delta}(\mathbf{t-}) + \Delta(t), \tag{7.5.1}$$

where $\boldsymbol{\varphi}$ and $\boldsymbol{\psi}$ represent the respective effects of $\mathbf{e}^2_\mathbf{R}(\mathbf{t-}) = \mathrm{e}_R(t-j)^2$ and $\mathbf{\Delta}(\mathbf{t-}) = \Delta(t-j^*)$, and $\Delta(t)$ denotes the model error. A possible shortcoming of model (7.5.1) is that if volatility is time varying, the volatility model may also be time varying. To address this shortcoming, we allow for time varying coefficients generated in terms of elements of $\mathbf{\Delta}(\mathbf{t-})$; for example,

$$e_R(t)^2 = \mu(t) + \boldsymbol{\varphi}(\mathbf{t})'\mathbf{e}_\mathbf{R}^2(\mathbf{t}-) + \boldsymbol{\psi}(\mathbf{t})'\boldsymbol{\Delta}(\mathbf{t}-) + \Delta(t), \tag{7.5.2}$$

where

$$\begin{aligned} \mu(\mathrm{t}) &= \mu + \boldsymbol{\mu}_1{}'\boldsymbol{\Delta}(\mathbf{t}-) + \delta_\mu(t), \\ \boldsymbol{\varphi}(\mathbf{t}) &= \boldsymbol{\varphi} + \boldsymbol{\varphi}_1'\boldsymbol{\Delta}(\mathbf{t}-) + \delta_\varphi(t), \\ \boldsymbol{\psi}(\mathbf{t}) &= \boldsymbol{\psi} + \boldsymbol{\psi}_1'\boldsymbol{\Delta}(\mathbf{t}-) + \delta_\psi(t), \end{aligned} \tag{7.5.3}$$

where $\boldsymbol{\mu}_1, \boldsymbol{\varphi}_1$, and $\boldsymbol{\psi}_1$ are coefficient vectors denoting the effects of elements of $\boldsymbol{\Delta}(\mathbf{t})$. Substituting expressions for $\mu(t), \varphi(t)$, and $\psi(t)$ in (7.5.3) into (7.5.2), we have a reduced, second-order GARCH model:

$$\begin{aligned} e_R(t)^2 = \mu &+ \boldsymbol{\varphi}'\mathbf{e}_\mathbf{R}^2(\mathbf{t}-) + \boldsymbol{\psi}'\boldsymbol{\Delta}(\mathbf{t}-) + \boldsymbol{\mu}_1{}'\boldsymbol{\Delta}(\mathbf{t}-) + \mathbf{e}_\mathbf{R}^2(\mathbf{t}-)'\boldsymbol{\varphi}_1\boldsymbol{\Delta}(\mathbf{t}-) \\ &+ \boldsymbol{\Delta}(\mathbf{t}-)'\boldsymbol{\psi}_1\boldsymbol{\Delta}(\mathbf{t}-) + \Delta_R(t). \end{aligned} \tag{7.5.4}$$

Alternative modeling schemes include replacing model (7.5.2) with a bilinear GARCH model [which allows for the effects of interactions between elements of $\mathbf{e}_\mathbf{R}^2(\mathbf{t}-)$ and $\boldsymbol{\Delta}(\mathbf{t}-)$ on $e_R(t)^2$] and/or generating time-varying coefficients in terms of ARMA or bilinear processes. Under each modeling scheme, estimations procedures are described in Section 7.3. As discussed in Section 4.4, the foregoing volatility forecasting procedures may be misleading when the model for $D(t)$ is specified incorrectly; see Section 9.4 for model diagnostics and alternative modeling procedures.

7.6 MODELING MONETARY GROWTH DATA

Kim and Nelson (1989) present an analysis of U.S. quarterly data (1964:1–1985:4) on monetary growth. The following time-varying parameter model was applied under the assumption that regression coefficients follow a random walk:

$$\begin{aligned} DM1(t) = \beta_0(t) &+ \beta_1(t)DINT(t-1) + \beta_2(t)INF(t-1) \\ &+ \beta_3(t)SURP(t-1) + \beta_4(t)\mathrm{DM1}(t-1) + \varepsilon(t), \end{aligned} \tag{7.6.1}$$

$$\beta_i(t) = \beta_i(t-1) + \delta_i(t), \qquad i = 0, 1, \ldots, 4, \tag{7.6.2}$$

where *DM1, DINT, INF*, and *SURP* denote, respectively, the change in M1 growth rate, the change in the interest rates on three-month Treasury bills, inflation measured by the Consumer Price Index, and the full employment budget surplus. Kalman filtering estimation (Schneider, 1988) is applied in obtaining the following variance estimates for $\varepsilon(t)$ and $\delta_i(t)$:

$$s_\varepsilon^2 = 0.126284, \quad s_{\delta 0}^{20} = 0.012133, \quad s_{\delta 1}^2 = 0.000896, \quad s_{\delta 2}^2 = 0.074544,$$
$$s_{\delta 3}^2 = 0.000683, \quad \text{and} \quad s_{\delta 4}^2 = 0.001184. \qquad (7.6.3)$$

Assuming constant coefficients and applying ordinary least squares estimation in (7.6.1), we have

$$\begin{aligned} DM1^\wedge(t) &= 0.496 + 0.300_{3.71} DINT(t-1) - 0.484_{7.18} INF(t-1) \\ &\quad + 0.174_{2.44} SURP(t-1) - 0.694_{3.80} DM1(t-1). \qquad (7.6.4) \end{aligned}$$

Here $DM1^\wedge(t)$ denotes the predicted value of $DM1(t)$; the coefficient subscripts are corresponding values of $|t|$, where t denotes Student's t statistic.

As in (7.1.2), the time-varying coefficients in (7.6.1) are assumed generated in terms of lagged shocks:

$$\beta_i(t) = \beta_i + \beta_{i1}\varepsilon(t-1) + \cdots + \beta_{iq}\varepsilon(t-q) + \delta_i(t). \qquad (7.6.5)$$

Substitution of (7.6.5) into (7.6.1) yields a reduced equation with numerous second-order terms. Following the variable selection and reduced model estimation procedures discussed in Sections 7.2 and 7.3, we obtain the following bilinear equation:

$$\begin{aligned} DM1^\wedge(t) &= 0.703 + 0.383_{4.69} DINT(t-1) - 0.538_{8.26} INF(t-1) \\ &\quad - 0.505_{2.96} DM1(t-1) + 0.221_{3.30} INF(t-1)e_R(t-1) \\ &\quad + 0.066_{2.42} INF(t-1)e_R(t-2), \qquad (7.6.6) \end{aligned}$$

where $e_R(t-j)$ denotes the estimate of $\varepsilon(t-j)$. Note that $SURP(t-1)$ in (7.6.1) has been replaced by the interactions $INF(t-1)e_R(t-1)$ and $INF(t-1)e_R(t-2)$ in (7.6.6), which may explain the relatively large value of the $INF(t-1)$ variance estimate ($s_{\delta 2}^2$) in (7.6.3).

Finally, (7.6.6) is examined for time-varying variation in terms of GARCH modeling, with the following result:

$$e_R(t)^{2\wedge} = 0.141 + 0.210_{2.26} e_R(t-1)^2 + 0.349_{3.84} e_R(t-3)^2. \qquad (7.6.7)$$

For the data under study, there is no evidence of time-varying coefficients for the volatility model.

7.7 MODELING GNP DEFLATOR GROWTH

In introducing GARCH processes, Bollerslev (1986) models quarterly data (1948.2–1983.4) on the rate of growth in the implicit GNP deflator in the United States. With $GD(t)$ denoting the implicit price deflator for GNP, $u(t) = 100 \log_e(GD(t)/GD(t-1))$ is modeled in terms of an AR(4) process complemented by a GARCH(1, 1) process:

$$u^\wedge(t) = 0.141 + 0.433_{5.34}u(t-1)0 + 0.229_{2.08}u(t-2) + 0.349_{4.33}u(t-3) - 0.162_{1.56}u(t-4); \quad (7.7.1)$$

$$e_R(t)^{2\wedge} = 0.007 + 0.135_{1.93}e_R(t-1)^2 + 0.829_{12.2}E[e_R(t-1)^2]. \quad (7.7.2)$$

Note: In Bollerslev's analysis, the GARCH(1, 1) equation, given by

$$E_c(e_R(t)^2) = \mu + \varphi e_R^2(t-1) + \psi\{e_R^2(t-1) - E_c[e_R^2(t-1)]\},$$

is written

$$E_c(e_R(t)^2) = \mu + [\varphi + \psi][e_R^2(t-1)] - \psi E_c[e_R^2(t-1)],$$

where E_c denotes conditional expectation.

Analysis of comparable quarterly data from 1955.1 to 1988.4 gives the following result:

$$u^\wedge(t) = 0.0020 + 0.365_{4.23}u(t-1)0 + 0.318_{3.62}u(t-2) + 0.183_{2.19}u(t-3), \quad (7.7.3)$$

$$E_c(e_R(t)^2) = 0.2(10^{-4}) + 0.233_{2.67}e_R(t-3)^2. \quad (7.7.4)$$

Comparison of (7.7.3) with (7.7.1) indicates changes in the $u(t)$ model between the two overlapping time segments, at least when the $u(t)$ models are limited to first-order AR processes with nonvarying coefficients. Given a lack of justification for nondynamic modeling, the model structure is reevaluated to include consideration of ARMA-type processes for $u(t)$ with time-varying coefficients.

Variable selection procedures (discussed in Section 7.2) uncovered significant effects of three-factor interactions involving lagged shocks. If such interactions are credible, several modeling options are indicated. For the option that is chosen, $u(t)$ is conjectured to follow an ARMA process with time-varying coefficients generated in terms of second-order MA processes.

(For the latter processes, the first-order MA processes were augmented to include interactions between the lagged shocks.) The following third-order reduced-form equation resulted from the variable selection and estimation procedures:

$$\begin{aligned} u^{\wedge}(t) &= 0.002 + 0.370_{4.31} u(t-1) + 0.340_{3.87} u(t-2) + 0.169_{1.97} u(t-3) \\ &\quad + 0.584_{2.85} e(t-3)e(t-5) - 1.33_{2.21} e(t-1)e(t-2)e(t-3). \end{aligned} \tag{7.7.5}$$

In view of the second- and third-order interactions in (7.5.5), additional studies are required to evaluate forecasts based on fitted models that are biased.

The accompanying GARCH modeling residuals show no evidence of heteroskedasticity. One possible implication of model (7.7.5) is that the effect of $e(t-3)^2$ in model (7.7.4) reflects the direct effect of the two- and three-factor shock interactions in model (7.7.5).

8

Single-Equation Modeling: Sports Gambling Markets

8.1 EFFECTS OF INTERACTIVE GAMBLING SHOCKS

In this book emphasis is on adaptive drift modeling in single time series and in simultaneous time series that are cointegrated. Earlier (Mallios, 2000), the writer ignored cointegration and focused on modeling single time series in the markets under study. This was not a serious omission for modeling game outcomes since lagged gambling shocks for each of the opposing teams have significant effects on game outcomes and also partially reflect the effects of cointegration when an equation is part of a system of equations. In this chapter we illustrate adaptive drift modeling in single equations prior to its application to cointegrated series in Chapters 9 and 10.

For modeling outcomes in sports gambling markets, known gambling shocks should always be distinguished from unknown statistical shocks. Recall from Section 1.2 that the gambling shock for the difference in points scored by opposing teams is given by *GSD* = (difference in scores) – (the line on the difference), while the gambling shock for the total points scored is given by *GST* = (total points scored by both teams) – (the line on the total).

Since the known lagged gambling shocks are correlated with the unknown lagged statistical shocks, the former may largely (although not entirely) reflect the effects of the latter in single-equation sports forecasting

Forecasting in Financial and Sports Gambling Markets: Adaptive Drift Modeling, By William S. Mallios

models. As such, in the initial stages of model building, the models (7.1.1) and (7.1.2) may be simplified by approximating the MA terms (the lagged statistical shocks) in terms of known lagged gambling shocks. Specifically, $\boldsymbol{\varepsilon}(\mathbf{t-})$ in (7.1.1)–(7.1.2) is replaced by $\mathbf{GSD(t-)}$ and $\mathbf{GST(t)}$:

$$E[D(t)] \approx \alpha_{\mathbf{D}}(\mathbf{t})'\mathbf{D(t-)} + \gamma_{\mathbf{GSD}}(\mathbf{t})'\mathbf{GSD(t-)} + \gamma_{\mathbf{GST}}(\mathbf{t})'\mathbf{GST(t-)} + \boldsymbol{\beta}(\mathbf{t})'\mathbf{x(t)}, \tag{8.1.1}$$

$$E[\boldsymbol{\psi}(\mathbf{t})] \approx \boldsymbol{\psi} + \boldsymbol{\psi}_1'\mathbf{GSD(t-)} + \boldsymbol{\psi}_2'\mathbf{GST(t-)}. \tag{8.1.2}$$

The coefficient vectors $\boldsymbol{\psi}(\mathbf{t}), \psi, \psi_1,$ and,$\boldsymbol{\psi}_2$ are each replaced, in turn, by $\boldsymbol{\psi} = \alpha_{\mathbf{D}}, \gamma_{\mathbf{GSD}}, \gamma_{\mathbf{GST}}$, and $\boldsymbol{\beta}$ [i.e., $E[\alpha_{\mathbf{D}}(\mathbf{t})] \approx \alpha_{\mathbf{D}} + \alpha_{\mathbf{D1}}'\mathbf{GSD(t-)} + \alpha_{\mathbf{D2}}'\mathbf{GST(t-)}, E[\gamma_{\mathbf{GSD}}(\mathbf{t})] \approx \gamma_{\mathbf{GSD}} + \gamma_{\mathbf{GSD1}}'\mathbf{GSD(t-)} + \gamma_{\mathbf{GSD2}}'\mathbf{GST(t-)}$, etc.]. The symbol $\approx$ denotes *approximated by* since added effects of lagged statistical shocks have been excluded.

The reduced equation for $E[D(t)]$ is obtained by substituting coefficient expressions for coefficients in (8.1.2) into (8.1.1):

$$E[D_R(t)] \approx \alpha'\mathbf{D(t-)} + \gamma_{\mathbf{GSD}}'\mathbf{GSD(t-)} + \gamma_{\mathbf{GST}}'\mathbf{GST(t-)} + \boldsymbol{\beta}'\mathbf{x(t)} + \mathbf{w}_1\mathbf{(t-)}'M\,\mathbf{w}_2\mathbf{(t-)}. \tag{8.1.3}$$

Effects of cross products between elements of $\mathbf{w}_1\mathbf{(t-)}' = (\mathbf{GSD(t-)}', \mathbf{GST(t-)}')$ and elements of $\mathbf{w}_2\mathbf{(t-)} = (\mathbf{D(t-)}', \mathbf{GSD(t-)}', \mathbf{GST(t-)}')'$ are represented by elements of the matrix M. All variables on the right-hand side of (8.1.3) are known and can be scanned for significance as in step 2 of Section 7.2. The entry of significant predictors involving lagged gambling shocks indicates a modeling simplification relative to financial modeling since they are known and partially reflect the unknown lagged statistical shocks. Added effects of the latter shocks may then be recovered through steps 3 through 6 in Section 7.2. Following model identification, the nonlinear estimation procedure in Section 7.3 is then applied if lagged statistical shocks and bilinear terms are found to be significant.

8.2 END OF AN ERA: MODELING PROFILE OF THE 1988–1989 LOS ANGELES LAKERS

The 1980s were the Los Angeles Lakers' (LAL) decade. The time had come for Kareem Abdul-Jabbar's farewell. Would LAL "three-peat" as NBA champions? In the Western Conference playoffs, LAL swept Portland, Seattle, and then Phoenix. In the Eastern Conference, the Detroit Pistons swept Boston, Milwaukee, and then dismembered the Chicago Bulls in six games.

Los Angeles Times sportswriter Mike Downey provided LAL coach Pat Riley with a scouting report on Detroit:

- *Dennis Rodman, alias "the worm." Leading characteristics: Great rebounder, serious defender, and major showboat.*
- *William (Bill) Laimbeer, alias unprintable. Leading characteristics: Thug, gangster, cheap shot artist.*
- *Isiah Lord Thomas III, alias "Zeke." Leading characteristics: Perennial all-star, explosive scorer. Has a tendency to kiss certain opponents on face before tipoff.*
- *Joe Dumars III, no alias. Leading characteristics: Steady as guards come, underrated in every phase of the game. Gentleman on and off court.*
- *Vincent Johnson, alias V.J. Leading characteristics: Shoots lights out. Shots have no arc whatsoever.*
- *If officials permit Pistons to play as physically as they did against Bulls without calling fouls, the Lakers will find themselves in a war.... A Piston's idea of no-harm, no-foul is if opponent regains consciousness within 5 minutes.* (*Los Angeles Times,* June 4, 1989)

This was the group that Kareem Abdul-Jabbar, Magic Johnson, James Worthy, Michael Cooper, and Byron Scott had to face. It was the Hollywood good guys versus the Motor City bad guys. And, of course, the good guys lost. Fifteen years later, a similar scenario was repeated when Detroit ended the Shaquille O'Neal–Kobe Bryant LAL saga in the 2004 NBA finals.

The LAL forecasting model for the 1989 play-offs was based on regular-season games for 1988–1989 and 1987–1988 games for both the regular season and play-offs. By play-off time, the LAL forecasting model had stabilized, as was evidenced by only marginal changes in updates of the reduced forecasting equation between successive play-off games. In the model (8.1.1)–(8.1.2), effects of **GST(t−)** were excluded since lines on point totals were not easily accessible (at least not through the news media) in the late 1980s. Based on the modeling procedure discussed in Section 8.1, forecasting equation (8.2.1) is based on nonlinear, weighted least squares estimation. Relative to the line on the difference, the equation provided correct forecasts in 75 % of the LAL play-off games.

$$
\begin{aligned}
D^{\wedge}(i = \text{LAL}, t) = {} & 0.977_{7.47} LD(i,t) - 1.126_{3.23} WS(i^*,t^*) \\
& + 0.378_{3.41} H(i,t) GSD(i^*,t-1) \\
& + 0.043_{2.25} LD(i^*,t^*-1) GSD(i^*,t^*-2)
\end{aligned}
$$

$$+0.025_{2.15} GSD(i^*, t^*-1) GSD(i^*, t^*-2)$$

$$+0.020_{2.35} H(i,t) GSD(i^*, t^*-1) e_R(i, t-2)^\wedge. \quad (8.2.1)$$

Individual variables are defined as follows:

$D^\wedge(i = \text{LAL}, t)$ = difference predicted in final score between LAL (team i)and its opponent(team i^*)in what was game t for LAL and game $t^*$for its opponent

$LD(i = \text{LAL}, t)$ = line on the difference for the Lakers

$WS(i^*, t^*)$ = number of consecutive games won or lost by LA's opponent just prior to the present game

$H(i = \text{LAL}, t)$ = 1if a home game for LAL and−1if an away game

$GSD(i = \text{LAL}, t - k)$ = lagged gambling shock on the score difference for LAL in game $t - k, k > 0$

$GSD(i^*, t^* - k)$ = lagged gambling shock on the difference for the LAL opponent in game $t^* - k$

$e(i = \text{LAL}, t - k)^\wedge$ = estimated statistical shock for LAL in game $t - k$

The coefficient of $L(i = \text{LAL}, t)$ is nearly 1, so that LAL forecasts amount to adjustments of the line based mostly on lagged shock interactions (which indicate that past volatility affects current expectations directly). The negative coefficient of $WS(i^*, t^*)$ indicates that $D^\wedge(i = \text{LAL}, t)$ is reduced (increased) by 1.126 points for each game of LA opponent's winning (losing) streak. Collecting terms involving $GSD(i^*, t^*-2)$ and substituting $D(i^*, t^*-1) - GSD(i^*, t^*-1)$ for $L(i^*, t^*-1)$, we have a stability condition: $0.043D(i^*, t^*-1) - 0.018GSD(i^*, t^*-1)GSD(i^*, t^*-2)$; that is, an opponent coming off a win $[D(i^*, t^*-1) > 0]$ tended to enhance LAL performance unless the opponent was on a roll $[GSD(i^*, t^*-1) > 0$ and $GSD(i^*, t^*-2) > 0]$.

Note that the lagged statistical shock $e_R(i = \text{LAL}, t-2)^\wedge$, as part of a three-factor interaction, is delayed, in the sense that $e_R(i = \text{LAL}, t-2)^\wedge$ enters as a predictor in place of $e_R(i = \text{LAL}, t-1)^\wedge$. In explanation, recall that a statistical shock is a deviation from *what should have happened*, which usually differs from the line and tends to reflect a deviation from the coach's perception of how his team should have performed. With so many games in the NBA season, the coach's means of effecting change in response to his team's performance in a particular game may not be immediate but, instead, may have delayed effects.

Collecting terms involving $H(i = \text{LAL}, t)$ interactions, we have

$$H(i = \text{LAL}, t)GSD(i^*, t-1)[0.378 + 0.020e_R(i = \text{LAL}, t-2)^\wedge].$$

At home $[H(i = \text{LAL}, t) = 1]$, where the Lakers lost very few games, their performance was enhanced when their opponent's previous outing was better than expected $[GSD(i^*, t-1) > 0]$ and their own past performance was better than expected $[e_R(i = \text{LAL}, t-2)^\wedge > 0]$. On the road, where peak performance is more difficult to maintain, these effects are often negative.

As will be illustrated by the analyses results in Chapter 10, current-day modeling profiles of NBA teams differ from those of 20 years ago. During the Pat Riley years in Los Angeles, lagged gambling shocks tended to dominate modeling results for Western Conference teams, whose run-and-shoot offenses were geared to produce fan-pleasing, high-scoring games. With such volatility, the line often went by the wayside. As such, gambling shocks were often large and their effects in subsequent games tended to be more pronounced, as reflected in equation (8.2.1).

During the LAL reign in the 1980s, volatility modeling in financial markets was in its developmental stages and applications of such modeling did not appear in the literature on sports modeling. However, it was soon recognized that volatility in financial markets provided a gambling venue for greater profits—accompanied by commensurate risk. This sparked evolutionary changes in the sports gambling markets in terms of spread betting.

8.3 SPREAD BETTING

Spread betting provides greater profit opportunities, accompanied by greater risks. In the mid-1970s, the Investors' Gold Index (later the IG Index) was founded to offer spread betting on gold prices. By 1985 the transition was to sports spread betting in terms of horse racing. For example, suppose you think that horse A is far superior to horse B. Your bookmaker quotes you odds of *2 lengths at 3 lengths*. You place a *buy* at $100 per length. Even though both horse A and horse B may finish out of the money, suppose that horse A finishes 8 lengths ahead of horse B. You win 8 less 3 = 4($100) = $400; that is, for every whole length greater than 3 lengths that horse A beats horse B, you receive $100. However, suppose that horse B beats horse A by 5 lengths. Then you lose 5 less 2 = 3($100) = $300; that is, for every whole length greater than 2 lengths that horse B beats horse A, you lose $100.

The bookmaker's spread line on the total number of points scored in the New York Giants vs. New England Patriots title game was [52 to 55].

Your choices: (1) *Buy* total points at 55. (To win, the total points must exceed 55.) (2) *Sell* total points at 52. (To win, the total points must be less than 52.) The final score: Giants 17, Patriots 14 = 31 total points. If you sold at 52 for $100, you won (52–31)($100) = $2100. If you bought at 55 for $100, you lost (58–31)($100) = $2700. Risk assessment is obviously of greater importance in spread betting than in conventional betting since losses are not fixed.

Risk assessment, discussed in Chapter 11, may be approached as follows. For a particular game of interest, let $T(t)$ and $LT(t)$ denote, respectively, the total points scored by opposing teams and the line on the total points scored. The procedure for modeling T(t) with time-varying coefficients is the same as that given for modeling $D(t)$ in (8.1.1)–(8.1.3).

Let $T(t) = E[T(t)] + \varepsilon_T(t) = T(t)^\wedge + e_T(t)$, where $E[T(t)]$ denotes the expected value of $T, \varepsilon(t)$ the model error, $T(t)^\wedge$ the forecast, and $e_T(t)$ the residual corresponding to $\varepsilon_T(t)$. Volatility modeling for $e_T(t)^2$ may be approached through either (7.5.1) or (7.5.4), whichever is appropriate.

Let $e_T(t)^{2\wedge}$ denote the variance forecast based on either (7.5.1) or (7.5.4). Then $T(t)^\wedge$ and $e_T(t)^{2\wedge}$ are estimates of the first two moments of some suitable probability distribution for $T(t) \geq 0$. Of the nonsymmetric gamma and exponential-type distributions that are probably suitable, the following Weibull density function is perhaps the most convenient to apply:

$$f(T) = cb^{-c}T^{c-1}\exp[-(x/b)^c], \tag{8.3.1}$$

where $T(t)$ is replaced by T. The cumulative Weibull distribution is given by

$$P(T < T_0) = 1 - \exp[-(T_0/b)^c], \tag{8.3.2}$$

$$E(T) = (b/c)\Gamma(1/c); \qquad \text{variance } (T)$$
$$= (b2/c)\{2\Gamma(2/c) - (1/c)[\Gamma(1/c)]^2\}. \tag{8.3.3}$$

Equating the estimates $T(t)^\wedge$ and $e_T(t)^{2\wedge}$ to the expressions for $E(T)$ and variance(T) in (8.3.3), we obtain method-of-moments estimates of b and c, say, $b^\wedge$ and $c^\wedge$. Substituting these estimates into (8.3.2), we may estimate the probabilities in (8.3.2), which are then used to evaluate the risks associated with either spread betting or conventional betting.

For the Giants vs. Patriots spread of [52 to 55], the critical tail probabilities used to model risk are given by $P(T(t) > 55)$ and $P(T(t) < 52)$. These tail probabilities are inversely related to risk; that is, the higher a particular tail probability, the lower the risk and the greater the amount of the wager.

8.4 MODELING PROFILE OF A DREAM TEAM: THE 1989–1990 SAN FRANCISCO 49ERS

In Super Bowl XXIV, the 1989–1990 49ers were seeking consecutive titles. The opponent was Denver. Bill Walsh had retired the previous year following the 49ers Super Bowl win over Cincinnati. Rookie coach George Seifert proceeded to guide the 49ers to a 14–2 season. The 49ers offense was led by Joe Montana (NFL most valuable player award and offensive player of the year award), wide receivers Jerry Rice and John Taylor, tight end Bret Jones, and running backs Roger Craig and Tom Rathman. Even backup quarterback Steve Young had an exceptional year. The 49ers defense, ranked third in the league, was led by defensive ends Charles Haley and Pierce Holt, linebackers Kenna Turner, Matt Millen, and Bill Romanowski, safety Ronnie Lott, and defensive backs Eric Wright and Chet Brooks.

Denver was trying to avoid becoming the second team, after the Minnesota Vikings, to lose a fourth Super Bowl. It wasn't to be. The 49ers won 55–10. Dan Reeves, coach of the losing Bronchos, commented: *We made some people, some experts, awful smart today*, alluding to the point spread that favored the 49ers by 11.5 to 12 points.

The 49er forecasting model was based on the same procedure used in obtaining the Los Angeles Laker model in Section 8.2. Forecasts were based on the 1988–1989 and 1989–1990 season; the estimation procedure employed nonlinear weighted least squares [as in LAL model (8.2.1)] to account for the increased importance of more recent games in their effect on subsequent games. For the 1989–1990 season, weekly updates of the 49er model provided exceptionally reliable forecasts, beating the line on the difference in 84% of the games. (On average, drift model forecasts in sports are reliable about 66% of the time, although forecast reliability is team specific. Some teams are far more predictable than others.)

Contributing to the forecasting effectiveness is the fact that weekly updates remained remarkably stable over the course of the season, dating back to the 1988–1989 season. As a summary of model updates, the year-end 49er forecasting model for $D(i = \mathrm{SF}, t)$ is as follows:

$$
\begin{aligned}
D^{\wedge}(i = \mathrm{SF}, t) = {} & 0.044_{3.80} LD(i,t) D(i,i^*,t_p) \\
& - 0.286_{5.82} LD(i,t) GSD(i,t-1) \\
& + 1.050_{12.2} L(i,t) WC(i,t) \\
& + 0.192_{2.83} WT(i,t) GSD(i,t-1) \\
& + 0.229_{3.88} WT(i^*,t^*) GSD(i^*,t^*-1). \qquad (8.4.1)
\end{aligned}
$$

Notice the absence of lagged statistical shocks. The implication is that they are largely reflected by the lagged gambling shocks.

Individual variables, in order of their appearance on the left-hand side of (8.4.2), are identified as follows:

$LD(i = \mathrm{SF}, t)$ = line on the difference

$D(i = \mathrm{SF}, i^*, t_p)$ = outcome of the game between the 49ers and their opponent in their previous encounter

$GSD(i = \mathrm{SF}, t-1)$ = gambling shock for the 49ers in their previous game

$WC(i = \mathrm{SF}, t)$ = number of consecutive 49er wins just prior to game t

$WT(i = \mathrm{SF}, t)$ and $WT(i^*, t^*)$ = number of wins in their previous five games for the 49ers and their opponent, respectively

$GSD(i^*, t^*-1)$ = gambling shock for the 49ers' opponent in their previous game

Notice that contrary to Los Angeles Laker model (8.2.1), no lagged statistical shocks entered the forecasting model for the 49ers. For single-equation sports modeling, the lack of MA terms in sports usually reflects exceptional forecasting effectiveness (since the gambling shocks reflect the statistical shocks).

During the 1989–1990, the 49ers were favored in all games except at Philadelphia (in the third week), where they were 1.5-point underdogs. (The 49ers won 38–20.) The 49ers were least favored (by 1 point) against the Los Angeles Rams in an away game. (The 49ers won 30–27.) These two games illustrate exceptions to the more typical modeling result that the lagged gambling shock $GSD(i = \mathrm{SF}, t-1)$ affects $D(i = \mathrm{SF}, t)$ negatively. Collecting terms with $GSD(i = \mathrm{SF}, t-1)$ on the right-hand side of (8.4.2), we have

$$-GSD(i = \mathrm{SF}, t-1)[0.286LD(i = \mathrm{SF}, t) - 0.192WT(i = \mathrm{SF}, t)].$$

For the 49ers, the quantity in brackets was always positive [and hence the effect of $G(i = \mathrm{SF}, t-1)$ on $D(i = \mathrm{SF}, t)$ is negative] except for the two games just mentioned. The implication is that a better/worse-than-expected 49er performance in week $t-1$ [in terms of a positive/negative value of $GSD(i = \mathrm{SF}, t-1)$] tended to degrade/improve 49er performance in week

t [in terms of the negative coefficient of $G(i = \text{SF}, t-1)$] unless the 49ers were underdogs or slight favorites in week t. The 49ers tended to forgo the letdown associated with a better-than-expected win when their next opponent was expected to be formidable.

Note that the positive effect of the interaction $WT(i^*, t^*)GSD(i^*, t^*-1)$ on $D(i = \text{SF}, t)$ indicates an enhanced 49er performance when they played a winning team [as indicated by a larger value of $WT(i^*, t^*)$] that performed better than expected the previous week [as indicated by a positive value of $G(i^*, t^*-1)$]. Positive effects of the interactions $LD(i = \text{SF}, t)D(i = \text{SF}, i^*, t_p)$ and $LD(i = \text{SF}, t)WC(i = \text{SF}, t)$ on $D(i = \text{SF}, t)$ indicate that the point spread should have been increased when the 49ers had beaten their opponent in their previous encounter or when they were in the midst of a winning streak.

The 51–10 Denver loss to the 49ers capped a season of extremes, many of which were suitable for spread betting, discussed in Section 8.3. Regarding the other 1989–1990 play-off teams, Pittsburgh started the season by losing 51–0 to Cleveland, then 41–10 to Cincinnati. Following these losses, Pittsburgh coach Chuck Noll regrouped the Steelers, made the play-offs, beat Houston, and then lost by 1 point to Denver. Cincinnati faded after their Super Bowl appearance a year earlier. However, in week 15, Cincinnati coach Sam Wynche made a statement to the black-bedecked Houston coach Jerry Glanville—his Bengels trashed the Oilers 61–7. The Los Angeles Rams (now the St. Louis Rams) started the season strong, faded, and then finished strong. It was to be the last successful season for John Robinson as Rams coach.

In round one of the play-offs, the 3-point underdog Rams beat Philadelphia 21–7 in a soggy Veterans Stadium. Ram quarterback Jim Everett in a postgame interview stated: *We're a better . . . team because we had to put up with a lot of pregame abuse. . . . I guess that comes with playing Philly and* [Philadelphia head coach] *Buddy Ryan*. Next, the 5.5-point underdog Rams (with a defense that finished the regular season 21st in the league) beat Bill Parcell's favored Giants 19–13 in New York. (In winning the Super Bowl the following year, Parcell was to become the first coach to take three different teams to the finale.) The Cinderella Rams (a 7.5-point underdog) reached midnight against the 49ers 41–3—as did Minnesota and Denver in subsequent games.

8.5 MAJOR LEAGUE BASEBALL: A DATA-INTENSIVE GAME

Formula 1 racing is the most data intensive of all sports. Cars transmit between 15 and 20 gigabytes of data on everything from tire wear to hydraulic pressures and down draft during the two hours or so that a race lasts. Major League Baseball is probably the second-most data-intensive sport. As such, MLB games are more amenable to effective forecasting than are NBA and NFL games, assuming that data on appropriate variables are available and are analyzed properly.

The odds provided by the money line on MLB games are determined largely by the starting pitchers. Late in the 2007 season, the money line on a key Red Sox–Yankees game was NY(+120) at Boston(−130). The starters were Josh Beckett (Boston) and Andy Pettitte (NY). Without the bookmaker's commission, the *true* money line was NY(+125) at Boston(−125); that is, the true odds of Boston's winning were 1 to 1.25. Since *odds to $1* can be converted to probability in terms of the relation

$$\text{probability} = [(\textit{odds to } \$1)+1]^{-1}, \tag{8.5.1}$$

the probability of a Boston (New York) win was 0.556 (0.444) based on the true money line; that is, if this was game t for team i = Boston, the probability of a Boston win is denoted by

$$P(i = \text{Boston}, t) = 0.556. \tag{8.5.2}$$

Similar to NFL and NBA games, a second line, denoted by $LTOT(i,t)$, is on the total runs scored by both teams. Typically, $LTOT(i,t)$ ranges from 7 to 9.

A win probability of 0.556 for Boston has a number of interpretations. The following five interpretations are not exhaustive.

Interpretation 1 If Boston and New York were to play many games under exactly the same conditions, Boston would win 55.6 % of the games. (The fallacy of this interpretation is that, in contrast to tossing a coin, no baseball game is replayed under precisely the same conditions.)

Interpretation 2 55.6 % of the gambling public thinks that Boston will win the game.

Interpretation 3 On a [0,1] scale, 0.556 is a subjective judgment that Boston will win.

Interpretation 4 Boston will score 55.6 % of the total runs scored in the game.

Interpretation 5 In an empirical Bayesian context, 0.556 is the expectation of the distribution of Boston win probabilities in games against the Yankees in recent times.

When *odds* are converted to *probabilities* and *probabilities* are converted to *expected winning margins* (Mallios, 2000), it is often the case that a favorite may win and have a negative gambling shock or an underdog may lose and have a positive shock. However, shocks defined in this manner are meaningless—in the sense of having no significant effects on subsequent game outcomes—since such expected winning margins are mathematical artifices that are typically unknown to opposing team players and the gambling public. In contrast, such gambling shocks have meaning in basketball and football, given that a point spread on the difference is known and understood by all concerned.

However, gambling shocks corresponding to the money line may have meaning when favorites lose or underdogs win and when favorites (underdogs) win (lose) by much more than anticipated. Accordingly, the baseball gambling shock corresponding to the money line for team i in the team(i,t) versus team (i^*,t^*) encounter is defined as follows:

$$
\begin{aligned}
GSD(i,t) &= TRA(i^*,t^*) - P(i,t)TOT(i,t),\\
&= 0 \qquad \text{if } P(i,t) > 0.5 \text{ and } TRA(i^*,t^*)\\
&\qquad\quad > TOT(i,t)/2 \text{ and } G(i,t) < 0,\\
&= 0 \qquad \text{if } P(i,t) < 0.5 \text{ and } TRA(i^*,t^*)\\
&\qquad\quad < TOT(i,t)/2 \text{ and } G(I,t) > 0,
\end{aligned}
\tag{8.5.3}
$$

where $TRA(i^*,t^*)$ is the total runs allowed by team i^* pitchers (starting and relief) in game $t^*, P(i,t)$ is the probability that team i wins, and $TOT(i,t) = TRA(i,t) + TRA(i^*,t^*)$ is the total runs scored by both teams. Under this definition, a favorite (underdog) who wins (loses) cannot have a negative (positive) gambling shock. For example, if $P(i,t) = 0.6$, team i is expected to score 60 % of the runs in that game. If team i scores in excess of 60 % of the total runs, their gambling shock is positive; if they win but score 60 % or less of total runs, their gambling shock is zero. Although heuristic, definition (8.5.3), which follows from interpretation 4, appears to be more useful in modeling relative to competing definitions that were evaluated. The effects of this definition are illustrated in the MLB modeling examples.

Tables 8.5.1–8.5.14 present an extensive but not comprehensive listing of variables associated with the outcome of the game between team(i,t)

TABLE 8.5.1 Major League Baseball: General Variables

$L(i, t) = P(i, t)$; defined in (8.5.2)
Lagged values of $L(i, t)$ and $L(i^*, t^*)$
Lagged values of $D(i, t)$ and $D(i^*, t^*)$
$D(i, i^*, t_p)$: the outcome of the most recent encounter (in game t_p) between teams i and i^*
$LTOT(i, t)$: the bookmaker's line on the total runs scored by both teams
Lagged values of $TOT(i, t)$ and $TOT(i^*, t^*)$
Lagged values of $GSD(i, t)$ and $GSD(i^*, t^*)$; see (8.5.3)
Lagged values of $GST(i, t) = TOT(i, t) - LTOT(i, t)$ and $GTOT(i^*, t^*)$
$W(i, t) = 1$ for home game
$\quad = -1$ for away game
$DN(i, t) = 1$ for day game
$\quad = -1$ otherwise
$NA(i, t) = 1$ for natural turf
$\quad = -1$ for artificial turf
$DE(i, t) = 1$ for intra-division game
$\quad = -1$ for inter-division game
$RL(i, t) = 1$ if starting pitcher for team i^* is right-hander
$\quad = -1$ otherwise
$WT(i, t), WT(i^*, t^*)$: percentage of games won in previous 10 games
$WT(i, i^*, t)$: percentage of games won by team i in games with team i^* prior to game t
$WC(i, t), WC(i^*, t^*)$: consecutive wins ($WC \geqslant 1$) or losses ($WC \leqslant -1$) prior to forthcoming game
$WC(i, i^*, t)$: consecutive wins or losses by team i in encounters with team i^* just prior to game t
$WH(i, t), WH(i^*, t^*)$: winning percentage at home
$WR(i, t), WR(i^*, t^*)$: winning percentage on the road
$WWHO(i, t) = WH(i, t) - WR(i^*, t^*)$ if $W(i, t) = 1$
$\quad = WR(i, t) - WH(i^*, t^*)$ if $W(i, t) = -1$
$WWHO(i^*, t^*) = WH(i^*, t^*) - WR(i, t)$ if $W(i^*, t^*) = 1$
$\quad = WR(i^*, t^*) - WH(i, t)$ if $W(i^*, t^*) = -1$

TABLE 8.5.2 Major League Baseball: Variables Relating to Runs Scored

$D(i, t) = -D(i^*, t^*)$
$RA(i, t)$: total runs allowed by team i starting pitcher in game t
$RRA(i, t)$: total runs allowed by team i relief pitchers in game t
$RS(i, t) = RA(i^*, t^*)$: (total runs scored by team i off team i^* starting pitcher) = (total runs allowed by team i^* starting pitcher)
$RRS(i, t) = RRA(i^*, t^*)$: (runs scored by team i off team i^* relief pitchers) = (total runs allowed by team i^* relief pitchers)
$TRA(i, t) = RA(i, t) + RRA(i, t) = RS(i^*, t^*) + RRS(i^*, t^*)$, $[D(i, t) = TRA(i^*, t^*) - TRA(i, t)]$
$TOT(i, t) = TRA(i, t) + TRA(i^*, t^*)$: total runs scored by opposing teams

and its opponent, team(i^*, t^*). Variables defined in these tables begin with Table 8.5.1: *General Variables* and end with Table 8.5.14: *Shocks Pertaining to Team Offense in Support of the Starting Pitcher in His Previous Start*.

Game-specific variables appear in Tables 8.5.1–8.5.5, while team averages begin in Table 8.5.6 with *Variables Pertaining to Average Performances by All Starting Pitchers*. For example, $IPA(i, t-1)$ in Table 8.5.5 denotes the average innings pitched by all team i pitchers in the last 40 games. If fewer than 40 games have been played, the deficit is made up by games in the latter portion of the previous season.

TABLE 8.5.3 Major League Baseball: Variables Relating to Team *i* Starting Pitcher in Game *t*

$IP(i, t)$: total innings pitched
$HT(i, t)$: total hits allowed
$ER(i, t)$: total earned runs
$BB(i, t)$: total bases on balls
$SO(i, t)$: total strikeouts

TABLE 8.5.4 Major League Baseball: Variables Relating to Team *i* Aggregate Relief Pitching in Game *t*

$RIP(i, t)$: total innings pitched by aggregate relief pitchers
$RHT(i, t)$: total hits allowed
$RER(i, t)$: total earned runs
$RBB(i, t)$: total bases on balls
$RSO(i, t)$: total strikeouts

TABLE 8.5.5 Major League Baseball: Variables Relating to Team *i* Offense in Game *t*

$OAB(i, t)$: total at bats by team i
$OHT(i, t)$: total hits $[OHT(i, t) = HT(i^*, t^*) + RHT(i^*, t^*)]$
$O2b(i, t)$: total two-base hits
$O3B(i, t)$: total three-base hits
$OHR(i, t)$: total home runs
$ORBI(i, t)$: total runs batted in
$OBB(i, t)$: total bases on balls $[OBB(i, t) = BB(i^*, t^*) + RBB(i^*, t^*)]$

TABLE 8.5.6 Major League Baseball: Variables Pertaining to Average Performances by All Starting Pitchers

$IPA(i, t-1), IPA(i^*, t^*-1)$: average innings pitched by all starting pitchers
$HTA(i, t-1), HTA(i^*, t^*-1)$: average hits allowed
$RNA(i, t-1), RNA(i^*, t^*-1)$: average runs allowed
$ERA(i, t-1), ERA(i^*, t^*-1$ earned run average
$BBA(i, t-1), BBA(i^*, t^*-1)$: average bases on balls
$SOA(i, t-1), SOA(i^*, t^*-1)$: average strikeouts

TABLE 8.5.7 Major League Baseball: Variables Relating to Average Performances by All Relief Pitchers

$RIPA(i, t-1), IRPA(i^*, t^*-1)$: average innings pitched by all relief pitchers
$RHTA(i, t-1), RHTA(i^*, t^*-1)$: average hits allowed
$RRNA(i, t-1), RRNA(i^*, t^*-1)$: average runs allowed
$RERA(i, t-1), RERA(i^*, t^*-1)$: earned run average
$RBBA(i, t-1), RBBA(i^*, t^*-1)$: average bases on balls
$RSOA(i, t-1), RSOA(i^*, t^*-1)$: average strikeouts

TABLE 8.5.8 Major League Baseball: Variables Pertaining to Average Performances by a Specific Starting Pitcher

$IPa(i, t_p), IPa(i^*, t_p^*)$: average innings pitched by specific starting pitcher
$HTa(i, t_p), HTa(i^*, t_p^*)$: average hits allowed
$RNa(i, t_p), RNa(i^*, t_p^*)$: average runs allowed
$ERa(i, t_p), ERa(i^*, t_p^*)$: earned run average
$BBa(i, t_p), BBa(i^*, t_p^*)$: average bases on balls
$SOa(i, t_p), SOa(i^*, t_p^*)$: average strikeouts

TABLE 8.5.9 Major League Baseball: Variables Pertaining to Average Performances by Relief Pitchers in Support of a Specific Starting Pitcher

$RIPa(i, t_p), RIPa(i^*, t_p^*)$: average innings pitched by all relief pitchers in support of starter
$RHTa(i, t_p), RHTa(i^*, t_p^*)$: average hits allowed
$RRNa(i, t_p), RRNa(i^*, t_p^*)$: average runs allowed
$RERa(i, t_p), RERa(i^*, t_p^*)$: earned run average
$RBBa(i, t_p), RBBa(i^*, t_p^*)$: average bases on balls
$RSOa(i, t_p), RSOa(i^*, t_p^*)$: average strikeouts

TABLE 8.5.10 Major League Baseball: Variables Pertaining to Average per Team Offense in Support of All Starting Pitchers

$OABA(i, t-1), OABA(i^*, t^*-1)$: average number of at-bats in support of all starting pitchers
$OHTA(i, t-1), OHTA(i^*, t^*-1)$: average number of hits
$O2BA(i, t-1), O2BA(i^*, t^*-1)$: average number of doubles
$O3BA(i, t-1), O3BA(i^*, t^*-1)$: average number of triples
$OHRA(i, t-1), OHRA(i^*, t^*-1)$: average number of home runs
$ORBIA(i, t-1), ORBIA(i^*, t^*-1)$: average number of runs batted in
$OBBA(i, t-1), OBBA(i,^*, t^*-1)$: average number of bases on balls
$OBA(i, t-1), OBA(i^*, t^*-1)$: batting average

TABLE 8.5.11 Major League Baseball: Variables Pertaining to Average per Team Offense in Support of a Specific Starting Pitcher

$OABa(i, t_p), OABa(i^*, t_p^*)$: average number of at-bats in support of starting pitcher
$OHTa(i, t_p), OHTa(i^*, t_p^*)$: average number of hits
$O2Ba(i, t_p), O2Ba(i^*, t_p^*)$: average number of doubles
$O3Ba(i, t_p), O3Ba(i^*, t_p^*)$: average number of triples
$OHRa(i, t_p), OHRa(i^*, t_p^*)$: average number of home runs
$ORBIa(i, t_p), ORBIa(i^*, t_p^*)$: average number of runs batted in
$OBBa(i, t_p), OBBa(i,^*, t_p^*)$: average number of bases on balls
$OBa(i, t_p), OBa(i^*, t_p^*)$: batting average

TABLE 8.5.12 Major League Baseball: Shocks Pertaining to the Starting Pitcher in His Previous Start (in game t_p)[a]

$SIP(i, t_p) = IP(i, t_p) - IPa(i, t_p - 1), SIP(i^*, t_p^*)$: shock for innings pitched by starting pitcher in his last start
$SHT(i, t_p) = HT(i, t_p) - HTa(i, t_p - 1), SHT(i^*, t_p^*)$
$SRN(i, t_p) = RN(i, t_p) - RNa(i, t_p - 1), SRN(i^*, t_p^*)$
$SER(i, t_p) = ER(i, t_p) - ERa(i, t_p - 1), SER(i^*, t_p^*)$
$SBB(i, t_p) = BB(i, t_p) - BBa(i, t_p - 1), SBB(i^*, t_p^*)$
$SSO(i, t_p) = SO(i, t_p) - SOa(i, t_p - 1), SSO(i^*, t_p^*)$

[a] See Table 7.5.3 for variable definitions.

TABLE 8.5.13 Major League Baseball: Shocks Pertaining to Aggregate Relief Pitching in Support of the Starting Pitcher in His Previous Start[a]

$SRIP(i, t_p) = RIP(i, t_p) - RIPa(i, t_p - 1), SRIP(i^*, t_p^*)$: shock for innings pitched by aggregate relief pitching in support of starting pitcher in his last start
$SRHT(i, t_p) = RHT(i, t_p) - RHTa(i, t_p - 1), SRHT(i^*, t_p^*)$
$SRRN(i, t_p) = RRN(i, t_p) - RRNa(i, t_p - 1), SRRN(i^*, t_p^*)$
$SRER(i, t_p) = RER(i, t_p) - RERa(i, t_p - 1), SRER(i^*, t_p^*)$
$SRBB(i, t_p) = RBB(i, t_p) - RBBa(i, t_p - 1), RSBB(i^*, t_p^*)$
$SRSO(i, t_p) = RSO(i, t_p) - RSOa(i, t_p - 1), RSSO(i^*, t_p^*)$

[a] See Table 7.5.4 for variable definitions.

TABLE 8.5.14 Major League Baseball: Shocks Pertaining to Team Offense in Support of the Starting Pitcher in His Previous Start[a]

$SOAB(i, t_p) = OAB(i, t_p) - OABa(i, t_p - 1), SOAB(i^*, t_p^*)$: shock for team offense in support of starting pitcher in his last start
$SOHT(i, t_p) = OHT(i, t_p) - OHTAa(i, t_p - 1), SOHT(i^*, t_p^*)$
$SO2B(i, t_p) = O2B(i, t_p) - O2Ba(i, t_p - 1), SO2B(i^*, t_p^*)$
$SO3B(i, t_p) = O3B(i, t_p) - O3Ba(i, t_p - 1), SO3B(i^*, t_p^*)$
$SOHR(i, t_p) = OHR(i, t_p) - OHRa(i, t_p - 1), SOHR(i^*, t_p^*)$
$SORBI(i, t_p) = ORBI(i, t_p) - ORBIa(i, t_p - 1), SORBI(i^*, t_p^*)$
$SOBB(i, t_p) = OBB(i, t_p) - OBBa(i, t_p - 1), SOBB(i, ^*, t_p^*)$

[a] See Table 7.5.5 for variable definitions.

Averages for individual players begin in Table 8.5.8 with *Variables Pertaining to Average Performances by a Specific Starting Pitcher*. For example, $IPA(i, t_p)$ in Table 8.5.8 denotes the average innings pitched by a starter in his last five games through his previous start in game t_p. In the early stages of the regular season, the average dates back to later starts in the previous season.

Shocks for individual players begin in Table 8.5.12 with *Shocks Pertaining to the Starting Pitcher in His Previous Start (in Game t_p)*. With $IP(i, t_p)$ defined earlier in Table 8.5.3, $SIP(i, t_p)$ in Table 8.5.12 denotes the shock associated with the number of innings pitched by the starting pitcher in his previous start in game t_p. For example, if $SIP(i, t_p)$ is a large positive (negative) value, the starter pitched far more (fewer) innings in his previous start than he had in the past. Such shocks associated with his previous start may affect his effort in the current game by reflecting physiological–psychological–sociological variables that have not been measured.

8.6 WHILE STILL UNDER THE CURSE: MODELING PROFILE OF THE 1990 BOSTON RED SOX

Moralists in the 19th century placed an extraordinary emphasis on probity. When congratulated on his baseball team, Harvard President Charles Eliot responded, "I'm told the team did well because one pitcher had a fine curve ball. I understand a curve ball is thrown with a deliberate attempt to deceive. Surely that is not an ability we should want to foster at Harvard." (Brooks, 6/8/99)

Boston won the World Series in 1918, the year in which the Kaiser surrendered. In 1920, Boston owner Harry Frazee financed the Broadway play *No, No, Nanette* by selling Babe Ruth to the Yankees for $100,000. Their next World Series appearance was in 1946, a year after Hirohito surrendered. They lost to the Cardinals: Ted Williams couldn't hit, and Red Sox third baseman Johnny Pesky managed to freeze when Enos Slaughter rounded third.

There were other memorable Red Sox World Series losses, but none more so in the minds of Red Sox faithful than the 1986 World Series against the Mets. Mookie Wilson's slow-hop grounder skipped between the legs of Red Sox first baseman Bill Buckner to end game six. The Bambino's Curse finally ended when the Red Sox, the symbol of heartbreak and human foible to their faithful, swept the Cardinals in 2004 and the Colorado Rockies in 2006.

The following modeling illustrations are intended to provide added support to *sports quant modeling* that has contributed to the success of general managers such as Billy Bean and Theo Epstein (see Section 2.2). Team/game-specific models for MLB are based on a structural system involving the following endogenous variables defined in Table 8.5.2: $D(i,t)$, $TOT(i,t)$, $RA(i,t)$, $RS(i,t)$, and $RRS(i,t)$. Based on these variables, a proposed structural system contains two equalities:

$$\begin{aligned} \mathrm{D}(i,t) &= RS(i,t) + RRS(i,t) - RA(i,t) - RRA(i,t), \\ TOT(i,t) &= RS(i,t) + RRS(i,t) + RA(i,t) + RRA(i,t), \end{aligned} \tag{8.6.1}$$

and four stochastic equations, one for each of $RS(i,t)$, $RRS(i,t)$, $RA(i,t)$, and $RRA(i,t)$. From the four reduced-form equations that result from a specification of the four stochastic equations, we may obtain reduced-form equations for $D(i,t)$ (*Who will win the game and by how much?*) and $\mathrm{TOT}(i,t)$ (*How many total runs will be scored?*).

Consider the reduced-form model for $D(i = \text{Boston}, t)$ under the assumption of time-varying coefficients. As in (8.1.1), the modeling process is initiated by replacing lagged statistical shocks with lagged gambling shocks:

$$\begin{aligned} E[D(i,t)] \approx\ & \boldsymbol{\alpha}(\mathbf{t})'\mathbf{D}(\mathbf{t}-) + \boldsymbol{\gamma}_{\mathbf{GSD}}(\mathbf{t})'\mathbf{GSD}(\mathbf{t}-) \\ & + \boldsymbol{\gamma}_{\mathbf{GST}}(\mathbf{t})'\mathbf{GST}(\mathbf{t}-) + \boldsymbol{\beta}(\mathbf{t})'\mathbf{x}(\mathbf{t}). \end{aligned} \tag{8.6.2}$$

The vector $\mathbf{x(t)}$ now includes all variables defined in Table 8.5.1 and Tables 8.5.6–8.5.14. Let $\mathbf{S}(t-)$ denote the shocks defined in Tables 8.5.12–8.5.14. Aside from $\mathbf{GSD(t-)}$ and $\mathbf{GST(t-)}$, elements of $\mathbf{S(t-)}$ may also contribute to the time-varying coefficients in (8.6.1). As such, in contrast to (8.1.2), expressions for the time-varying coefficients are augmented to include the effects of $\mathbf{S(t-)}$:

$$E[\boldsymbol{\psi}_t] \approx \boldsymbol{\psi} + \boldsymbol{\psi}_1'\mathbf{GSD(t-)} + \boldsymbol{\psi}_2'\mathbf{GST(t-)} + \boldsymbol{\psi}_3'\mathbf{S(t-)}. \tag{8.6.3}$$

Substitution of (8.6.2) into (8.6.1) leads to a reduced equation for $E[D(i,t)]$ with a multitude of possible predictor variables—all of which are known.

As presented in Section 8.2, model identification in (8.6.2) may be approached in several ways. The first is to reduce the dimensionality of the predictor variables through a factor analysis and to scan, through stepwise regression, factor scores associated with eigenvector with roots sufficiently large. A second approach is simply to scan all the possible predictors directly and retain those with significant effects. For purposes of simplification, we present results based on the second approach and temporarily forgo the factor analysis approach. With a few exceptions, both approaches lead to noncontradictory forecasts.

Following the model identification in (8.6.2), we then substitute expressions in (8.6.3) into (8.6.2) to obtain a higher-order reduced model for $D(i,t)$ that contains MA terms. The MA terms are scanned for significance through the nonlinear weighted least squares estimation procedure described in Section 7.3. With the presence of MA terms in bilinear-type or higher-order ARMA-type models, the model identification and estimation process is tedious.

The fitted forecasting model for $D(i = \text{Boston}, t)$ is based on the latter part of the 1989 season and all 1990 regular season games (excluding the 1990 play-offs). Data from the latter part of the 1989 season are necessary for purposes of determining lags for the earlier games of the 1990 season. The reduced forecasting model for $D(i = \text{Boston}, t)$ is estimated by

$$\begin{aligned}
D(i = \text{Boston}, t)^{\wedge} = {} & 1.136_{4.28} DE(i,t) + 0.136_{2.57} GTOT(i^*, t^*-1) \\
& - 0.413_{2.19} SRBB(i, t_p) \\
& - 0.453_{2.98} SO2B(i, t_p) + 0.628_{2.09} SOHR(i, t_p) \\
& + 0.346_{2.86} SOBB(i, t_p) \\
& - 1.426_{4.49} RIPA(i^*, t_p^*) - 0.194_{2.71} SOHT(i^*, t_p^*) \\
& - 0.362_{3.04} SOBB(i^*, t_p^*) \\
& + 0.879_{2.34} [RRNA(i, t-1) - RRNA(i^*, t-1)]
\end{aligned}$$

$$
\begin{aligned}
&+ 3.287_{2.94}[RSOA(i,t-1)/RIPA(i,t-1)] \\
&+ 3.058_{6.25}[RBBa(i,t_p) - RBBa(i,t_p^*)] \\
&- 0.659_{4.40}SER(i,t_p) + 0.156_{2.56}e_R(i,t-2)SER(i,t_p) \\
&- 1.020_{2.56}e_R(i,t-1)SRHT(i,t_p) \\
&- 0.160_{2.07}e_R(i,t-2)SBB(i^*,t_p^*) \\
&+ 0.097_{2.62}e_R(i,t-2)RIPA(i,t_p).
\end{aligned} \tag{8.6.4}
$$

Model updates during the last weeks of the regular season showed only minor changes in coefficients and model structure.

The Boston model is clearly dominated by variables reflecting pitching performances by starters and relief pitchers. There is every reason to expect the same to hold for current-day MLB games. However, given the evolution of the pitching game over the past decade, quant forecasting models are likely to differ considerably from (8.6.4). The pitching evolution has been attributed to the following factors:

1. *Relievers are more effective than starters because of the way they are used (so now they are being used more).*
2. *All pitchers have become less effective the longer they are in games (so all pitchers are being used for shorter outings).*
3. *The starting rotation could be shortened to four pitchers, even without shortening their average outing.*
4. *Relievers are also not used to their limit.* (A study on the evolution of pitching staff usage, www.livewild.org/bb//pitchingstaff/index.htlm)

In (8.6.4), the positive effect of $DE(i = \text{Boston}, t)$ reflects the relatively poor performance of Boston against American League West teams during that period; that is, a game against an ALW team subtracted 1.136 points, on the average, from $D^\wedge(i = \text{Boston}, t)$. The positive effect of $GTOT(i^*, t^* - 1)$, the gambling shock on the total line for Boston's opponent in their previous game, reflected heightened preparation on Boston's part and/or stability on the part of their opponent. *Stability* is defined as the tendency for larger positive (negative) shocks associated with the outcome of the current game to have negative (positive) effects on the outcome of a subsequent game. The negative effect of $SRBB(i = \text{Boston}, t_p)$, the shock pertaining to aggregate bases on balls by relief pitchers in support of the starting pitcher in his last outing, probably reflects stability; that is, poorer (better) than expected support for the starter in his last outing tended to be compensated by better (poorer) than expected support for the starter in his next start.

Similar to the effect of $SRBB(i = \text{Boston}, t_p)$, effects of the shocks $SO2B(i = \text{Boston}, t_p), SOHR(i = \text{Boston}, t_p)$, and $SOBB(i = \text{Boston}, t_p)$ indicate that Boston's offensive support for its starting pitcher was not so much in terms of their average support for that pitcher but rather, in terms of how well or poorly, relatively speaking, they supported him in his previous start.

These variables also illustrate that $D^\wedge(i = \text{Boston}, t)$ was more influenced by events from the starting pitcher's last start in game t_p than from their last game. The positive effects of the shocks $SOHR(i = \text{Boston}, t_p)$, and $SOBB(i = \text{Boston}, t_p)$ may indicate streaks in Boston's offensive over- or underperformance in terms of home runs and bases on balls in support of the starting pitcher. The negative effect of $SO2B(i = \text{Boston}, t_p)$ may indicate a stability. Or possibly $SO2B(i = \text{Boston}, t_p)$ reflects variables such as the effect of Fenway Park's configuration in enhancing two-base hits.

$RIPA(i^*, t_p^*)$ measures the average innings pitched by relief pitchers in support of the opposition's starting pitcher through his previous start. The negative effect of this variable has led some observers to conclude that relief pitchers should start games and starting pitchers should end games. This strategy fails to recognize that large values of $RIPA(i^*, t_p^*)$ do not necessarily indicate an ineffective starting pitcher. Rather, it may indicate that the starter's effectiveness may end relatively early and that the opposition's manager knows when to pull his starter (based on, say, the number of pitches thrown, pitching velocity and changes in velocity, number of strikes/balls thrown, etc.—all of which should be considered in more extensive model development).

The negative effects of $SOHT(i^*, t_p^*)$ and $SOBB(i^*, t_p^*)$, shocks pertaining to team offense in support of the opposition's starting pitcher in his previous start may reflect stability; for example, the performance of Boston's opponent in game t was enhanced (degraded) if their starting pitcher received better than average (poorer than average) offensive support in his previous start.

The positive effect of $RRNA(i = \text{Boston}, t-1) - RRNA(i^*, t-1)$, the difference in average, aggregate runs allowed by relief pitchers between opposing teams, is counterintuitive—unless it is argued that Boston's offense tended to compensate for its relief pitchers when their aggregate performance was inferior to that of Boston's opponent. There is a logical, positive effect of pitching effectiveness, measured in terms of $[RSOA(i = \text{Boston}, t-1)/RIPA(i = \text{Boston}, t-1)]$, the ratio of the average strikeouts to the average innings pitched by Boston relief pitchers through game $t-1$. There is also a logical, negative effect of $[RBBa(i = \text{Boston}, t_p) - RBBa(i, t_p^*)]$, the average bases on ball by relievers in support of Boston's starting pitcher through his previous start minus the comparable figure for the opponent's starting pitcher.

The negative effect $SER(i = \text{Boston}, t_p) = ER(i = \text{Boston}, t_p) - \text{ERA}(i = \text{Boston}, t_{p-1})$, the earned run shock for Boston's starting pitcher in his previous outing, is counterbalanced by the positive effect of its interaction with $e_R(i = \text{Boston}, t{-}2)$, Boston's statistical shock as of game $t{-}2$. Collecting terms containing $SER(i = \text{Boston}, t_p)$, we have $SER(i = \text{Boston}, t_p)[-0.659 + 0.156e_R(i = \text{Boston}, t{-}2)]$. In cases where $SER(i = \text{Boston}, t_p)$ and $e_R(i = \text{Boston}, t{-}2)$ are opposite in sign, stability is reinforced; for opposite signs, stability is lessened.

The final three interactions involve lagged statistical shocks dating back to either game $t{-}1$ or $t{-}2$. The effects of $SRHT(i = \text{Boston}, t_p)$, the shock in the relief support for the starting in his previous start; $SBB(i^*, t_p^*)$, the shock for the opposition's starting pitcher in his last start; and $RIPA(i = \text{Boston}, t_p)$, the average innings pitched by all relief pitchers in support of Boston's starting pitcher, are affected by values of either $e_R(i = \text{Boston}, t{-}1)$ and $e_R(i = \text{Boston}, t{-}2)$.

There are both interpretational and conceptual difficulties in modeling $D(i = \text{Boston}, t)$ in terms of $D^\wedge(i = \text{Boston}, t)$ in (8.6.3). First, many of the variables appearing in (8.6.3) appear to be aliases for other variables. Second, many of the variables deal with pitching performances. As such, it would be preferable to model individual pitching performances in terms of the number of runs allowed, as illustrated in the following section.

8.7 PORTRAIT OF CONTROVERSY: MODELING PROFILE OF ROGER CLEMENS WITH THE 1990 RED SOX

For effective intragame decision making, modeling of pitcher performance should proceed on an inning-by-inning basis or, preferably, on an out-by-out basis, or even on a pitch-by-pitch basis. To avoid the accompanying task of compiling a database that would allow such microgame modeling, a simplified, aggregate modeling approach is illustrated in terms of Roger Clemens' per game performances with the 1990 Red Sox. A modeling portrait of Clemens' nemesis during that time period, Dave Stewart of the Oakland A's, is discussed in Section 8.8.

Table 8.7.1 presents modeling results for the following four variables associated with Clemens' successive per game performances in 1990: $RA(i = \text{Boston}, t)$, the number of runs allowed by Clemens; $RRA(i = \text{Boston}, t)$, the number of runs allowed by relief pitchers in support of Clemens; $RA(i^*, t^*)$, the number of runs in support of Clemens; and $RRA(i^*, t^*)$, the number of runs in support of Clemens' relief pitchers.

TABLE 8.7.1 Year-end 1990 Pitcher-Specific Models for[a]

$RA(i,t)$ Model		$RRA(i,t)$ Model	
Predictor	Coefficient ($\|t\|$)	Predictor	Coefficient ($\|t\|$)
$WT(i, t)$	4.943 (10.2)	$NA(i, t)$	1.162 (5.82)
$SER(I, t_p)$	−0.294 (2.92)	$WC(i^*, t^*)$	−0.202 (3.79)
$SO3B(i, t_p)$	−1.354 (5.96)	$SRSO(i, t_p)$	0.369 (5.85)
$SOHR(i, t_p)$	−0.886 (4.76)	$SO3B(i, t_p)$	−1.889 (7.00)
$OHRa(i^*, t_p^*)$	−0.986 (3.06)	$O2Ba(i^*, t_p^*)$	0.661 (3.14)
$BBA(i, t-1)$ - $BBA(i^*, t^*-1)$	−1.065 (6.69)	$SOAB(i^*, t_p^*)$	0.125 (3.54)
$BBa(i, t_p)$ - $BBA(i^*, t_p^*)$	0.919 (3.94)	$TOT(i, t-1)$	0.128 (3.75)
$[e(i, t-2)][RL(i,t)]$	0.150 (4.52)	$[GSD(i, t-1)][RL(i, t)]$	−0.333 (4.71)
$[e(i, t-2)][O3Ba(i^*,t_p^*)]$	−0.019 (2.43)	$[e(i, t-1)][SRSO(i^*, t_p^*)]$[b]	−0.110 (4.20)
$RA(i^*,t^*)$ Model		$RRA(i^*,t^*)$ Model	
Predictor	Coefficient ($\|t\|$)	Predictor	Coefficient ($\|t\|$)
$NA(i, t)$	−1.100 (8.88)	$GTOT(i, t-2)$	0.165 (3.79)
$O3Ba(i, t_p)$	−13.150 (7.35)	$SOBB(i, t_p)$	0.479 (4.20)
$SHT(i, t_p)$	0.401 (7.65)	$SO2B(i^*, t_p^*)$	−0.508 (2.48)
$SOBB(i, t_p)$	0.283 (6.05)	$RNA(i, t-1)$ - $RNA(i^*,t^*-1)$	−1.659 (5.31)
$RNa(i^*, t_p^*)$	0.877 (8.30)	$RHTA(i, t-1)$ - $RHTA(i^*,t^*-1)$	1.106 (4.78)
$SBB(i^*, t_p^*)$	−0.244 (4.25)	$O3BA(i, t-1)$ - $O3BA(i^*,t^*-1)$	−3.394 (3.91)
$SSO(i^*, t_p^*)$	−0.164 (3.33)	$RNA(i, t-1)$ - $RNA(i^*,t^*-1)$	−1.124 (4.56)
$O3BA(i, t-1)$ - $O3BA(i^*,t^*-1)$	−3.656 (8.89)	$[GTOT(i, t-1)][GTOT(i, t-2)]$	−0.066 (6.76)
$D1ERA(i, t-1)$[c]	−2.285 (4.39)	$[GSD(i, t-1)][SRRN(i, t-1)]$	−0.308 (3.97)

[a] $RA(i$ = Clemens, $t)$, $RRA(i$ = aggregate relief pitching in support of Clemens, $t)$, $RA(i^*$ = opposition starter, $t^*)$, and $RRA(i^*$ = aggregate relief pitching in support of opposition starter, $t^*)$.
[b] Lagged statistical shocks are based on $D^\wedge(i$ = Boston, $t)$ in (8.6.4).
[c] $D1ERA(i, t-1) = ERA(i, t-1) - ERA(i^*, t^*-1)$ when $ERA(i, t-1) > ERA(i^*, t^*-1)$; = 0, otherwise.

Relative to $D^\wedge(i = \text{Boston}, t)$ in Table 8.6.1, the Clemens modeling profile in Table 8.7.1 provides an alternative and confirmatory prediction for $D(i = \text{Boston}, t)$ for games in which Clemens was the starting pitcher. As with the Boston team model in (8.6.4), lags for early 1990 season games are based on the latter portion of the 1989 season. Conceptually, the Clemens modeling profile is that first step to modeling all individual player performances.

Consider, first, the Clemens model for $RA(i,t)$. Clemens tended to pitch more effectively when recent Red Sox performances had been poor to mediocre. In terms of runs allowed, this effectiveness is reflected by the lead predictor, $WT(i,t)$, Boston's win percentage over the last 10 games, which has a positive coefficient of 4.943; that is, for each loss in Boston's last 10 games prior to game t, Clemens tended to allow nearly 0.5 fewer runs in game t. However, a caveat to the positive effect of $WT(i,t)$ on $RA(i,t)$ is the negative effect of the interaction $[e_R(i,t-2)][O3Ba(i^*,t_p^*)]$. Suppose that for the first term of the interaction, we have $e_R(i,t-2) \ll 0$. This indicates a poor past performance by Boston, in which case Clemens' game t performance should have been enhanced. However, larger values of the other member of the interaction, $O3Ba(i^*,t_p^*)$, indicates a *fast* opposition team (*fast* refers to team speed). Since Clemens did not perform as well against fast teams, the negative effect of this interaction indicates that fast teams tended to degrade his performance even in games where his performance was expected to be enhanced [such as in the aftermath of $e_R(i,t-2) \ll 0$].

Clemens also tended to pitch better against better teams, which is reflected by the negative coefficient of $BBA(i,t) - BBA(i^*,t^*)$; for example, if better teams can be characterized as having higher values of BBA (the average number of bases on ball by a team's offense), a positive value of $BBA(i,t) - BBA(i^*,t^*)$ contributes to the superiority of team i over team i^*.

Clemens tended not to fare as well when the opposition starter was a control pitcher, which is reflected in the positive coefficient of $BBa(i,t_p) - BBa(i^*,t_p^*)$; that is, since control pitchers have lower values of BBa (the average number of bases on ball by the specific pitcher in his previous starts), negative values of $BBa(i,t_p) - BBa(i^*,t_p^*)$ indicate that Clemens' opposition counterpart had better control—which tended to add to the number of runs allowed by Clemens.

The predictor $OHRa(i^*,t_p^*)$, the average per game home runs in support of the opposition starter, has a negative coefficient of −0.986. Large values of this predictor may indicate a mediocre pitcher—in the sense that better starters tend to receive less offensive support (because they usually don't require as much), whereas poorer starters tend to receive more offensive support (because they usually do require such support). Thus, Clemens tended to be more effective when facing a lineup that had to provide more offensive power for its starting pitcher.

The three shock predictors, $SER(i,t_p), SO3B(i,t_p)$, and $SOHR(i,t_p)$, reflect pitcher stability. The negative coefficient of $SER(i,t_p)$ indicates that when Clemens allowed fewer (more) earned runs in a previous start (relative to his earned run average up to that start), he tended to allow more (fewer) earned runs in a subsequent start. The other two shocks, $SO3B(i,t_p)$, and $SOHR(i,t_p)$, pertain to offensive support for Clemens. For example, if in Clemens' previous start, Boston's offense provided an excess of support—in terms of getting more than three base hits and home runs relative to other games in which Clemens was the starter, then Clemens tended to receive less than the expected support in his subsequent start.

The interaction $[e_R(i,t-2)][RL(i,t)]$—with a positive coefficient of 0.150—combines Boston's statistical shock of two games ago [i.e., $e_R(i,t-2) > 0$ if the performance was "good" and $e_R(i,t) < 0$ if it was "poor"] with whether the opposition starter in game t is a right-hander [$RL(i,t) = 1$] or a left-hander [$RL(i,t) = -1$]. In explaining the effect of this interaction, we first note that Boston's offense did better against left-handed pitching, a fact well known to all Boston pitchers. Suppose, in addition, that $e_R(i,t-2) \gg 0$, which indicates a poor Boston performance two games ago and presages an anticipated Boston turnaround. The combination of an anticipated Boston turnaround (due to stability, which may take two games to achieve) and a left-handed opposition starting pitcher in game t may indicate that Clemens' performance did not need to be as sharp (in which case he tended to allow more runs). If, under the same circumstances, the opposition starter was a right-hander, a sharper performance by Clemens was required (in which case he tended to allow fewer runs).

The Clemens model for $RA(i^*,t^*)$ describes how Boston's offense reacted with Clemens as the starter. The first point to be noted is that six of the nine predictors are pitcher specific, either in terms of shocks or of aggregate performance. Thus, Boston's offensive performance was attuned to the abilities of the opposing starter, to Clemens' known abilities, and to the most recent performances by Clemens and his opposition counterpart.

The negative coefficient of $NA(i,t)$ shows that Boston hitters were more productive on artificial turf $[NA(i,t) = -1]$ than on natural turf $[NA(i,t) = 1]$. For this particular equation, the predictor $O3Ba(i,t_p)$ reflects the shock $SO3B(i,t_p)$; that is, major changes in $O3Ba(i,t_p)$ are a result of large values of $SO3B(i,t_p)$. Thus, the negative coefficient of $O3Ba(i,t_p)$ may indicate that Clemens' game t performance was enhanced when increased offensive support (in terms of three base hits) was necessary in his previous start. Another possible explanation is that Clemens competed with his offense—in the sense that when $SO3B(i,t_p) \gg 0$, Clemens reacted by reducing the runs he allowed in his subsequent start.

The ability of the opposing starter is reflected by $RNa(i^*,t_p^*)$, which has a logical, positive coefficient of 0.877; that is, with Clemens pitching, Boston's offense tended to average roughly 0.9 run for every average run allowed by the opposition pitcher prior to game t^*. [*Note:* The coefficient of $RNa(i^*,t_p^*)$ is higher for other Boston starters; that is, other Boston starters tended to require more offensive support than did Clemens.]

The positive coefficient of the shock $SHT(i,t_p)$ indicates stronger (weaker) offensive support for Clemens in game t if he allowed an excess (a deficiency) of hits in his previous start.[1] The shock $SOBB(i,t_p)$,

[1] A lack of offensive support plagued Clemens throughout his career. In 2005, the 42-year-old Clemens was with the Houston Astros. In a midseason game against the Dodgers, *Clemens was victimized by poor run support again. The Rocket allowed two runs in seven strong innings but didn't get a decision as the Astros beat the Dodgers 3–2 . . . in the ninth. . . . His ERA, the best in the major leagues, rose to 1.48 from 1.41 as he settled for his eighth no-decision of the year* (Associated Press, 7/9/05).

pertaining to the excess or deficiency of bases on balls by Boston's offense in support of Clemens' previous start, has a positive coefficient, which may indicate a team streak effect. The negative coefficients of the shocks $SBB(i^*, t_p^*)$ and $SSO(i^*, t_p^*)$ reflect stability of performance on the part of the opposition starting pitcher; for example, an excess of bases on balls and/or strikeouts by the opposing starting pitcher in his previous start tended to result in a deficiency of bases on balls and/or strikeouts in his subsequent start.

The negative coefficient of the difference $O3BA(i, t-1) - O3BA(i^*, t^*-1) = D(O3BA)$ (say) is explained as follows. When higher values of *O3BA* reflect above-average team speed and $D(O3BA) < 0$, Clemens required more offensive support since he tended not to perform as well against fast teams. When higher values of *O3BA* reflect offensive power and $D(O3BA) > 0$, Clemens required less offensive support since he tended to do well against power teams. The negative effect (−2.285) of the difference $D1ERA(i, t-1) = ERA(i, t-1) - ERA(i^*, t^*-1)$, is not unreasonable; that is, the greater the value of $D1ERA(i, t)$, the greater the opposition pitching strength and the fewer the runs by Boston's offense.

Table 7.7.1 also presents modeling results for $RRA(i, t)$, the runs allowed by Boston relief pitchers when Clemens did not go the distance, and $RRA(i^*, t^*)$, the runs allowed by the opposition relief pitchers in support of Clemens' opposition counterpart. Predictors for $RRA(i^*, t^*)$ include lagged gambling shocks, corresponding to the line on the total number of runs scored: $GTOT(i, t-2)$, with a coefficient of 0.165, and the interaction $[GTOT(i, t-1)][GTOT(i, t-2)]$, with a coefficient of −0.066. Taken together, these two predictors reflect stability; for example, the positive contribution of a $GTOT(i, t-2) \gg 0$ is offset by the negative contribution of the interaction when $GTOT(i, t-1) \gg 0$.

Table 8.7.1 also presents modeling results for $RRA(i, t)$, the runs allowed by Boston relief pitchers when Clemens did not go the distance, and $RRA(i^*, t^*)$, the runs allowed by the opposition relief pitchers in support of their starting pitcher. The negative effect of $NA(i, t)$ on $RRA(i, t)$ indicates that Boston's relief pitchers allowed more (fewer) runs on natural (artificial) turf. Thus, Boston's hitters and relief pitchers both performed better on artificial turf. The positive effect of $SRSO(i, t_p)$, the shock corresponding to the total strikeouts by relievers in support of Clemens, indicates a carryover or streak effect on the part of the relievers.

$GSD(i, t)$, the gambling shock on the difference in scores defined in (7.5.2), enters as an interaction in both the $RRA(i, t)$ and $RRA(i^*, t^*)$ models. For the former model, the negative effect of $[GSD(i, t-1)][RL(i, t)]$ indicates that Boston's relievers were more (less) effective when the opposition starter was a right-hander and when Boston performed better (worse) than expected in their previous game. For the $RRA(i^*, t^*)$ model, the negative effect of $[GSD(i, t-1)][SRRN(i, t-1)]$ indicates that the opposition relievers were less effective if the runs allowed by Boston's relievers in game $t-1$ exceeded the norm and if Boston's expected performance in that game was subpar.

Predictors for $RRA(i^*, t^*)$ also include lagged gambling shocks corresponding to the line on the total number of runs scored: $GTOT(i, t-2)$, with a coefficient of 0.165, and the interaction $[GTOT(i, t-1)][GTOT(i, t-2)]$, with a coefficient of −0.066. Taken together, these two predictors reflect stability [e.g., the positive contribution of a $GTOT(i, t-2) \gg 0$ is offset by the negative contribution of the interaction when $GTOT(i, t-1) \gg 0$].

8.8 PITCHER OF THE YEAR IN 1990: MODELING PROFILE OF THE OAKLAND'S BOB GIBSON

The Oakland A's found the San Francisco earthquake far more troublesome than did their opponent, the San Francisco Giants, in the 1989 World Series. And as the favorite in 1990, they played according to form—except for the last four games. The A's had baseball's best pitcher in Bob Gibson and the winningest pitcher in Dave Stewart. With Stewart pitching, the A's usually won, but not as easily. It was often the first game of a series and Stewart's counterpart was typically the opposition's best pitcher. With Welch pitching, the A's also tended to win, but with greater ease. Welch's counterpart tended not to be the opponent's best pitcher, and Oakland's hitters reacted accordingly.

In the American League play-offs, the A's swept the Red Sox to win the pennant. In the last game of the play-offs, it was Stewart vs. Clemens. In the second inning, Clemens showed demonic behavior and was ejected by umpire Terry Clooney. Said Bill Rigney, the A's senior advisor: "*The thing I find amazing about Stew* [Stewart] *is that he always finds a way to win the big game. He's always pitching against a Clemens or* [Chuck] *Finley or* [Dave] *Steib but he always finds a way to win. The only guy I can think of with the same consistency was* [Sandy] *Koufax.*" *A's manager Tony LaRusso added: "When you get right down to it, Stew was the difference in this game, not Clemens or Clooney.*" In 2009, Bob Gibson was elected to the Baseball Hall of Fame, while Clemens was in the process of denying charges of using performance-enhancing drugs.

In the World Series, it was the A's muscle of Mark McGwire (a user of performance enhancing drugs) and José Canseco and the pitching of Stewart and Welch vs. Cincinnati's Billy Hatcher, a 0.265 hitter prior to the series, and José Rijo, a pretty good pitcher. Hatcher got nine hits in 12 at-bats

TABLE 8.8.1 Year-end Pitcher-Specific Models[a]

RA(*i*,*t*) Model		*RA*(*i*,*t*) Model	
Predictor	Coefficient ($\|t\|$)	Predictor	Coefficient ($\|t\|$)
$WC(i,t)$	−0.399 (4.22)	$TOT(i^*,t^*-2)$	−0.170 (4.48)
$WT(i^*,t^*)$	4.087 (4.51)	$SRHT(i,t_p)$	0.673 (8.34)
$G(i^*,t^*-1)$	−0.290 (3.31)	$SO3B(i,t_p)$	1.020 (3.11)
$SRHT(i,t_p)$	−0.307 (2.89)	$SORBI(i,t_p)$	−0.357 (6.06)
$SO3B(i,t_p)$	2.102 (5.49)	$IPa(i^*,t_p)$	0.704 (5.17)
$WZ(i,t)$[b]	−10.248 (4.51)	$RERA(i,t-1)-RERA(i^*,t^*-1)$	−0.707 (4.02)
$RNa(i,t_p)-RNa(i^*,t_p^*)$	−0.683 (2.80)	$OHRA(i,t-1)-OHRA(i^*,t^*-1)$	−0.235 (6.63)
$[G(i,t-1)][SRRN(i,t_p)]$	−0.909 (3.24)	$[G(i,t-1)][SRSO(i^*,t_p^*)]$	0.485 (10.5)
$D2ERA(i,t-1)$[c]	−3.854 (10.80)	$[G(i,t-1)][RL(i,t)]$	0.495 (6.58)
		$D2ERA(i,t-1)$[c]	0.888 (2.95)

[a] $RA(i,t)$ and $RA(i^*,t^*)$; i = Dave Stewart, and i^* = opposing pitcher.
[b] $WZ(i,t) = WR(i,t) - WH(i^*,t^*)$ if $W(i,t) = -1$; $= 0$, otherwise.
[c] $D2ERA(i,t-1) = ERA(i,t-1) - ERA(i^*,t^*)$ when $ERA(i,t-1) < ERA(i^*,t^*)$; $= 0$, otherwise.

during the series and Rijo pitched two very good games. The outcome: *The Oakland lambs were led to slaughter as the Reds swept the A's.*

For the Stewart RA(i,t) model in Table 8.8.1, the negative coefficient of $WC(i,t)$ indicates that the longer the Oakland win steak, the fewer the runs he allowed; that is, in game t, Stewart allowed an average of 0.4 fewer runs for each game of an Oakland win streak through game $t-1$. Conversely, the better Oakland's opponent [in terms of $WT(i^*,t^*)$, the percentage of games won by the opponent over their last 10 games], the more runs he allowed; that is, in game t, Stewart tended to allow an average of 0.4 run more for each game won by Oakland's opponent in their last 10 games.

The negative coefficient of the lagged opposition gambling shock, $G(i^*,t^*-1)$, indicates stability between performances on the part of the opposition (i.e., under stability without caveats, abnormal performance in one direction tends to be followed by subsequent performance in the opposite direction) and/or anticipation on the part of Stewart [i.e., Stewart tended to be more (less) effective against a team whose most recent performance was better (worse) than expected].

The shock corresponding to hits allowed by relief pitchers in support of Stewart in his previous appearance, $SRHT(i,t_p)$, has a negative coefficient of −0.307. The implication is that Stewart compensated for his most recent relief support by pitching better in game t if the relief support in game t_p was relatively weak and "easing up" if that support was relatively strong. The effect of $SRHT(i,t_p)$ appears in both the Stewart RA(i,t) and $RA(i^*,t^*)$ models, though with opposite signs. An implication is that both Stewart and the Oakland offense adjusted following abnormal relief performances in support of Stewart's previous start. The opposing effects of $SRHT(i,t_p)$ in the $RA(i,t)$ and $RA(i^*,t^*)$ models, are another indication of the team synergism of the 1990 A's.

The predictor

$$WZ(i,t) = WR(i,t) - WR(i^*,t^*) \text{ for Oakland road games,}$$
$$= 0 \text{ for Oakland home games}$$

has a negative coefficient, which indicates that the better the opposition on the road, the weaker the Stewart performance. On the other hand, the negative coefficient of $RNa(i,t_p) - RNa(i^*,t_p^*)$ indicates that Stewart's performance was enhanced against better pitchers (as measured in terms of average runs allowed per game).

The following explanation is given for the negative effect of the interaction $[G(i,t-1)][SRRN(i,t_p)]$. If $G(i,t-1) \gg 0$ and $SRRN(i,t_p) \ll 0$, an implication is that the A's were winning—which would have enhanced Stewart's game t performance. [This tends to enforce the effect of $WC(i,t)$ on $RA(i,t)$, as discussed earlier.] If $G(i,t-1) \ll 0$ and $SRRN(i,t_p) \gg 0$, an implication is that the A's were playing poorly, which would have provided motivation for Stewart's game t performance. [In subsequent modeling, this interaction should be redefined—similar to (2.5.1) and (2.5.2)—as two variables, one for when $G(i,t-1)$ and $SRRN(i,t_p)$ are of opposite signs and one when they are of the same sign.]

The predictor $SO3B(i,t_p)$ is of special importance since it occurs in the Clemens $RA(i,t)$ model with a negative coefficient and in the corresponding Stewart and Welch models with positive coefficients. Recall that an excess of average three-base hits can reflect offensive power, team speed, or both. Thus, large values of $SO3B(i,t_p)$ reflect abnormal power or abnormal team speed in game t_p. In the case of the Clemens $RA(i,t)$ model, $SO3B(i,t_p)$ tends to reflect abnormal power, and its negative coefficient may indicate that Clemens required such support in his previous start. Hence, his game t performance tended to compensate for that required support. In the case of the Stewart $RA(i,t)$ model, $SO3B(i,t_p)$ is more a reflection of team speed, and its positive coefficient may indicate "an exceptionally good team effort" in Stewart's previous start. In such instances, there may have been a tendency for Stewart to "ease up" in his subsequent start. It is also likely that the positive effect of $SO3B(i,t_p)$ counteracted the negative effect of $WC(i,t)$—which is another indicator of stability.

Regarding the Stewart model for $RA(i^*,t^*)$, the predictor $SO3B(i,t_p)$ appears in both the $RA(i,t)$ and $RA(i^*,t^*)$ models with positive coefficients. Thus, if $SO3B(i,t_p) \gg 0$, the indication is that Stewart eased up in his subsequent start; it also indicates that Oakland's offense picked up the slack in that subsequent start.

The positive effect of the interaction $[G(i,t-1)][SRSO(i^*,t_p^*)]$ may be interpreted as follows. Recall that coefficients of lagged shocks tend to reflect stability. This interaction illustrates caveats to stability. If $G(i,t-1) > 0$, the effect of the interaction is positive if $SRSO(i^*,t_p^*) > 0$ (which means that Oakland momentum was more likely to be maintained if the opposition starter received abnormally strong relief support in his previous start). If $G(i,t-1) < 0$, the effect of the interaction is negative if $SRSO(i^*,t_p^*) < 0$ (which means that chances of an Oakland turnaround were lessened if the opposition starter received abnormally weak relief support in his previous start). The former case may reflect actions based on decisions by Oakland's manager, and the latter, actions based on decisions by the opposition manager.

The interaction $[G(i,t-1)]$ $[\mathrm{RL}(i,t)]$ has a positive effect in both the team-specific model for Oakland and the Stewart model for $\mathrm{RA}(i^*,t^*)$. The interpretation of this effect is the same for both models.

9

Simultaneous Financial Time Series

9.1 THE CURSE OF HIGHER DIMENSIONALITY

Too often, attention is limited to a single time series when the series under study is, in fact, one of a system of simultaneous time series with feedbacks between series (Quenouille, 1957). This situation is typical in financial and sports gambling markets. Published analyses of simultaneous time series, including allowances for dynamic processes, have tended toward oversimplified modeling assumptions. Such simplifications are understandable given the complexities and perplexities of effectively forecasting price changes or game outcomes (relative to bookmakers' lines) during periods of market inefficiency.

Consider a partial list of such modeling complexities.

1. A tentative model must be specified, typically in terms of a first-order ARMA-type system of simultaneous equations. In this writing, the ARMA-type model is in terms of a reduced regression system that results from an unspecified structural regression system—except for the case of the Major League Baseball structural system in (8.6.1). Estimation of the structural system parameters is not considered in this book. Instead, the focus is on effective forecasting through the reduced system.

2. For effective forecasting, one must, realistically, allow for time-varying coefficients, in the sense that the ARMA-type equations must be transitional to accommodate evolving market conditions. This requires the specification of a second system of equations that is assumed to generate each of the time-varying coefficients.
3. The substitution of expressions for the time-varying coefficients into the ARMA-type reduced system in complexity 1 leads to higher-order, ARMA-type equations. Due to the excessive numbers of possible predictors that comprise such reduced systems, variable selection procedures must be applied to identify significant predictor variables. Stepwise regression procedures for variable selection almost invariably involve trial and error.
4. Given that a tentative reduced system has been identified in item 3—in the sense that a model is chosen to be fitted to the data for purposes of forecasting—one is faced with choosing from a variety of estimation procedures which may or may not include Bayesian learning. Following estimation, the effectiveness of each fitted forecasting model must then be evaluated.
5. Risk assessment then comes into play. This involves the task of modeling volatility associated with forecasts for price changes or game outcomes. Volatility forecasting is typically in terms of GARCH-type systems of equations, where dependent variables are defined by residual squares and cross products that result from the fitted forecasting equations for price changes and game outcomes.
6. Finally, model effectiveness ultimately depends on frequent model updates that address gradual and abrupt drift scenarios. For real-time forecasting under such scenarios, continual model updates are the norm.

These complexities were aptly described by Jenkins and Alavi (1981) as *the curse of higher dimensionality*. The *curse* has many added facets. For example, if long-term relations exist between processes, disequilibria between such relations (termed *between-relation shocks*) are related but not redundant with disequilibria within processes (termed *within-relation shocks* and quantified by MA variables in ARMA processes). Both shock types usually have direct effects on subsequent outcomes within one or more of the simultaneous equations. Effects of the two shock types are typically confounded in the estimation stages.

There is yet another modeling complication. In the human behavioral markets under study, the situation arises periodically where sufficiently large shocks [as in Gould's *punctuated equilibrium* (Gould, 1984)] alter model structure, either temporarily or on a longer-term basis. During these periods, effective forecasting may not be possible. In such situations, the objectives are to anticipate excessive volatility (if at all possible), to recognize the increased forecasting risks, and then to reformulate an effective model structure with the apparent return of stability.

We begin by considering systems of simultaneous, nonstationary time series with non-time-varying coefficients. Long-term relations/associations between such series are said to be *cointegrated* under the following conditions:

Definition 9.1.1: A vector $\mathbf{z(t)}$ of m simultaneous, nonstationary time series is said to be *cointegrated* if each individual series must be differenced [e.g., $d\mathbf{z}(t) = \mathbf{z(t)} - \mathbf{z(t-1)}$] to make it stationary and if there exists a linear combination of $\mathbf{z(t)}$, say, $\boldsymbol{\ell}'\mathbf{z(t)}$, that defines a stationary process for some nonzero vector $\boldsymbol{\ell}$ (termed a *cointegrating vector*).[1]

Cointegration can be illustrated in terms of the simultaneous price and volume movements depicted in candlestick charts. One simplistic measure of price volatility is in terms of the differences between highs and lows, $H(t) - L(t)$. Suppose that $H(t)$ and $L(t)$ can be modeled individually as stationary AR processes:

$$dH(t) = \lambda_{11} dH(t-1) + \lambda_{12} dL(t-1) + \delta_1(t),$$
$$dL(t) = \lambda_{21} dL(t-1) + \lambda_{22} dH(t-1) + \delta_2(t), \qquad (9.1.1)$$

where $d(\cdot)$ denotes first differences, the λ's denote direct effects, and the δ's denote contemporaneous model errors. Aside from possible effects of MA terms, bilinear terms and time-varying coefficients, a shortcoming of (9.1.1) is that *valuable long-run information may be lost* if there exists a long-term relation between $H(t)$ and $L(t)$. Given such a relation, the between-relation shocks may affect subsequent changes in $dH(t)$ and $dL(t)$.

From Definition 9.1.1, the vector $(H(t), L(t))'[= \mathbf{z(t)}, \text{say}]$ is defined by two simultaneous, nonstationary time series, each of which have been differenced to make them stationary, as in (9.1.1). Then $H(t)$ and $L(t)$ are cointegrated if there exists a linear combination between the two variables: say,

$$H(t) = \ell_0 + \ell_1 L(t) + \varepsilon(t), \qquad (9.1.2)$$

[1]Recall that a *stationary* time series is a stochastic process whose joint probability distribution does not change when shifted in time or space. As a result, parameters such as the mean and variance, if they exist, also do not change over time or position.

that defines a stationary process. To test for stationarity in (9.1.2), the customary procedure is to apply ordinary least squares (OLS) in (9.1.2), estimate $\varepsilon(t)$ in terms of the residuals $e(t)$, and then reapply OLS in estimating the coefficient ρ in the regression equation

$$de(t) = \rho e(t-1) + \delta(t); \tag{9.1.3}$$

$$de(t) = e(t) - e(t-1). \quad \text{If}$$

$$H_0 : \rho = 0 \text{ is rejected in favor of } H_a : \rho < 0, \tag{9.1.4}$$

$H(t)$ and $L(t)$ are cointegrated. If H_0 is not rejected, $H(t)$ and $L(t)$ are not cointegrated (Granger, 1983). MacKinnon (1991) provides tables for testing $H_0 : \rho = 0$ for any sample size and also when the number of regressors in (9.1.2) is increased—as is the case when the $e(t)$ display serial correlation:

$$de(t) = \rho e(t-1) + \boldsymbol{\rho}'\mathbf{de(t-)} + \delta(t), \tag{9.1.5}$$

where $\mathbf{de(t-)}$ contains lags of $de(t)$ with effects denoted by elements of the vector $\boldsymbol{\rho}$.

Equations (9.1.1) to (9.1.5) are illustrated in terms of weekly data from 1/5/99 to 10/29/01 for the NASDAQ Composite index ($COMPX). This time period encompasses the peak of the high-tech bubble; see Figure 9.1.1 for the corresponding candlestick chart that excludes volume trends. Application of OLS in (9.1.2) yields

$$H(t)^\wedge = -21.1 + 1.095_{72.1}L(t), \quad R^2 = 0.973. \tag{9.1.6}$$

Letting $H(t) = H(t)^\wedge + e(t)$, we obtain the following result in fitting (9.1.5) through OLS estimation:

$$\begin{aligned} de(t)^\wedge = 0 &- 0.275_{3.39}e(t-1) - 0.208_{2.29}de(t-1) \\ &- 0.166_{2.00}de(t-2), \quad R^2 = 0.226. \end{aligned} \tag{9.1.7}$$

(Absolute Student t values are given by the coefficient subscripts.) Under (9.1.4)–(9.1.5), $H(t)$ and $L(t)$ are cointegrated.

As a second illustration of cointegration, consider the daily candlestick chart in Figure 9.1.2 for Exxon Mobile (XOM) from 9/1/07 to 2/2/09. Fitting (9.1.2) through OLS, we have

$$H(t)^\wedge = 14.620_{21.9} + 0.854_{106.3}L(t), \quad R^2 = 0.969. \tag{9.1.8}$$

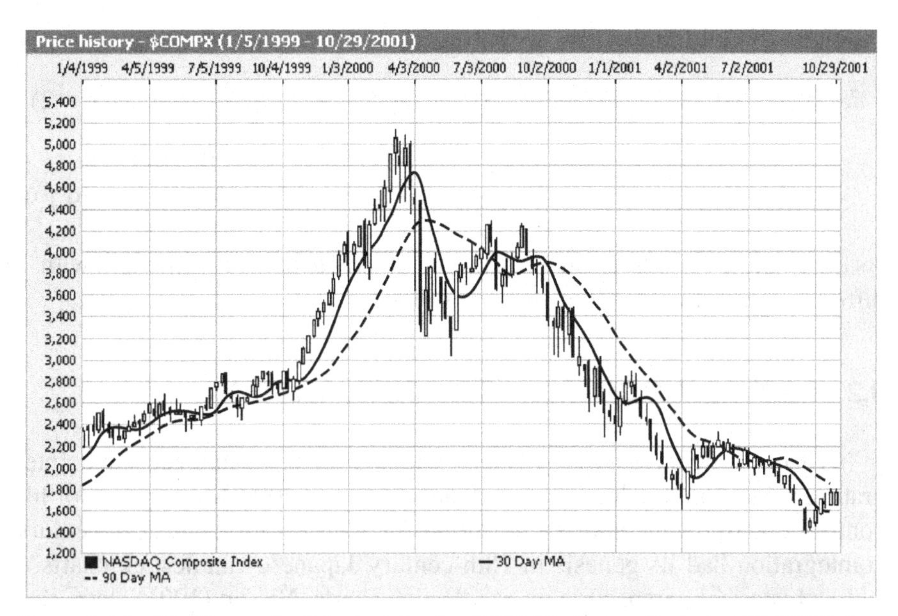

Figure 9.1.1 *Weekly candlestick chart for the NASDAQ Composite Index ($COMPX) from 1/5/99 to 10/29/01.*

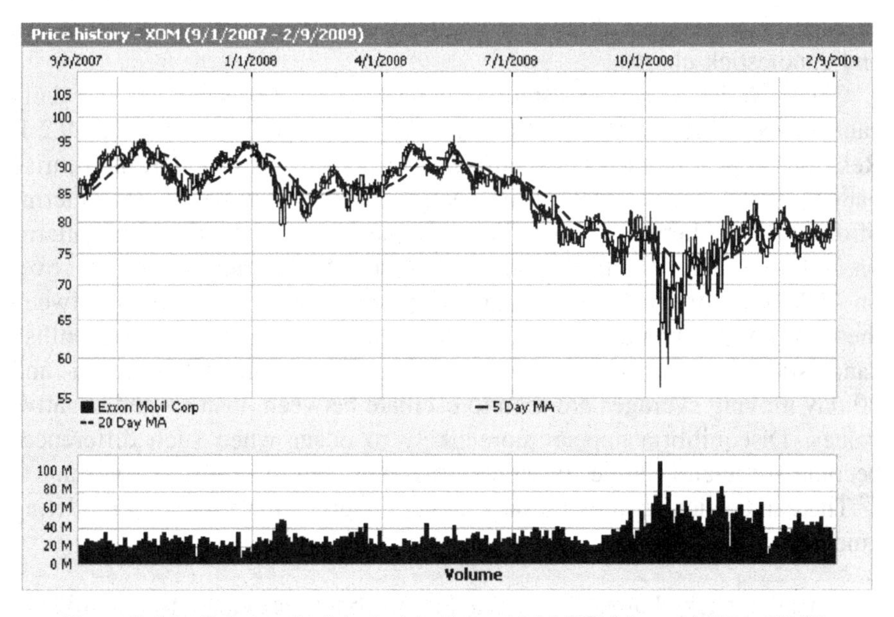

Figure 9.1.2 *Daily candlestick chart for Exxon Mobil (XOM) from 9/1/07 to 2/2/09.*

Fitting (9.1.5) through OLS, we have

$$de(t)^\wedge = -0.233_{4.5.1}e(t-1) - 0.354_{5.52}de(t-1) - 0.155_{2.44}de(t-2)$$
$$- 0.163_{2.71}de(t-3) - 0.050_{0.96}de(t-4), \quad R^2 = 0.273. \tag{9.1.9}$$

As in (9.1.7), (9.1.9) is an illustration of cointegration, although with a different structure regarding serial correlations.

9.2 FROM CANDLESTICKS TO COINTEGRATION

In his 1987 paper with Robert Engle, Clive Granger formalized the cointegrating vector approach and subsequently shared a 2003 Nobel prize for his contribution to the technique's development. Conceivably, late 20th-century cointegration had its genesis in 18th-century Japanese candlestick charts.

In a historical perspective on candlestick charts, Nisson (1991) states that the charts provided their originator with profitable insights into the market psychology of the Osaka Rice Exchange in feudal Japan (see Section 5.1). Since market psychology may be quantified at least conceptually in terms of cointegrated time processes that oscillate between poles of rational and irrational behavior, parallels may be drawn between cointegrated processes and candlestick charts.

Consider the $m = 7$ nonstationary time series described by the daily candlestick chart for the NASDAQ-100 Index (QQQQ) in Figure 9.2.1. Relations appear to exist between the seven series. In particular, the bearish patterns that appear at each of the relative maxima may be viewed in terms of disequilibria between series. For example, the bearish engulfing patterns on 3/20, 4/7, and 5/13 (see boxes 1, 2, 4, and 5) and the dark cloud cover on 4/27 (see box 3) are accompanied by (1) increasing distances between the two moving averages and (2) abrupt changes from consecutive bullish candlesticks to a bearish candlestick. Moreover, distances between 5- and 20-day moving averages are seen to oscillate between positive and negative values. Disequilibria appear more likely to occur when such differences become sufficiently large, in which case they tend to return to a norm.

The candlestick–cointegration analogy underlies the following modeling guidelines.

1. Candlestick charts, as presented in this book, provide a rough guide to near-term forecasting during periods of apparent market inefficiency; or, for the skeptics, they are an informative picture of what has happened in the past.

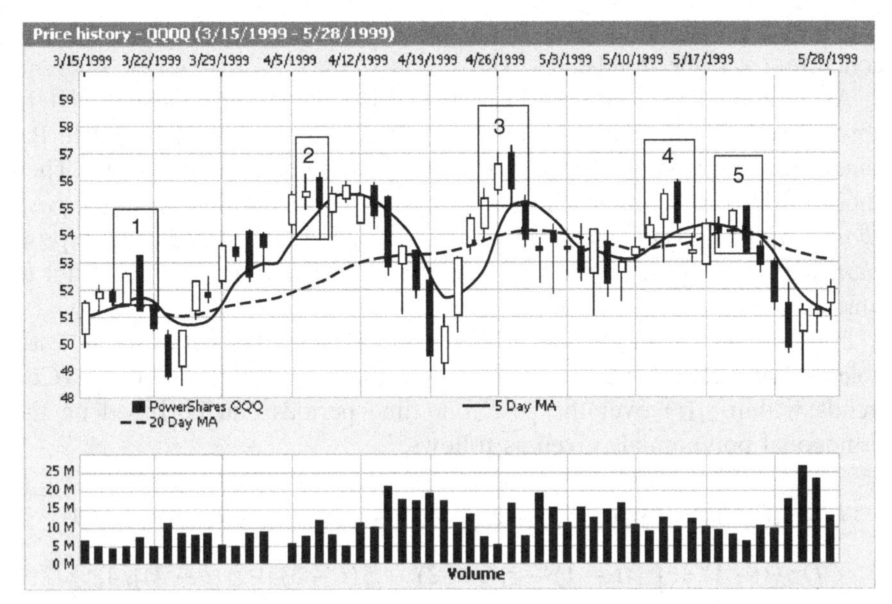

Figure 9.2.1 Daily candlestick chart for QQQQ, an exchange-traded fund related to the performance of the NASDAQ-100 Index.

2. Indeterminate trends may reflect periods of market efficiency or uncertainty (when effective forecasting may be difficult or not possible).
3. When trends are indeterminate, a common recourse is to reevaluate the trend in terms of different time periods; for example, if the candlestick trend defined by days is indeterminate, switch to candlesticks defined by hours or minutes or, more conservatively, possibly by months.
4. Modeling short-term movements in cointegrated time series should go hand-in-hand with accompanying candlestick graphics.
5. Major discrepancies between model forecasts and graphic projections are contentious and require investigation through model updates.
6. Finally, candlestick configurations that are said to be bullish, bearish, or otherwise should be evaluated for predictive validity in the modeling procedure either directly or indirectly.

Due to the large number of bullish and bearish patterns that have been documented in the numerous chartist Web sites, quantifying each individual pattern in modeling evaluations is a major task. Such quantification could begin with the specification of a dummy variable for each pattern—an approach that quickly becomes tedious, especially with the inclusion of candlestick positionings vis-à-vis the oscillating moving average trends and concurrent volume trends.

As an alternative to dummy variables, the following candlestick pattern approximations are explored. For each of the time series depicted in the candlestick chart, suppose that the focus is on the forecast for time t. Then short-term trends within each time series are quantified in terms of the *past, short-term* linear (i.e., straight line), quadratic, and cubic trends, where *past, short-term* is defined in terms of, say, the four or five time periods prior to time t.

For example, if, in Definition 9.1.1, $z_i(t)$ denotes the ith of the m variables in the vector **z(t)**, the linear (LIN), quadratic (QD), and cubic (CB) trends within $z_i(t)$ over the past four time periods can be based on the orthogonal polynomials given as follows:

$$
\begin{aligned}
LINz_i(t-1) &= [3z_i(t-1) + z_i(t-2) - z_i(t-3) - 3z_i(t-4)]/20, \\
QDz_i(t-1) &= [z_i(t-1) - z_i(t-2) - z_i(t-3) + z_i(t-4)]/4, \\
CBz_i(t-1) &= [z_i(t-1) - 3z_i(t-2) + 3z_i(t-3) - z_i(t-4)]/20.
\end{aligned} \tag{9.2.1}
$$

The $m = 8$ variables under study in **z(t)** are $O(t), H(t), L(t), C(t)$, the slower and faster moving averages for $C(t)$, $v(t) = \log_e V(t)$, and the moving average of $v(t)$—as discussed in (5.1.1) for daily candlesticks. The trends in (9.2.1) may be redefined so that they apply, for example, to longer periods of time (such as for five-period patterns 9 and 10 in Table 5.1.1).

The conjecture is that bullish and bearish patterns reflect disequilibria between and/or within cointegrated time series, and that these patterns can be quantified in terms of unusual values of individual short-term trends and/or in terms of disparities between and/or within such trends for at least two of the time series.

For example, with reference to (9.2.1), the bearish engulfing and dark cloud patterns (see patterns 2 and 4 in Table 5.1.1) may be described by the following trend disparities. Prior to time t, a large negative quadratic trend in the closing prices from $t-4$ to $t-1$—denoted by $QD(\text{close}) \ll 0$—is accompanied by positive linear and/or quadratic and/or cubic trends in opening prices—denoted, respectively, by $LIN(\text{open}) > 0$, $QD(\text{open}) > 0$, and $CB(\text{open}) > 0$. Alternative disparities for this pattern could be described in terms of (1) $QD(\text{close}) \ll 0$, accompanied by positive trends through $H(t-1)$ and/or $O(t-1)$ or (2) negative quadratic and/or cubic trends in the candlestick bodies through $B(t-1)$ [where $B(t) = O(t) - C(t)$] accompanied by positive trends in one or more of the other processes.

For the bullish engulfing and piercing patterns (see patterns 1 and 3 in Table 5.1.1), $QD(\text{close}) \gg 0$ is accompanied by negative linear and/or quadratic and/or cubic trends in the corresponding opening prices and highs. Significant quadratic or cubic parallelisms may also indicate changes in directions, such as the distinct quadratic trends in patterns 11 and 12.

Suppose that bullish and bearish configurations do, in fact, reflect disequilibria between and/or within cointegrated processes and are adequately quantified in terms of the trend disparities between and/or within trends comprising the configurations. If so, then disequilibria variables—the confounded within- and between-relation shocks—can then be approximated in terms of the disparity variables.

For larger systems of equations, conventional methods of estimating disequilibria variables (as discussed in Section 9.3) are far more cumbersome than estimating disparity variables. Modeling simultaneous time series is thus simplified when the former can be approximated by the latter. This modeling simplification will be shown to lead to effective forecasting during periods of market inefficiency.

9.3 COINTEGRATION IN TERMS OF AUTOREGRESSIVE PROCESSES

To begin the model identification stage, $\mathbf{z(t)}$, representing the m simultaneous time series in Definition 9.1.1, is modeled in terms of the vector AR(p) process:

$$\mathbf{z(t)} = A_1\mathbf{z(t-1)} + A_2\mathbf{z(t-2)} + \cdots + A_p\mathbf{z(t-p)} + \boldsymbol{\varepsilon}\mathbf{(t)}. \quad (9.3.1)$$

This process is rewritten in the following error correction form (Johansen, 1995):

$$\begin{aligned} d\mathbf{z(t)} &= B\mathbf{z(t-1)} + B_1\mathbf{dz(t-1)} + B_2\mathbf{z(t-2)} + \cdots \\ &\quad + B_{p-1}\mathbf{dz(t-p+1)} + \boldsymbol{\varepsilon}\mathbf{(t)}, \end{aligned} \quad (9.3.2)$$

$$\begin{aligned} d\mathbf{z(t)} &= (z_h(t) - z_h(t-1); \quad B_{i^*} = -(A_{i+1} + \cdots + A_p), \quad i^* \\ &= 1, \ldots, p-1, \\ B \ &= A_1 + \cdots + A_p - I_m. \end{aligned} \quad (9.3.3)$$

Here I_m denotes the $m \times m$ identity matrix. Regarding the model error $\boldsymbol{\varepsilon}\mathbf{(t)}$, it is assumed that $E(\boldsymbol{\varepsilon}\mathbf{(t)}) = \mathbf{0}$ and $E(\boldsymbol{\varepsilon}\mathbf{(t)}\boldsymbol{\varepsilon}\mathbf{(t)}') = \Sigma$:

$$\boldsymbol{\varepsilon}\mathbf{(t)} : (\mathbf{0}, \Sigma). \quad (9.3.4)$$

At a later modeling stage, Σ is assumed to be time varying to accommodate volatility modeling.

The $m \times m$ matrix B in (9.3.3) is rewritten as $B = CD$, where C is on the order $m \times m^*, D$ is of order $m^* \times m$, and $m^* < m$. Regarding

$$B\mathbf{z(t-1)} = CD\mathbf{z(t-1)}C\mathbf{u(t-1)}, \tag{9.3.5}$$

rows of D define the $m^* < m$ cointegrating vectors, while elements of $D\mathbf{z(t-1)} = \mathbf{u(t-1)}$ represent the between-relation disequilibria at time $t-1$; elements of C denote effects of $\mathbf{u(t-1)}$ on $d\mathbf{z(t)}$. The rank of B determines the value of m^*.

When $m = m^*$, (9.3.1) defines a system of stationary rather than non-stationary processes and is not of interest in the current context. When, in (9.3.3), $A_1 + \cdots + A_p - I_m = 0$, then $B = 0$ and there are no cointegrating relations.

Based on notation in (9.3.5), (9.3.2) is rewritten as

$$\begin{aligned} d\mathbf{z(t)} = C\mathbf{u(t-1)} + B\mathbf{z(t-1)} + B_1\mathbf{dz(t-1)} + B_2\mathbf{dz(t-2)} + \cdots \\ + B_{p-1}\mathbf{dz(t-p+1)} + \boldsymbol{\varepsilon}\mathbf{(t)}. \end{aligned} \tag{9.3.6}$$

Johansen's maximum likelihood procedure (1995) for estimating $\mathbf{u(t-1)}$ relaxes the assumption that the cointegrating vectors defined by D in (9.3.5) are unique. D and subsequently $\mathbf{u(t-1)}$ are then estimated indirectly; that is, OLS estimation in (9.3.2) provides the estimate $B^\wedge$ of B, whereupon $D^\wedge$, the estimate of D, is based on the estimated rank of $B^\wedge$; $\mathbf{u(t-1)}$ is then estimated in terms of $D^\wedge\mathbf{z(t-1)}$.

A simpler alternative to the Johansen procedure for estimating $\mathbf{u(t-1)}$ is to apply factor analysis (Afifi and Clark, 1996) as a means of reducing the dimensionality of $\mathbf{z(t-1)}$ in terms of $m^* < m$ linear combinations thereof. If the m^* linear combinations (termed *factors*) define stationary processes, the factor scores associated with the factors provide an estimate of $\mathbf{u(t-1)}$ directly. Suppose, however, that one or more of the factors do not define a stationary process. Aside from the implication that the variables involved in the particular factor are not cointegrated, it may be the case that the between-relation associations are time varying. If so, one alternative is to allow for time-varying effects of disparity variables associated with $\mathbf{z(t-1)}$, as discussed in Section 9.2 and illustrated in the following section.

Model (9.3.6) can be deduced intuitively as follows. Instead of initiating the modeling procedure with the assumption of an AR process in (9.3.1), suppose that we begin with an ARMA($p, q = 1$) process that includes a vector of exogenous variables denoted by $\mathbf{x(t)}$ with coefficient matrix B_x;

that is,

$$d\mathbf{z(t)} = B\mathbf{z(t-1)} + \Sigma_{i>0} B_i \mathbf{dz(t-i)} + B_x \mathbf{x(t)} + C\boldsymbol{\varepsilon}\mathbf{(t-j)} + \boldsymbol{\varepsilon}\mathbf{(t)}. \tag{9.3.7}$$

Then (9.3.2) becomes

$$d\mathbf{z(t)} = B\mathbf{z(t-1)} + \Sigma_{i>0} B_i \mathbf{dz(t-i)} + B_x \mathbf{x(t)} + C[\mathbf{z(t-1)}) - E(\mathbf{z(t-1)}] + \boldsymbol{\varepsilon}\mathbf{(t)}, \tag{9.3.8}$$

where $E(\cdot)$ denotes expectation. If $C[E(z(t-1)]$ is reflected in terms of the other predictor variables, then, approximately,

$$d\mathbf{z(t)} = B^*\mathbf{z(t-1)} + \Sigma_{i>0} B_i \mathbf{dz(t-i)} + B_x \mathbf{x(t)} + \boldsymbol{\varepsilon}\mathbf{(t)}, \tag{9.3.9}$$

where $B^* = B + C$. But (9.3.9) is the same as (9.3.2) except for the covariable vector. By this reasoning it becomes clearer that the $\mathbf{u(t-1)}$ in (9.3.6) is a reflection of both within- and between-relation shocks.

9.4 ESTIMATING DISEQUILIBRIA THROUGH FACTOR ANALYSIS

The factor analysis approach to estimating between-relation disequilibria is illustrated in terms of the data in Figure 9.4.1, a daily candlestick chart of price and volume trends (10/18/09 to 2/27/09) for Baidu (BIDU). (As of late 2009, Baidu held 70% of the Chinese Internet search market, while Google's share was estimated at 26%.)

The eight processes depicted in Figure 9.4.1 include the four time series comprising a candlestick $[O(t), H(t), L(t)$, and $C(t)]$ and two moving averages for $C(t)$, the 5- and 20-day moving averages based on successive days just prior to $C(t)$ (as denoted by *Cb5* and *Cb20*). The lower portion of the chart includes the volume $V(t)]$ and the 5-day moving average of $V(t)$ based on successive days prior to $V(t)$. In the analysis, $v(t) = \log_e V(t)$ replaces $V(t)$, and $vb5$, the corresponding 5-day moving average of $v(t)$, replaces the 5-day moving average of $V(t)$.

Linear combinations of $\mathbf{z(t-1)}$ in (9.3.5) are based on the correlation matrix of the $z_h(t-j), j = 1, \ldots, T$; $h = 1, \ldots, m$. Principal components analysis is applied to the correlation matrix in obtaining $m^* < m$ orthogonal eigenvectors associated with sufficiently large eigenroots. The m^* eigenvectors (or factors) are then rotated to preserve orthogonality and allow for

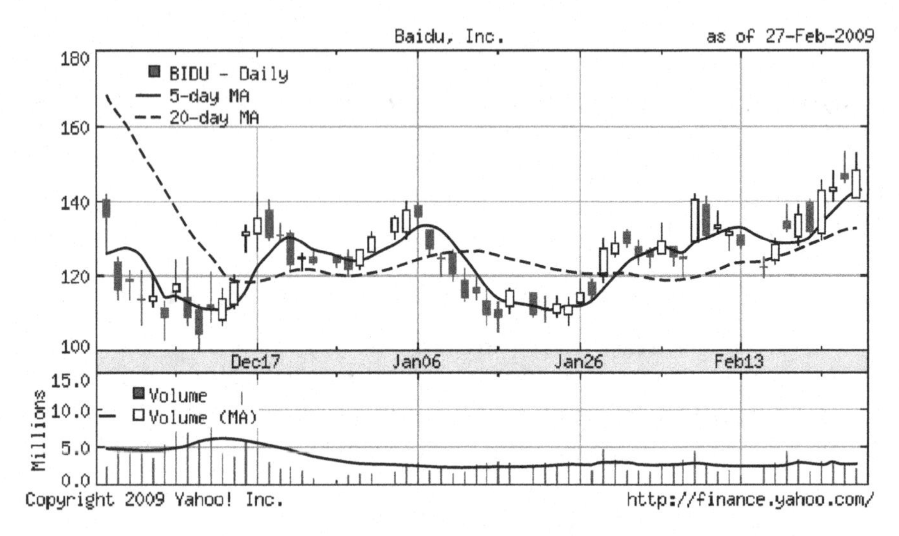

Figure 9.4.1 *Daily candlestick chart for Baidu (BIDU) from 10/18/09 to 2/27/09.*

easier interpretation. If each rotated factor defines a stationary process, the factors are estimates of the cointegrating vectors [given by the rows of D in (9.3.5)]. Moreover, factor scores corresponding to each of the m^* factors provide a direct estimate of elements of $\mathbf{u(t-1)} = D\mathbf{z(t-1)}$.

Table 9.4.1 presents results of principal components analysis applied to the 8×8 correlation matrix associated with the Baidu data. The analysis shows that the first three components (or eigenvectors) account for 90.31% of the information contained in the original eight variables. As a rule of thumb, only eigenvectors with eigenroots > 1 are retained. However, our applications will include eigenvectors with roots *somewhat less than 1* if such linear combinations make sense and can be shown to be stationary processes. For this illustration, the first three components are retained and rotated. Results of varimax rotation, given in Table 9.4.2, show that the first factor is the relation among $O(t), H(t), L(t)$, and $C(t)$, the second between the two moving averages for $C(t)$, and the third between $v(t)$ and its 5-day moving average.

The question remains as to whether the three linear combinations in Table 9.4.2 define stationary processes. To answer this question, the factor scores corresponding to each linear combination are analyzed in terms of model (9.1.4). Denoting the factor scores for the ith linear combination as $u_i(t), i = 1, 2, 3$, and setting $du_i(t) = u_i(t) - u_i(t-1)$, we follow the procedure in (9.1.5) and fit

$$du_i(t) = \rho_i u_i(t-1) + \boldsymbol{\rho}_i' \mathbf{du}_i\mathbf{(t-)} + \delta(t) \qquad (9.4.1)$$

TABLE 9.4.1 Principal Components Analysis Applied to the BIDU Correlation Matrix

	Initial Eigenvalues			Extraction Sums of Squared Loadings			Rotation Sums of Squared Loadings		
Component	Total	Percent of Variance	Cumulative %	Total	Percent of Variance	Cumulative %	Total	Percent of Variance	Cumulative %
1	4.035	50.432	50.432	4.035	50.432	50.432	3.727	46.585	46.585
2	2.369	29.615	80.047	2.369	29.615	80.047	1.972	24.649	71.234
3	0.821	10.263	90.309	0.821	10.263	90.309	1.526	19.075	90.309
4	0.403	5.039	95.349						
5	0.229	2.861	98.210						
6	0.093	1.164	99.374						
7	0.034	0.419	99.793						
8	0.017	0.207	100.000						

TABLE 9.4.2 Rotation of the First Three Eigen Vectors in Table 9.4.1[a]

	Component		
	1	2	3
O	0.906	−0.111	0.306
H	0.955	0.069	0.239
L	0.953	−0.264	−0.002
C	0.960	−0.121	−0.041
Cb5	0.359	0.021	0.900
Cb20	−0.108	0.573	0.730
Lnv	−0.074	0.858	0.172
Invb5	−0.129	0.898	0.016

[a] Rotation method: Varimax with Kaiser normalization.

through OLS estimation. Results are as follows:

$$du_1(t)^\wedge = -0.212_{2.99}u_1(t-1) + 0.323_{2.83}du_1(t-1), \qquad (9.4.2)$$

$$du_2(t)^\wedge = -0.102_{2.33}u_2(t-1) - 0.186_{2.24}du_2(t-4), \qquad (9.4.3)$$

$$du_3(t)^\wedge = -0.113_{2.12}u_3(t-1) - 0.234_{4.04}du_3(t-4), \qquad (9.4.4)$$

where $du_1(t-1), du_2(t-4)$, and $du_3(t-4)$ denote significant serial correlations.

Since results in (9.4.2) to (9.4.4) indicate that each of the three linear combinations defines a stationary process, the factor scores corresponding to the three factors at time $t-1$ provide a direct estimate of $\mathbf{u(t-1)} = D\mathbf{z(t-1)}$. However, alternative approaches should be considered when factor scores do not define stationary processes or have insignificant effects on $d\mathbf{z(t)}$. One alternative is to apply the Johansen maximum likelihood procedure. A second alternative is to quantify the confounded effects of the between- and within-relation shocks in terms of the disparity variables discussed in Section 9.2 and to evaluate whether effects of disparity variables are time varying. Through trial and error, the latter alternative appears to provide a promising, less-laborious, effective forecasting approach.

9.5 SIMULTANEOUS TIME SERIES: ADAPTIVE DRIFT MODELING

Model (9.3.6) is generalized in terms of an adaptive system by allowing for time-varying coefficients:

$$dz(\mathbf{t}) = C(t)\mathbf{u}(\mathbf{t}-\mathbf{1}) + \Sigma_i B_i(t)\mathbf{dz}(\mathbf{t}-\mathbf{i}) + \boldsymbol{\varepsilon}(\mathbf{t}), \qquad (9.5.1)$$

where $i = 1, \ldots, p-1$. The hth of the m equations in (9.5.1) is written

$$dz_h(t) = \mathbf{C_h}(\mathbf{t})'\mathbf{u}(\mathbf{t}-\mathbf{1}) + \Sigma_i \mathbf{B_{hi}}(\mathbf{t})'\mathbf{dz}(\mathbf{t}-\mathbf{i}) + \varepsilon_h(t). \qquad (9.5.2)$$

As discussed in Section 4.3, one approach to modeling $\mathbf{C_h}(\mathbf{t})$ and $\mathbf{B_{hi}}(\mathbf{t})$ is to assume that coefficients are generated by a random walk (West and Harrison, 1997):

$$\mathbf{C_h}(\mathbf{t}) = \mathbf{C_h}(\mathbf{t}-\mathbf{1}) + \boldsymbol{\delta}_{\mathbf{Ch}}(\mathbf{t})^*,$$
$$\mathbf{B_{hi}}(\mathbf{t}) = \mathbf{B_{hi}}(\mathbf{t}-\mathbf{1}) + \boldsymbol{\delta}_{\mathbf{Bh}}(\mathbf{t})^*,$$

where the $\boldsymbol{\delta}_{\mathbf{\cdot h}}(\mathbf{t})^*$ denote model errors. However, for the markets under study, a more realistic assumption is that changes, if any, in coefficients are due to shocks that occur at previous times. Accordingly, coefficients are assumed generated by lags of $\mathbf{u}(\mathbf{t})$:

$$\mathbf{C_h}(\mathbf{t}) = \mathbf{C_h} + \Sigma_k C_{hk}\mathbf{u}^*(\mathbf{t}-\mathbf{k}) + \boldsymbol{\delta}_{\mathbf{Ch}}(\mathbf{t}), \qquad (9.5.3)$$

$$\mathbf{B_{hi}}(\mathbf{t}) = \mathbf{B_{hi}} + \Sigma_k B_{hik}\mathbf{u}^*(\mathbf{t}-\mathbf{k}) + \boldsymbol{\delta}_{\mathbf{Bh}}(\mathbf{t}), \qquad (9.5.4)$$

where the C_{hk} and B_{hik} are matrices of coefficients for $k \geq 1$, and $\boldsymbol{\delta}_{\mathbf{Ch}}(\mathbf{t})$ and $\boldsymbol{\delta}_{\mathbf{Bh}}(\mathbf{t})$ denote model errors.

For gradual drift,

$$\mathbf{u}^*(\mathbf{t}-\mathbf{k}) = \mathbf{u}(\mathbf{t}-\mathbf{k}). \qquad (9.5.5)$$

For abrupt drift where elements of $\mathbf{u}(\mathbf{t}-\mathbf{k})$ and are sufficiently large in modulus,

$$\mathbf{u}^*(\mathbf{t}-\mathbf{k}) = \mathbf{u}(\mathbf{t}-\mathbf{k}) + A_u\mathbf{v_u}(\mathbf{t}-). \qquad (9.5.6)$$

Elements of A_u denote the effects of $\mathbf{v_u(t-)}$, a vector of variables having significant, direct effects on the particular abrupt drift scenario.

The vector $\mathbf{v_u(t-)}$ in (9.5.6) is of particular importance when cointegration is time varying and/or when the confounded effects of lagged between- and within-relation shocks are reflected inadequately by the $\mathbf{u(t-k)}$. In these situations, $\mathbf{v_u(t-)}$ is comprised of lagged disparity variables [or disparities between and/or within near-term trends in elements of $\mathbf{z(t-i)}$ for $i>0$], which are intended to compensate for shortcomings of the $\mathbf{u(t-k)}$.

Substitution of (9.5.3)–(9.54) into (9.5.2) results in a linear, second-order reduced model in the following variables: lags of $\mathbf{dz(t)}$ [denoted by $d\mathbf{z(t-)}$], lags of the $\mathbf{u(t-k)}$ [denoted by $\mathbf{u(t-)}), \mathbf{v_u(t-)}$], and corresponding interactions. In abbreviated notation, the second-order reduced model is written

$$dz_h(t) = f_h[d\mathbf{z(t-)},\ u\mathbf{(t-)},\ v_{\boldsymbol{u}}\mathbf{(t-)};\ \boldsymbol{\Theta}] + \varepsilon_{Rh}(t) = f_h + \varepsilon_{Rh}(t), \quad (9.5.7)$$

where $\boldsymbol{\Theta}$ denotes the parameter vector. Since all the variables are known, the stepwise regression procedures discussed in Section 7.3 may be applied in identifying appropriate predictor variables and estimating their effects. Thereafter, one may recover information provided by moving average variables in the event that they are not adequately reflected by $u\mathbf{(t-)}$ and $\mathbf{v_u(t-)}$.

Aside from these single-equation procedures, there are simultaneous, system-wide estimation procedures that provide increased efficiency when the reduced system of equations is restricted (Mallios, 1989; Zellner, 1971).

9.6 SIMULTANEOUS TIME SERIES: ADAPTIVE VOLATILITY MODELING

Population and sample forms of model (9.5.7) are written

$$dz_h(t) = f_h + \varepsilon_{Rh}(t) = f_h^{\wedge} + e_{Rh}(t), \quad (9.6.1)$$

where $f^{\wedge}$ is the sample form of f and $e_{Rh}(t)$ is the residual corresponding to $\varepsilon_{Rh}(t)$. Following procedures discussed in Section 4.4, assume that $e_{Rh}(t)^2$ follows a GARCH(p_h, q_h) process with time-varying coefficients:

$$\begin{aligned} e_{Rh}(t)^2 &= \mu_h(t) + \Sigma_{i^*\geq 1}\varphi_{hi^*}(t)e_{Rh}(t-i^*)^2 \\ &\quad + \Sigma_{i^{**}\geq 1}\psi_{hi^{**}}(t)\delta_h(t-i^{**}) + \delta_h(t), \end{aligned} \quad (9.6.2)$$

where $\varphi_{hi*}(t)$ denotes the effect of the autoregressive term $e_{Rh}(t-i^*)^2$, $\psi_{hi**}(t)$ the effect of the moving average term $\delta_h(t-i^{**}) = e_{Rh}(t)^2 - E[e_{Rh}(t)^2]$, and $\delta_h(t)$ the model error.

Analogous to (4.4.3), assume that each of $\mu_h(t), \varphi_{hi*}(t)$, and ψ_{hi**} are generated in terms of lags of $\delta_h(t)$ in first-order linear regression models. Substituting the latter expressions for $\mu_h(t), \phi_{hi*}(t)$, and ψ_{hi} into (9.6.2) leads to a bilinear reduced model for $e_{Rh}(t)^2$. Significant bilinear terms provide evidence that the GARCH(p_h, q_h) process is dynamic. Nonsignificant bilinear terms indicate that coefficients in (9.6.2) are non-time varying. Note, however, that since the model for $dz_h(t)$ in (9.5.1) allows for time-varying coefficients and since $dz_h(t)^\wedge$ generates $e_{Rh}(t)^2$, it is necessary to update model (9.6.2) with each update of model (9.5.1) to access whether (9.6.2) is or is not dynamic.

As a means of modeling volatility, GARCH processes have at least two shortcomings when applied individually (and not collectively) to equations that are part of a system of cointegrated equations. The first shortcoming, discussed earlier, is that $e_{Rh}(t)^2$ in (9.6.2) may be biased and misleading if the model for $dz_h(t)$ is ill specified. The second shortcoming is that model (9.6.2) does not consider information from the other fitted equations that comprise the system, such as the possible effects of $e_{Rh^*}(t-i)^2, h^* \neq h$, and residual cross products on $e_{Rh}(t-i)^2$. When ignored and significant, effects of $e_{Rh^*}(t-i)^2$ and the cross product $e_{Rh^*}(t-i)e_{Rh}(t-i)$ are pooled with model error in (9.6.2).

9.7 EXPLORATORY MODELING: MARATHON OIL COMPANY

Figure 9.7.1 presents the daily candlestick chart for Marathon Oil Company (MRO) from 10/24/09 to 1/9/09.

As in Section 9.4, factor analysis is applied to the correlation matrix associated with the seven time series in Figure 9.7.1: namely,

$$\mathbf{z(t)} = [O(t), H(t), L(t), C(t), cb5(t), cb(20), v = \ln V, \log_e vb5]'. \quad (9.7.1)$$

The analysis results given in Tables 9.7.1 and 9.7.2 show that rotated eigenvectors associated with the two largest eigenroots account for 86% of the total variation. The first factor is associated with prices and the second with volumes. Both factors define stationary processes, as shown through the following analysis of factor scores corresponding to each factor. Let $u_i(t), i = 1, 2$, denote the factor scores for the ith linear combination and set $du_i(t) = u_i(t) - u_i(t-1)$. Following (9.1.5), we fit

$$du_i(t) = \rho_i u_i(t-1) + \rho_i' \mathbf{du_i(t-)} + \delta(t) \quad (9.7.2)$$

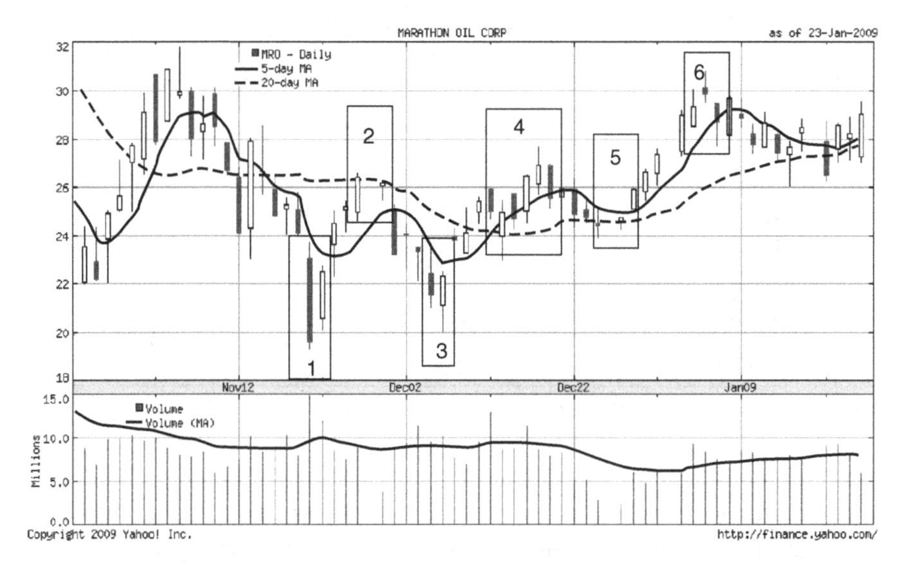

Figure 9.7.1 *Daily candlestick charts for Marathon Oil Company from 10/24/09 to 1/9/09.*

TABLE 9.7.1 Principal Components Analysis Applied to the MRO Correlation Matrix

	Initial Eigenvalues			Extraction Sums of Squared Loadings			Rotation Sums of Squared Loadings		
Component	Total	Percent of Variance	Cumulative %	Total	Percent of Variance	Cumulative %	Total	Percent of Variance	Cumulative %
1	5.143	64.287	64.287	5.143	64.287	64.287	5.141	64.263	64.263
2	1.744	21.805	86.092	1.744	21.805	86.092	1.746	21.829	86.092
3	0.515	6.432	92.524						
4	0.404	5.056	97.580						
5	0.095	1.186	98.766						
6	0.059	0.740	99.506						
7	0.028	0.355	99.861						
8	0.011	0.139	100.000						

through OLS estimation. Estimates of (9.7.2) are given by

$$du_1(t)^\wedge = -0.129_{4.32}u_1(t-1) + 0.157_{1.89}du_1(t-1) - 0.063_{0.75}du_2(t-1)$$

and

$$du_2(t)^\wedge = -0.102_{2.54}u_2(t-1) - 0.174_{1.92}du_2(t-4) + 0.167_{1.82}du_2(t-4). \tag{9.7.3}$$

These results indicate that both factors in Table 9.7.2 define stationary processes. As such, corresponding factors scores are measures of disequilibria.

TABLE 9.7.2 Rotation of the First Two Eigenvectors in Table 9.7.1[a]

	Factor	
	1	2
O	0.975	−0.062
H	0.987	0.012
L	0.952	−0.183
C	0.952	−0.103
lg1cb5	0.937	0.118
lg1cb20	0.721	0.519
Lnv	−0.018	0.836
lnvb5	−0.080	0.846

[a]Rotation method: Varimax with Kaiser normalization.

The factor analysis results are premised on the assumption that the correlation matrix remains the same over time. If not, cointegration may be time varying, in which case the factor analysis must be updated with each model update. To determine the extent to which the correlation matrix varies, a sensitivity analysis was performed as follows. Correlation matrices were determined for each of the daily three-month periods prior to and after the time period presented in Figure 9.7.1. Results show that when they occur, changes in the correlation matrix are minimal to very gradual. The implication is that model updates through drift modeling does not necessarily require a corresponding update of the factor analysis results, which may be required only periodically. This procedure is followed for all forthcoming modeling exercises.

For the BIDU analysis in Tables 9.4.1 and 9.4.2, the three eigenvectors associated with the three largest eigenroots are rotated, whereas for the MRO analysis in Tables 9.7.1 and 9.7.2, only the first two eigenvectors are rotated. In practice, eigenvectors with eigenvalues of less than 1 are typically discarded. However, in our analysis, eigenvectors with eigenvalues of less than 1 may be retained and rotated if the eigenroots are sufficiently close to 1. For the MRO analysis, only the first two eigenvectors are rotated, since the third eigenroot (with a value of 0.515) is well below 1. For the BIDU analysis, the first three eigenvectors are rotated since the third eigenroot (with a value of 0.821) is much closer to 1 and accounts for a significant 10% of the variation. The task of selecting the largest eigenvectors may involve trial and error, since it is necessary to show that all the rotated eigenvectors define stationary processes.

For MRO, the focus is on forecasting that $dz_h(t)$ corresponding to the daily change in per share closing prices, denoted by $dC(\mathrm{MRO}, t) = C(\mathrm{MRO}, t) - C(\mathrm{MRO}, t-1)$. To illustrate the modeling procedure, modeling is based on daily MRO data from 8/1/08 through 11/20/08 inclusive. Figure 9.7.1 depicts 11/20/08 in terms of the large dark body lying below the moving average band; see box 1. The low for the day is also the absolute minimum for all days depicted. (Events of 11/20/08 describe a *selling climax*, where a sharp drop in the closing price during a downward trend is accompanied by an excessive trading volume.)

From a chartist perspective, a bullish harami pattern (see pattern 9 in Table 5.1.1) is contained in box 1. The question at hand is whether the model forecast foretold the turning point *before* observing the white candlestick on 11/21/08. The answer is yes, but prior to discussing forecasting effectiveness, we present the modeling result for the model update through 11/20/08.

Based on model (9.5.7), the model forecast for $dC(\text{MRO}, t)$ is given by

$$\begin{aligned} dC(\text{MRO}, t)^{\wedge} = & -39.19_{5.49}[LIN\{LW(t-1)\}][LIN\{R(t-1)\}] \\ & - 5.51_{4.57}[LIN\{H(t-1)\}][QD\{R(t-1)\}] \\ & - 0.90_{2.78}[LIN\{H(t-1)\}][u_2(t-1)] \\ & + 5.02_{2.76}[QD\{R(t-2)\}][v(t-1) - vb5(t-1)]. \quad (9.7.4) \end{aligned}$$

Although coefficients and/or predictor variables are subject to either slight or significant change with each model update, (9.7.4) is discussed for purposes of interpreting the effects of disparity variables (usually appearing in terms of interactions), evaluating the effectiveness of forecasts over a short span of days following 11/20/08, comparing model forecasts with chartist forecasts, and assessing criteria for determining when model updates become necessary.

Based on the short-term trends defined in (9.2.1), individual variables comprising the four interactions in (9.7.4) are identified as follows: $LIN\{LW(t-1)\}$ is the linear (straight line) trend in the lower candlestick wick [defined by $LW(t) = \min[O(t), C(t)] - L(t)]$ through time $t-1$; $LIN\{R(t-1)\}$ is the linear trend in the range, $R(t) = H(t) - L(t)$, through time $t-1$; $LIN\{H(t-1)\}$ is the linear trend in the high, $H(t)$, through time $t-1$; $QD\{R(t-1)\}$ is the quadratic trend in $R(t)$ through time $t-1$; $u_2(t-1)$, the factor score associated with the second factor in Table 9.7.1, measures the shock at time $t-1$ that corresponds to the linear relation between the volume variables; $QD\{R(t-2)\}$ is the quadratic trend in the range through time $t-2$; and $\log_e V(t-1) - \log_e Vb5(t-1)$ is the difference between $\log_e$ volume and its 5-day moving average at time $t-1$.

Although the four interactions in (9.7.4) are interrelated, the negative effect of the first reflects bearish volatility when it is the product of linearly increasing ranges $LIN\{R(t-1)>0]$ and linearly increasing lower wicks ($LIN\{LW(t-1)>0]$. Note, however, that $LIN\{R(t-1)$ should be interpreted in light of $QD\{R(t-1)$, which appears in two of the other interactions. At issue is which of the two, $LIN\{R(t-1)$ or $QD\{R(t-1)$, dominates a particular scenario.

The effect of the second interaction is counterbalanced by the effect of the fourth interaction. Namely, the second interaction has a negative effect when $LIN\{H(t-1)\}>0$ (where the high's are increasing linearly) and $QD\{R(t-1)\}>0$ (which describes an abrupt, short-term upward surge in the ranges and volatility). This negative effect is counterbalanced by the positive effect of the fourth interaction when $QD\{R(t-2)\}>0$ (which describes a positive upward surge in the ranges through time $t-2$) that is followed by a supporting surge in the volume [as measured in terms of $\log_e V(t-1) - \log_e Vb5(t-1)>0$]. However, the latter positive effect may be counterbalanced by the third interaction when $\log_e V(t-1) - \ln Vb5(t-1)>0$ is accompanied by $u_2(t-1)>0$ (or when the surge in volume is excessive, as measured by a larger positive value of the shock associated with the linear combination of the volume variables in Table 9.7.1).

Runs that accompany an uptrend or downtrend are common indicators of forthcoming relative maxima or minima. (In Figure 9.7.1, note the run of six consecutive white bodies beginning with the first white body in box 5.) Overviews of candlestick charts indicate that successions of bullish bodies [$B(t) = C(t) - O(t)>0$] greater than 4 or 5 often portend short-term profit taking and that comparable successions of bearish bodies [$B(t)<0$] give rise to short-term buying and/or covering short positions. However, *runs* are often prolonged in the case of price bubbles (see Section 9.8). (*Runs* also reflect disparity variables that are predictive indicators in other gaming situations.[2])

Consider next a comparison of candlestick forecasts (CF) with the model forecasts (MF) for $dC(t)$ in (9.7.4). MF forecasts for 11/21/08 though12/05/09 are correct (in terms of predicting positive or negative price changes) for all except for 12/2/08 (the forecast is negative) and 12/5/08 (the forecast is negative for the white body in box 3). The correct MF for the downturn on 11/28/08 (see box 2) is due primarily to the

[2]The effect of runs has interesting parallels with blackjack strategies when a single deck is not shuffled after each hand but is used until the deck runs out. When an excess of "tens" remain in the second and, particularly, the third hands, the strategy is to bet high, since the probability of winning is greatly increased. A paucity of "tens" results in a lower win probability. Longer series of runs in candlestick bodies and a paucity or excess of "tens" tend to be effective short-term predictors.)

second interaction in model (9.7.4). Although the model (9.7.4) forecast is negative for the white body upturn in box 3 (on 12/5/08), a model update through12/5/08 correctly forecasts positive price changes through the next three trading days. With continuing model updates, especially at or near key turning points such as in boxes 3, 4, and 5, the explanatory interactions in forecasting models such as model (9.7.4) usually change.

As mentioned earlier, the candlestick forecasts do not usually anticipate upturns or downturns above or below the moving average bands. However, once these turns are observed and associated with bullish or bearish patterns, short-term forecasts will often have merit. Note, for example, the positive harami pattern in box 1, the negative harami pattern in box 2, the positive piercing pattern in box 3, the final bearish engulfing pattern in box 4, the bullish breakaway pattern in box 5, and the bearish evening star in box 6.

Beginning in mid-December with the bearish piercing pattern (the first dark body in box 4), daily model updates are hit and miss. However, given the bearish engulfing pattern in box 4 (the last dark body in box 4), updated model forecasts are mostly correct through the bullish breakout pattern in box 5. Given a model update following box 5, forecasts are correct through the bearish evening star pattern in box 6. Thereafter, price movements again become erratic, so that model updates are again mostly hit and miss. As mentioned earlier, for such periods of erratic price movements (on a daily basis), intraday candlestick charts, combined with intraday adaptive modeling, may provide more definitive forecasting. For selected hedge funds, the duration of a trade is said to last an average of less than 2 days. As such, intraday trading (based on intraday) modeling would appear to be the norm, as opposed to our interday modeling of MRO.

In Figure 9.7.1, the data beyond 11/21/08 (the first box) show little evidence of variance heterogeneity in terms of GARCH modeling. This is not the case for data prior to 11/21/08, as is evident from an inspection of the volatile daily price ranges and price changes prior to this date.

9.8 THE HIGH-TECH BUBBLE OF 2000

Figures 9.8.1 and 9.8.2 depict the formation and deflation of the high-tech bubble in terms of weekly candlestick charts for the Nasdaq-100 index (COMPX) and Yahoo (YHOO), one of the COMPX components. *Irrational exuberance* is said to have driven YHOO share prices from $6 in June 1998 to a high of $125 during the week of 12/27/99. Share prices then collapsed to $4 by October 2001.

The massive price increases during the second half of 1999 occurred during enactment of the Gram–Leach–Bliley Act, signed by President Clinton

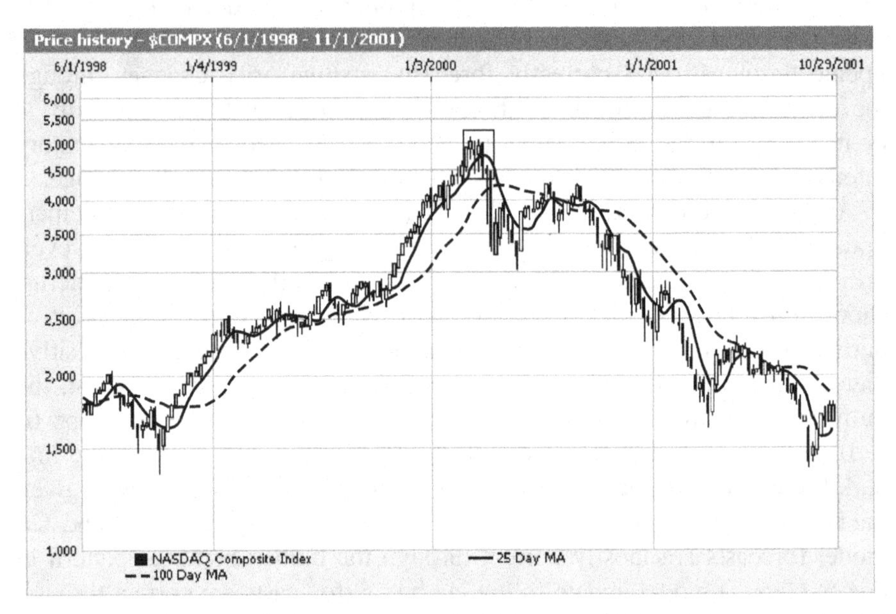

Figure 9.8.1 *Weekly candlestick chart for the NASDAQ Index ($COMPX) during formation and deflation of the high-tech bubble.* (Source: *MSN Money*)

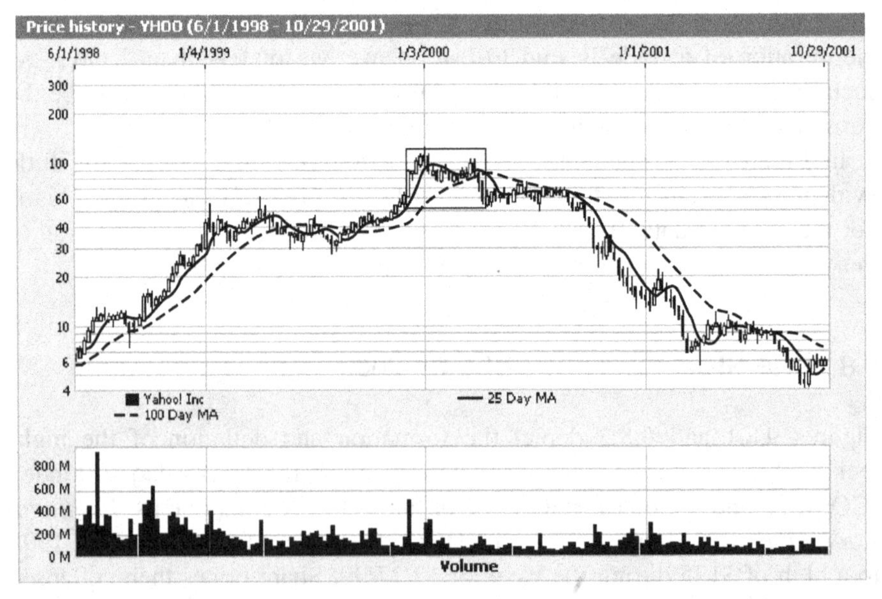

Figure 9.8.2 *Weekly candlestick chart for Yahoo (YHOO) during formation and deflation of the high-tech bubble.* (Source: *MSN Money*)

on 11/12/99. (In signing the bill, Clinton was counseled by his Treasury Secretary Larry Summers, former Treasury Secretary Robert Rubin, and Fed Chairman Alan Greenspan.) The Act repealed the 1933 Glass–Steagall Act, which had kept commercial banking and insurance separate from investment banking. A year later, in 2000, Clinton, again on the advice of his counsel, signed legislation to shield derivatives from federal regulation. The deregulation of derivatives was particularly surprizing since it followed the meltdown of the hedge fund Long-Term Capital Management, a meltdown that resulted from speculation in derivatives.[3]

It has been suggested that the bursting of the high-tech bubble and subsequent bear market may have been partially triggered by the adverse findings of fact in the *United States* v. *Microsoft* case, which at the time was being heard in federal court. The findings, which declared Microsoft a monopoly, were widely expected in the weeks before their release on 4/3/00.

Peak portions of the high-tech bubble are presented for the NASDAQ index (Figure 9.8.3), Yahoo (Figure 9.8.4), and two other NASDAQ components, Microsoft (Figure 9.8.5) and Citrix (Figure 9.8.6). Cross-sectional views of NASDAQ components are often useful, since price movements in one issue often portend and confirm price movements in like-kind issues, especially when related issues are cointegrated, at least periodically, and thus tend to move in harmony. (*Note:* Disappointing quarterly earning reports for a particular company is a caveat to such harmonious price movements; see Figure 5.5.1 regarding a disappointing earnings report for CTXS.)

Except for the MSFT, weekly price increases from early November 1999 through the end of December 1999 are in terms of an unusually large number of consecutive bullish candlestick bodies where $C(t) > O(t)$ and $C(t) > C(t-1)$. (These are the positive or *bullish runs* of candlestick bodies disussed in Section 9.7.) Following these bullish runs, bearish or nearly bearish patterns are indicated on 1/3/00; see box 1 in each figure. For COMPX and CTXS the negative patterns are misleading, whereas for YHOO and MSFT the indicators lead correctly to short-term price declines. Except for CTXS, the decisive bearish indicators that precede major price declines occur the trading day prior

[3]During the fiscal meltdown in Greece in 2010, James Richards, the former general counsel of Long Term Capital Management, stated: *This dynamic of pushing out spreads and calling in margin is the same one that played out at Long Term Capital Management in 1998 and AIG in 2008 and it is happening again this time in Europe.... Credit default swaps let anyone bet on anything. We have given Wall Street huge incentives to burn down your house.... Your neighbor cannot buy insurance on your house because they have no insurable interest in it. Such insurance is considered unhealthy because it would cause the neighbor to want your house to burn down—and maybe even light the match* (Richards, 2/12/10).

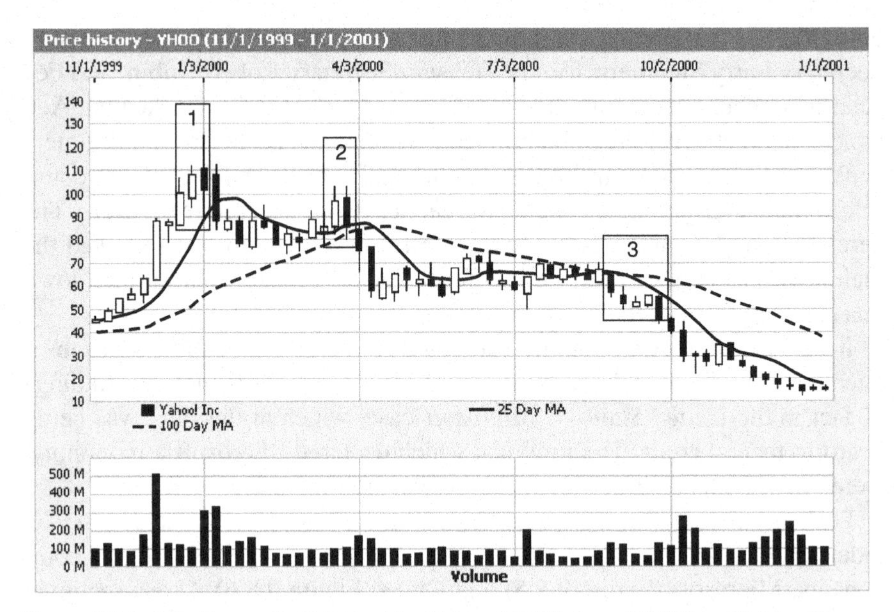

Figure 9.8.3 *Weekly candlestick chart for Yahoo (YHOO) from 11/1/99 to 1/1/01.* (Source: *MSN Money)*

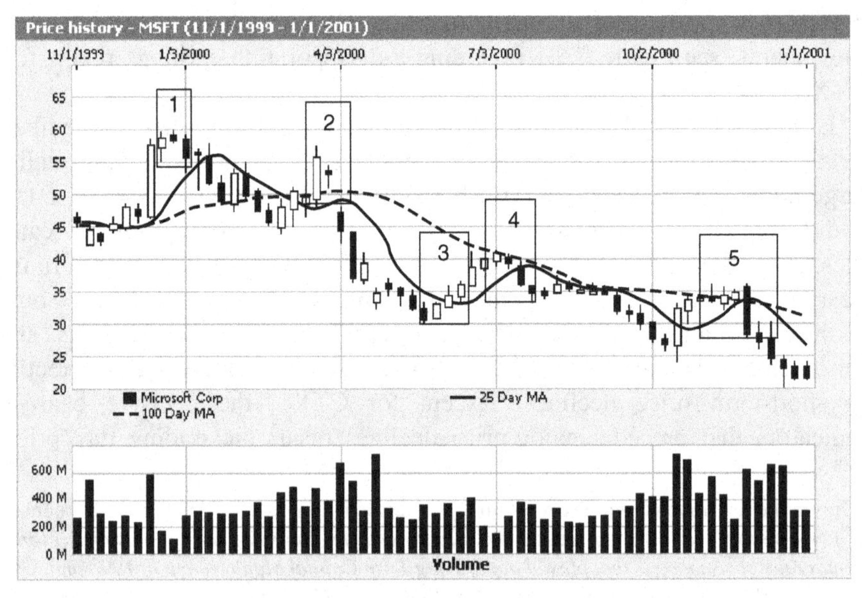

Figure 9.8.4 *Weekly candlestick chart for Microsoft (MSFT) from 11/1/99 to 1/1/01.* (Source: *MSN Money)*

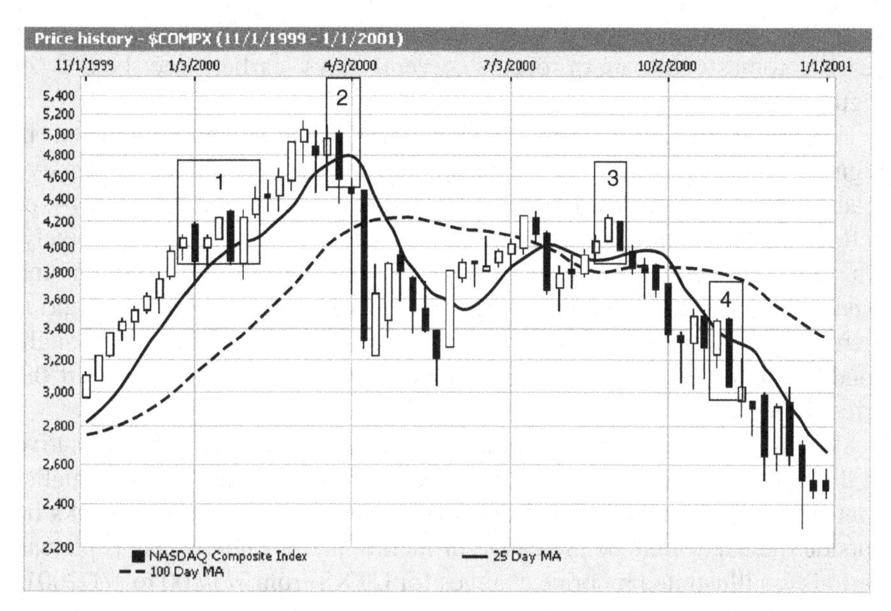

Figure 9.8.5 *Weekly candlestick chart for the NASDAQ Index ($COMPX) from 11/1/99 to 1/1/02. (Source: MSN Money)*

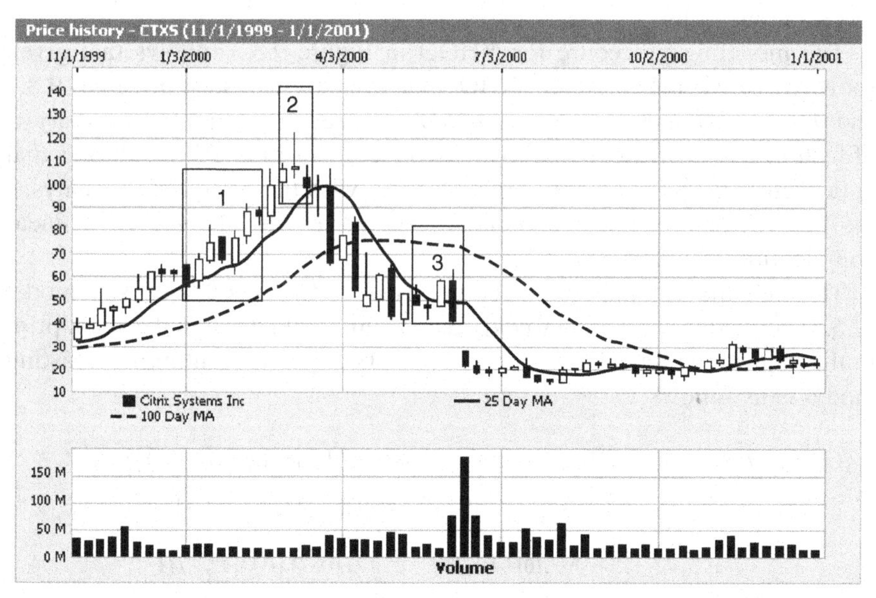

Figure 9.8.6 *Weekly candlestick chart for Citrix (CTXS) from 11/1/99 to 1/1/01. (Source: MSN Money)*

to 4/3/00; see box 2 in Figures 9.8.3–9.8.5. For CTXS, the decisive bearish tombstone pattern occurs several days earlier; see box 2 in Figure 9.8.6.

For major market upturns or downturns such as those depicted in Figures 9.8.3–9.8.6, it is not unusual that two or even three successive relative maxima (minima) occur before a major downturn (upturn)—or before the two moving average curves cross one another decisively. On the other hand, a sustained series of candlesticks that lie on and especially above the moving average band—when the width of the band is increasing—is often indicative of an overbought equity. The overbought condition is reinforced if trading volumes are not sufficient to support the price increases.

The moving average band and the placement of the candlesticks relative to the band may be viewed as a financial variation of time-varying quality control charts. Overbought and oversold conditions (where candlesticks lie outside the band) may be indicative of market inefficiency, whereas plateau periods (as illustrated by price changes for CTXS from 7/3/200 to 1/1/2001) may be indicative of greater market efficiency. Clearly, traders who sold and/or shorted very near the peak of the bubble capitalized on the irrational behavior of those who bought. On the other hand, those who sold too soon missed out on major profits, while those who shorted too soon may have faced margin calls.

The modeling procedure for YHOO in Figure 9.8.3 follows that given for MRO in Section 9.7. The YAHOO factor analysis results in Tables 9.8.1 and 9.8.2 are similar to those for MRO in Tables 9.7.1 and 9.7.2. Analysis of factor scores corresponding to the two factors in Table 9.8.2 indicates that both factors define stationary processes. As with the MRO factor analysis, the first factor is a linear combination of prices and the second a linear combination of volumes.

Through the estimation procedure used in the MRO analysis, model (9.5.7) is fitted to the Yahoo weekly data from 3/2/98 through the week prior to the bearish configuration on 1/3/00; see box 1. The resulting forecasting model is as follows:

$$
\begin{aligned}
dC(\text{YHOO}, t)^{\wedge} = 2.70_{3.48} &- 2.54_{4.04}[u_1(t-1)][\text{run}\{B(t-1)\}] \\
&- 0.39_{6.18}[dLW(t-1)][\text{run}\{\{B(t-1)\}] \\
&- 8.60_{7.10}[LIN\{v(t-1)\}][CB\{R(t-1)\}] \\
&- 1.66_{3.38}[CB\{UW(t-1)\}][LIN\{B(t-1)\}] \\
&- 0.02_{3.87}[dL(t-1)][dR(t-1)] \\
&+ 1.73_{3.23}[LIN\{LW(t-1)\}][QD\{UW(t-2)\}]. \quad (9.8.1)
\end{aligned}
$$

TABLE 9.8.1 Principal Components Analysis Applied to the Yahoo Correlation Matrix

	Initial Eigenvalues			Extraction Sums of Squared Loadings			Rotation Sums of Squared Loadings		
Component	Total	Percent of Variance	Cumulative %	Total	Percent of Variance	Cumulative %	Total	Percent of Variance	Cumulative %
1	5.798	72.473	72.473	5.798	72.473	72.473	5.633	70.416	70.416
2	1.614	20.173	92.646	1.614	20.173	92.646	1.778	22.230	92.646
3	0.337	4.211	96.857						
4	0.201	2.508	99.365						
5	0.030	0.371	99.737						
6	0.013	0.168	99.905						
7	0.006	0.069	99.974						
8	0.002	0.026	100.000						

TABLE 9.8.2 Rotation of the First Two Vectors of Table 9.8.1[a]

	Component	
	1	2
O	0.990	−0.069
H	0.991	−0.044
L	0.987	−0.083
C	0.986	−0.057
lg1cb5	0.979	−0.131
lg1cb20	0.865	−0.306
Lnv	−0.073	0.908
lnvb5	−0.114	0.909

[a]Rotation method: Varimax with Kaiser normalization.

Components of each interaction in (9.8.1) are identified as follows. The notation in (9.7.1) is used throughout to define the eight time series presented in the candlestick charts. The only difference is that for weekly charts—as opposed to daily charts—the two volume trends are defined in terms of weeks rather than days. The shock corresponding to the first factor score at time $t-1$ is denoted by $u_1(t-1)$. The variables $LIN\{\cdot\}, QD\{\cdot\}$, and $CB\{\cdot\}$ denote the linear, quadratic, and cubic trends within processes, as defined in (9.2.1), [e.g., $LIN(v(t-1)$ is the linear trend in $v = \log_e V$ from $t-4$ to $t-1$. Changes in the values of the LW and L from $t-2$ to $t-1$ are denoted by $dLW(t-1)$ and $dL(t-1)$, respectively. *Runs in candlestick bodies* (i.e., the number of successive positive or negative bodies) through time $t-1$ are denoted by $STR\{B(t-1)\}$.

The first two interactions indicate that an increasing number of positive bodies through time $t-1$ has an increasingly negative effect on $dC(\text{YHOO}, t)^\wedge$ when the lagged shock corresponding to the first factor (the linear combination of prices) is greater than expected and $dLW(t-1)>0$.

This implies that there has been an uninterrupted increase in prices, that the general increase is above expectations, and that the lower wick has just increased, which may indicate profit taking (and the possible formation of bearish hammer).

The combined negative effects of the interactions $[LIN\{v(t-1)\}][CB\{R(t-1)\}]$ and $[dL(t-1)][dR(t-1)]$ are a bearish reflection of volatility; that is, bearish volatility can be described by larger, short-term increases in the volumes and ranges that are finalized or reinforced by increases in both the low and the range. The negative effect of the interaction $[CB\{UW(t-1)\}][LIN\{B(t-1)\}]$ reflects the formation of a bearish tombstone where the upper wick increases abruptly while prices are increasing.

Model (9.8.1) correctly forecasts a decrease in $dC(\text{YHOO}, t)$ for the bearish configuration on 1/3/00 as well as price decreases over the next three weeks and a price increase associated with the positive engulfing candlestick four weeks following 1/3/00. However, a price plateau occurs over the next six weeks wherein daily updated model forecasts are hit and miss through the week ending with the larger white body on 3/20/00—which is just one week prior to the major downturn on 3/27/00; see box 2.

A model update through 3/20/00 correctly forecasts a price decrease associated with the bearish dark cover on 3/27/00 as well as the decreases the following two weeks. Thereafter, there is a plateau in prices from 4/17/00 through 8/28/00, during which time updated weekly forecasts are again hit and miss. The model update through the bearish pattern on 9/25/00 (see box 3) provides reliable forecasts for the downturn beginning the week of 10/2/00 and nearly all the weeks thereafter.

Since results of GARCH-type modeling may change in response to changes in the model for $dC(\text{YHOO}, t)$, the time-varying model in (9.6.1) is applied. Based on the model update for $dC(t)$ through 4/17/00, results of volatility modeling based on a time-varying GARCH process is as follows:

$$\begin{aligned} e_R(\text{YHOO}, t)^{2\wedge} &= 9.24 + 0.14_{1.76} e_R(\text{YHOO}, t-1)^2 \\ &\quad + 0.73_{2.86} e_R(\text{YHOO}, t-3)^2 \\ &\quad -0.55_{2.05} \Delta_R(\text{YHOO}, t-3)^\wedge, \end{aligned} \tag{9.8.2}$$

where $e_R(\text{YHOO}, t)^{2\wedge}$ denotes the GARCH estimate of $e_R(\text{YHOO}, t)^2$; $e_R(\text{YHOO}, t) = dC(\text{YHOO}, t) - dC(\text{YHOO}, t)^\wedge$; and $\Delta_R(\text{YHOO}, t-3)^\wedge = e_R(\text{YHOO}, t-3)^{2\wedge} - E[e_R(\text{YHOO}, t-3)^2]$ denotes a moving average term.

Model (9.8.2) implies that volatility during period t is affected positively affected by volatility lags in period $t-1$ and $t-3$. The positive effect of

$e_R(\text{YHOO}, t-3)^2$ on $e_R(\text{YHOO}, t)^2$ is counterbalanced by the negative effect of $\Delta_R(\text{YHOO}, t-3)$, the shock associated with $e_R(\text{YHOO}, t-3)^2$. Since there are no significant interactions between AR and MA terms in (9.8.2), GARCH coefficients appear to be non-time varying for the time period under study.

In general, if the equation for $dC(t)$ is ill specified, the GARCH variance forecast [as in (9.8.2)] is also likely to be biased or ineffective [since $e_R(t)^{2\wedge}$ is based on the model for the mean]. As such, price volatility should also be evaluated in terms of forecasts for two related variables in **z(t)** that quantify volatility: The first is the range $R(T) = H(t) - L(t)$ and the second is the candlestick body, $B(t) = O(t) - C(t)$. Forecasts of $dC(t)$ and the GARCH-type process associated with $dC(t), R(t)$, and $B(t)$ should be evaluated for consistencies in interpretation.

It is of interest to note that as of November 2009, none of the high-tech issues discussed in this section had recovered to achieve their dotcom bubble peaks—with one exception. Amazon.com reached an all-time high following announcement of their third-quarter 2009 earnings; see Figures 9.8.7 and 9.8.8. *No company symbolized the dotcom boom better than Amazon, the first public retailer to go public. By the time it peaked in December 1999, it had gained 6000% in the 30 months since its flotation. It did so without*

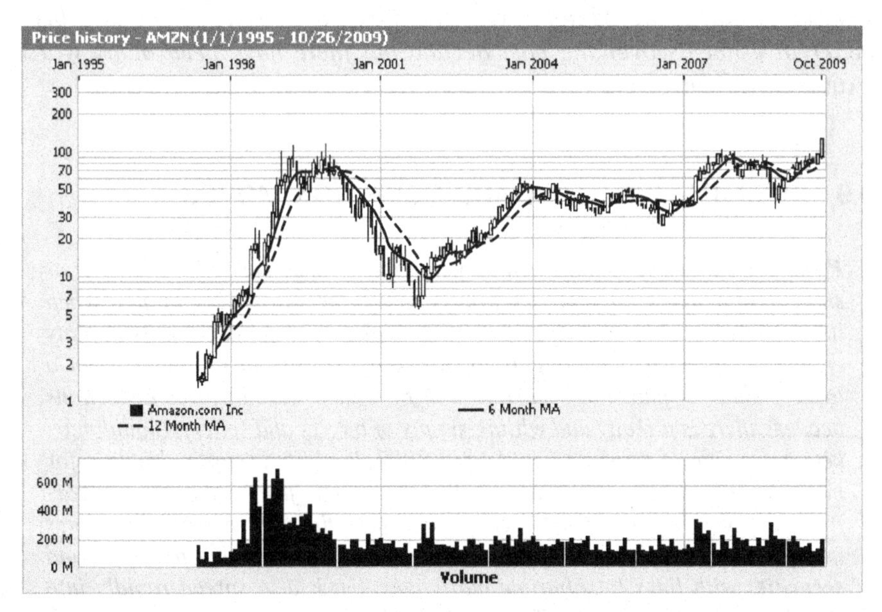

Figure 9.8.7 *Monthly candlestick chart for Amazon.com (AMZN) from flotation in May 1997 through October 2009. (Source: MSN Money)*

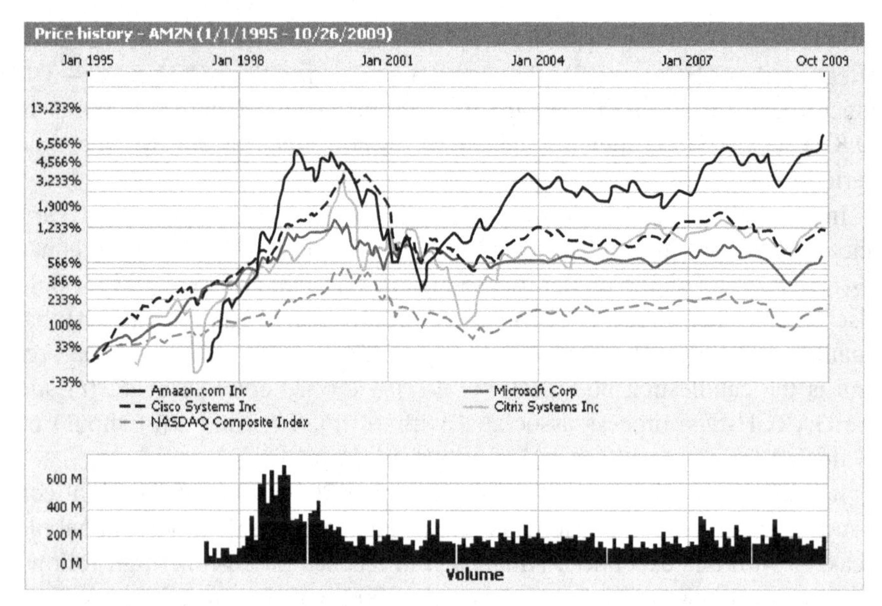

Figure 9.8.8 *Comparison of AMZN price trend with those of CSCO, CTXS, MSFT, and the NASDAQ Composite Index.* (Source: *MSN Money)*

making a profit. The insanity of this became evident as Amazon's stock subsequently fell 94%. But incredibly, it turns out that in the long term, buying Amazon at the top would not have worked out so badly.... There have been better investments over the past decade but there have been many worse (Authers, 10/27/09).

9.9 TWENTY-FIVE STANDARD DEVIATION MOVES

Firms are said to be liquid when they are able to meet current obligations or short-term demand for funds. A firm is said to be solvent but illiquid when its assets exceed its liabilities but it is unable to liquidate assets rapidly enough to meet current obligations. Markets are said to be liquid when a large volume of financial securities can be traded without price distortions because there is a ready and willing supply of buyers and sellers. Liquid markets are a sign of normalcy. In August 2007, liquidity abruptly dried up for many firms and securities markets. Suddenly some firms were able to borrow and investors were able to sell certain securities only at prohibitive rates and prices, if at all. The liquidity crunch was most extreme for firms and securities with links to subprime mortgages, but it also spread rapidly into seemingly unrelated areas. The stock market experienced unusual volatility and investors rushed to buy the safest of all investments, U.S. Treasury securities. On August 31, 2000, Federal Reserve Chairman Ben Bernanke noted that "although this episode appears to have been triggered largely by heightened concerns about subprime mortgages, global financial losses have far exceeded even the most pessimistic projections of credit losses on those loans." The spread of disruptions from housing into other debt markets is an example of financial contagion, or systemic risk. Contagion spread among non-bank institutions: mortgage lenders, hedge funds, and issuers of various types of securities, including commercial paper, asset-backed securities, structured products, and debt supporting leveraged buyouts and takeovers. As fear of risk has increased, these institutions saw sources of credit vanish and struggled to meet existing financing commitments, to post additional collateral, and to cope with portfolio losses. Some financial institutions, primarily mortgage lenders and hedge funds, have been unable to resolve liquidity problems and have closed. (opencrs.com/document/RL34182)

David Viniar, the chief financial officer of Goldman Sachs, and a man widely regarded as having done more than most to guard his institution against the credit crunch, complained of "25-standard deviation moves, several days in a row" back in August 2007. Such events are so impossibly unlikely that writing down the probability that they would occur would fill the rest of this paragraph with zeros. The implication should have been clear instantly: the mathematical model on which so many trillions of dollars depended was simply wrong. The financial sector will never see a real "25-standard deviation move," but it will see plenty of modeling errors. (*Financial Times* editorial, 12/23/08, Fixing cracks in the crystal ball)

Regarding a *25-standard move* for $dC(t)$, recall that the standard deviation (σ) is equal, approximately, to the range (R) divided by six: $R \approx 6\sigma$. If $\sigma = 25$, then $R \approx 150$. With reference to the daily candlestick chart for Dow Jones in Figure 9.9.1, a *25-standard move* for $dC(t)$—assuming normality for $dC(t)$—means that in a period of one trading day, the closing price increased or decreased by \$150. During the August 2007 period in question, price changes of \$150 were more or less common. Inspection of Figure 9.9.1, as well as other candlestick charts presented thus far, indicates that $dC(t)$ may be distributed according to nonnormal, fat-tailed distributions.

Given that price changes [the $dC(t)$] are periodically volatile and follow nonnormal-type distributions, commentaries such as "*the mathematical model* (i.e., the assumption of a normal distribution for price changes) *on which so many trillions of dollars depended were simply wrong*" apparently appeals to doomsayers and those who are illiterate in statistical modeling.

A basic overview of statistical modeling and its assumptions would seem to be in order. In modeling $dC(t)$, the objective is to explain, as much as

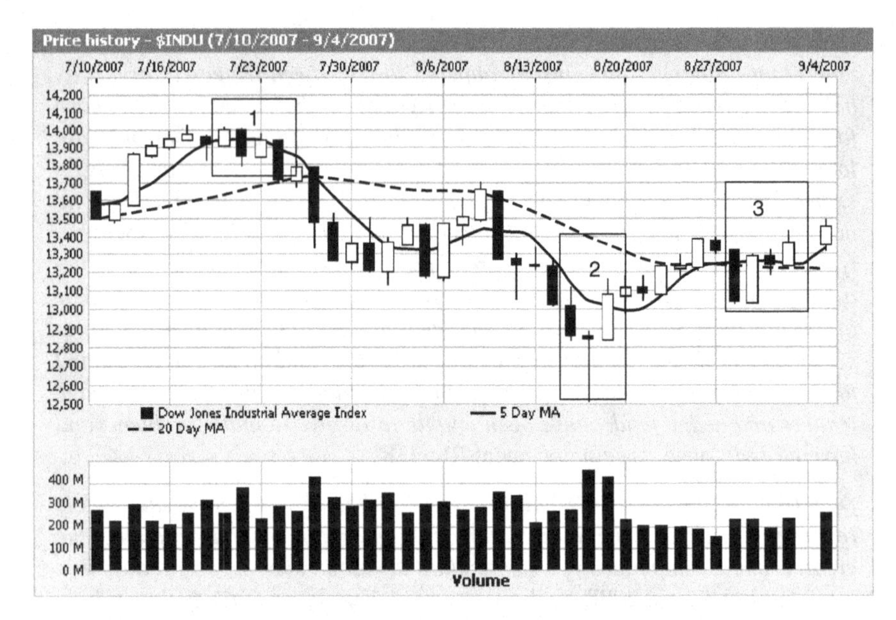

Figure 9.9.1 Daily candlestick chart for Dow Jones ($INDU): 7/30/07 to 9/4/07.

possible, variations in $dC(t)$ in terms of explanatory variables—whatever the distribution of $dC(t)$. The model expresses $dC(t)$ as a function of such variables. This function also includes a contemporaneous model error component denoting that portion of $dC(t)$ which is not explained by the explanatory variables.

The critical distributional assumption in modeling is in regards to the model error, such as $\varepsilon_{Rh}(t)$ in model (9.5.7)—not the distribution of $dC(t)$. Whatever the distribution of $dC(t)$, the model error tends to be symmetrically distributed—with either homogeneous or heterogeneous variance—when the model effectively explains a reasonable portion of the variation in $dC(t)$. GARCH-type modeling is a means of forecasting the variability associated with heterogeneous error variance. Consequently, if it is thought that *the mathematical model is simply wrong*, such commentary should be directed at the model's explanatory variables, not at the distribution of $dC(t)$.

The daily candlestick chart for Goldman Sachs (GS) in Figure 9.9.2 covers the same time period as that for $INDU in Figure 9.9.1. GS and $INDU are seen to move in tandem during the *25-standard deviation moves* of August 2007. From a chartist perspective, the bearish indicators for GS in the second and third trading days prior to 7/23/07 (see box 1) were reinforced by the bearish indicator for INDU the day prior to 7/23/07; see box 1.

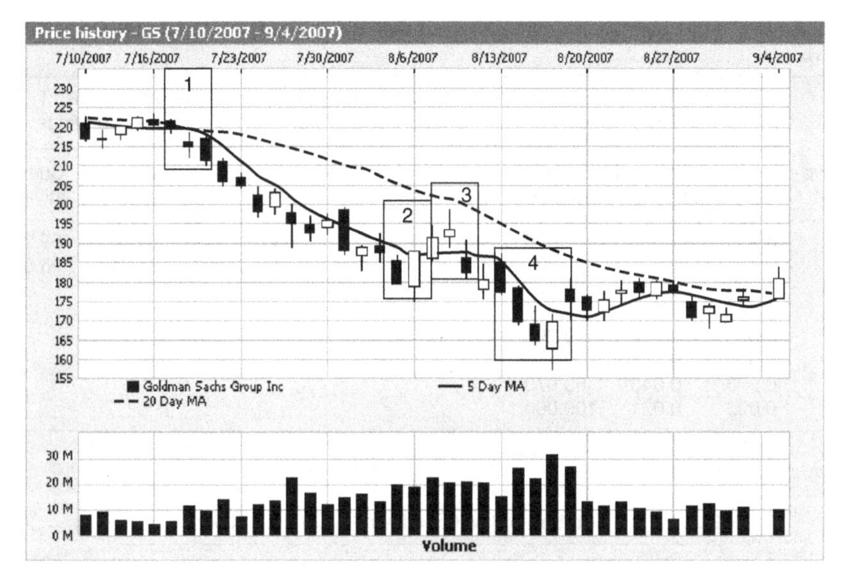

Figure 9.9.2 *Daily candlestick chart for Goldman Sachs (GS): 7/30/07 to 9/4/07.* (Source: *MSN Money*)

Both charts display bullish engulfing indicators on 8/6/07; see box 2 for GS. However, the bullish trend is unexpectedly short-lived. During the week of 8/6/07, a number of high-profile quant funds reported unprecedented losses. The speed and price impact of these initial losses put pressure on a broader set of quant fund portfolios, resulting in a substantial downturn on 8/9/07; see box 3 for GS. The downturn continued until the selling climax on 8/16/07, at which time bullish patterns appear in both Figures 9.9.1 (box 2) and 9.9.2 box 4). The question at hand is whether adaptive drift modeling could have enhanced chartist forecasts.

Results of factor analysis for GS are presented in Tables 9.9.1 and 9.9.2. The second factor (with an associated eigenroot of less than 1) is included to allow the factors to account for at least 90% of the total variation. Factor scores defined by both factors can be shown to define stationary processes.

The first model update for $dC(\text{GS},t)$ is based on daily data from 2/7/07 through the bearish engulfing pattern on 7/19/07, two trading days prior to 7/23/07; see box 1. Model forecasts support the chartist bearish forecasts. However, the model predicted a continued downward price movement for bullish engulfing pattern on 8/6/07, which mandated a model update then if not before.

Compared to the model update through 7/19/07, the model update through 8/6/07 results in changes in most of the explanatory variables and

TABLE 9.9.1 Principal Components Analysis Applied to the Goldman Sachs Correlation Matrix

	Initial Eigenvalues			Extraction Sums of Squared Loadings			Rotation Sums of Squared Loadings		
Component	Total	Percent of Variance	Cumulative %	Total	Percent of Variance	Cumulative %	Total	Percent of Variance	Cumulative %
1	6.948	86.850	86.850	6.948	86.850	86.850	4.742	59.281	59.281
2	0.694	8.677	95.527	0.694	8.677	95.527	2.900	36.246	95.527
3	0.202	2.519	98.046						
4	0.117	1.457	99.504						
5	0.023	0.292	99.796						
6	0.012	0.144	99.939						
7	0.003	0.040	99.979						
8	0.002	0.021	100.000						

TABLE 9.9.2 Rotation of the First Two Factors in Table 9.9.1[a]

	Component	
	1	2
O	0.992	0.043
H	0.993	0.081
L	0.994	−0.050
C	0.989	0.001
Lg1cb5	0.977	0.144
Lg1cb20	0.884	0.384
Inv	0.105	0.907
Invb5	0.020	0.925

[a] Rotation method: Varimax with Kaiser normalization.

is given as follows:

$$\begin{aligned} dC(\text{GS},t)^\wedge = {} & -1.85_{5.75} + 4.37_{5.56}[QD\{R(t-1)\}][\text{CB}\{R(t-1)\}] \\ & + 10.84_{6.45}[u_2(t-1)][QD\{B(t-1)\}] \\ & + 3.33_{5.55}[QD\{LW(t-1)\}][\text{run}\{B(t-1)\}] \\ & + 0.60_{3.96}[QD\{L(t-1)\}][QD\{B(t-1)\}] \\ & - 33.60_{3.53}[CB\{v(t-1)][LIN\{B(t-1)\}] \\ & - 1.46_{2.75}[QD\{LW(t-1)\}][u_2(t-1)]. \end{aligned} \tag{9.9.1}$$

Beginning with 8/7/07, model (9.9.1) correctly forecast all price movements up to, but not including, the next bullish engulfing pattern that occurred two trading days prior to 8/20/07; see box 4. For the white body in box 4, the model forecast is for a continued downward price movement—at which time the model is again updated if not before.

Regarding interpretations of interactions in model (9.9.1), the positive effect of $[QD\{R(t-1)\}][CB\{R(t-1)\}]$ indicates that this effect is negative if the quadratic and cubic trends in the price ranges do not reinforce one another. The variable $u_2(t-1)$—the lagged shock associated with volumes—appears in two interactions, the first with $[QD\{B(t-1)\}]$ and the second with $[QD\{LW(t-1)\}]$. The first $u_2(t-1)$ interaction has a positive (negative) effect when the volume is sufficiently large to support a quadratic upswing (downswing) in bullish (bearish) body sizes. The second $u_2(t-1)$ interaction has a negative effect when a sufficiently large volume accompanies an upswing in the trend of the lower wicks of the candlesticks.

The interaction $[QD\{LW(t-1)\}][\text{run}\{B(t-1)\}]$ has a negative effect when a larger run in bullish bodies is accompanied by an upswing in lower wick trends. The effect of the interaction $[CB\{v(t-1)][LIN\{B(t-1)\}]$ is negative when the linear trend in bullish bodies is accompanied by a downswing in the volumes.

Based on the model update for $dC(\text{GS}, t)$ in (9.9.1), the result of volatility modeling based on a GARCH process is as follows:

$$\begin{aligned} e_C(\text{GS}, t)^{2\wedge} = {} & 12.68_{2.21} + 0.54_{2.69} e_C(t-1)^2 - 0.43_{2.18} e_C(t-2)^2 \\ & + 0.39_{4.11} e_C(t-3)^2 \\ & + 0.48_{2.23} \Delta_C(t-1)^{\wedge} - 0.44_{2.16} \Delta_C(t-2)^{\wedge}. \end{aligned} \quad (9.9.2)$$

Model notation follows that given for model (9.6.2). Variance heterogeneity in (9.9.2) is seen to be affected by the first three autoregressive terms and the first two moving average terms. Since there is no evidence of interactions between autoregressive and moving average terms, the implication is that the GARCH process is not time varying, at least for the time period under study.

9.10 THE MARCH 2009 NADIR

Optimists declared early March 2009 as the nadir for 2008–2009 financial market convulsions, while pessimists tended to view the "V" recovery as the first stage of a "W" recovery. Others, particularly active traders, were content to view that period as a near-term buying opportunity. Figure 9.10.1 presents the relative price changes for the daily NASDAQ Index (COMPX) from 2/2/09 to 5/14/09 and superimposes comparable changes in two other

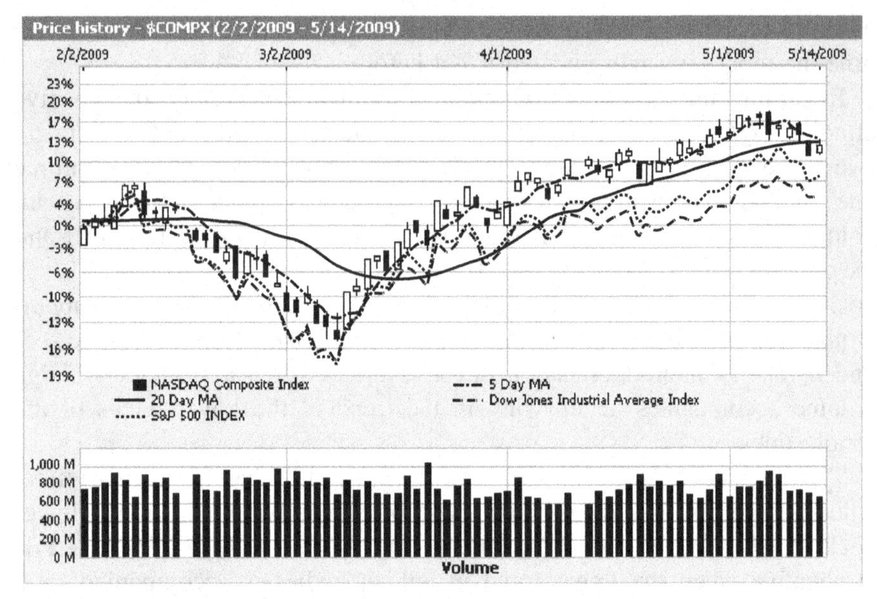

Figure 9.10.1 *Comparison of three major indices during the March 2009 nadir.* (Source: *MSN Money)*

major indexes, the Dow Jones Industrial Index and the Standard and Poor 500 Index. All the indexes clearly display pronounced minima on 3/9/09 and bullish patterns the following day.

Figure 9.10.1 also serves to illustrate discrepancies in recovery rates between the NASDAQ Composite Index (COMPX), the Dow Jones Industrial Index, and the S&P Index following the low point. *Some day the tech sector will be cyclical and silicon chips will obey the same economic rules as forklift trucks, rising and falling with the economy. But that moment has not yet arrived, and markets still rely on the tech sector not just as a vehicle for growth but as a defensive redoubt. Since hitting a bottom in November 2008, US tech stocks have outperformed the S&P 500 by 32.8% and beaten cyclical stocks by 32.8%* (Authers, 7/16/09).

However, while COMPX has consistently outperformed other popular indexes following the low point, Figure 9.10.2 shows that financial issues such as Goldman Sachs and Bank of America vastly outperformed COMPX.

> *Goldman, long the most prestigious bank on the Street and now plainly the biggest winner to emerge from the crisis, put aside $11.4bn to cover compensation for the first six months of the year. If current trends stay on course, Goldman will pay out an average of $770,000 to each of its 29,000 employees. . . . Joblessness is still rising across the western world. Losses on US credit cards hit a record 10.44% in June. Housing starts show signs of stabilizing but rising*

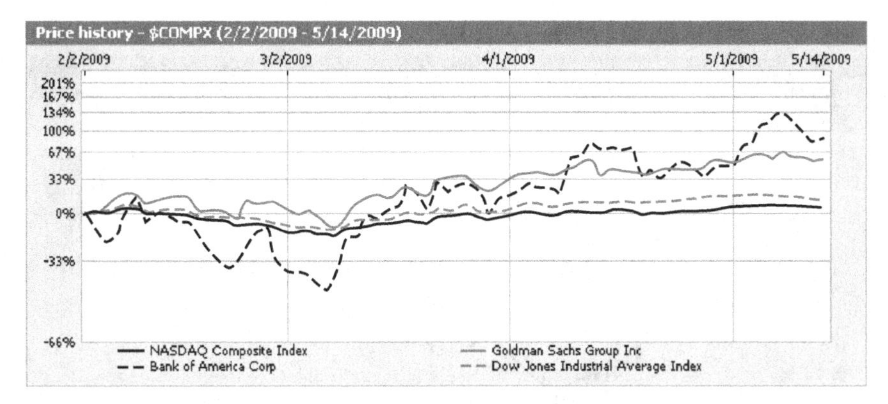

Figure 9.10.2 *Comparison of two major indices with Bank of America and Goldman Sachs during the March 2009 nadir.* (Source: *MSN Money*)

unemployment should force many more mortgages into default. . . . Goldman's results have triggered a backlash. A Rolling Stone *article that this month described Goldman as a "vampire squid sucking the face of humanity" seems to have struck a particular cord.*[4] (Authers, 7/18/09)

The daily candlestick charts for Radio Shack (RSH) in Figure 9.10.3 and for Polaris Industries (PII) in Figure 9.10.4 show a doubling in price by early May following the March nadir.

Figure 9.10.5 presents the weekly candlestick chart for the Chinese oil company CNOOC Limited (CEO) from 3/10/08 to 5/11/09. For weekly candlestick charts, the two closing price moving averages are based on the prior 25-day moving average (denoted *cb25*) and the prior 100-day moving average (denoted *cb100*). A selling climax occurred during the week of 10/20/08 with 6.3 million shares traded. The following week, a bullish engulfing pattern (see box 1) defined the absolute minimum five months before the *March 2009 nadir*; see box 1. From 10/20/08 to early May 2009, share prices doubled in value. Note that the weekly chart in Figure 9.10.5 is in accord with the daily charts in Figures 9.10.3 and 9.10.4 in that all three charts display bullish patterns during the 2009 nadir.

The March 2009 nadir may be compared to the August 1982 nadir. Figure 9.10.6 compares the NASDAQ Composite and S&P 500 Indexes

[4]*During the second quarter of 2009, Goldman Sachs made more than $100m in trading revenue on a record 46 separate days (or 71% of the time). This broke the previous high of 34 days during the first quarter of 2009. Its very counterintuitive to think that they'd be able to generate this much profit and this much revenue in the middle of an ongoing recession,* said William Cohan, author of *House of Cards* about the collapse of Bear Stearns Co. *But the fact that so many of their competitors are out of business or severely wounded has put them in a very strong position* (Harper, 8/5/09).)

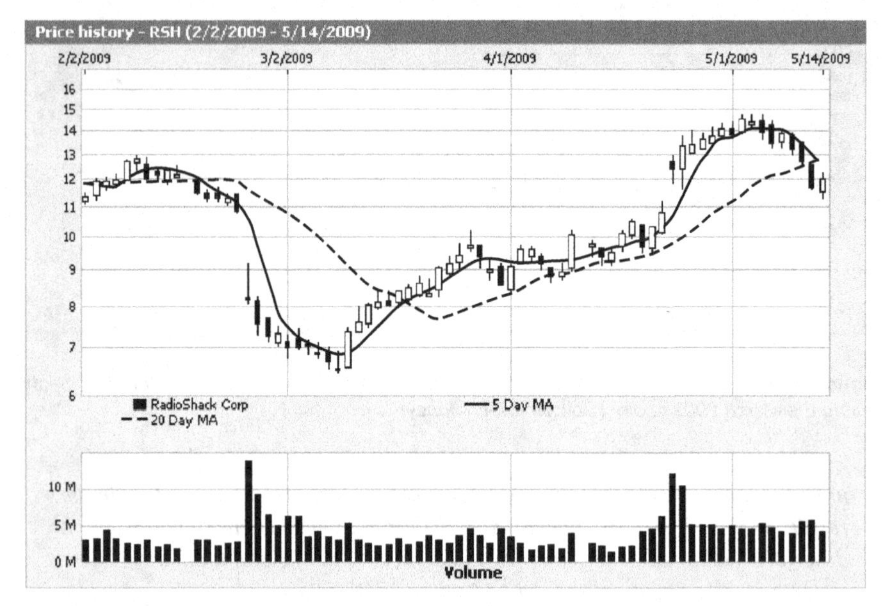

Figure 9.10.3 *The March 2009 nadir: daily candlestick chart for Radio Shack (RSH). (Source: MSN Money)*

Figure 9.10.4 *The March 2009 nadir: daily candlestick chart for Polaris Industries (PII). (Source: MSN Money)*

Figure 9.10.5 *Weekly candlestick charts for the Chinese oil company CNOOC (CEO).* (Source: *MSN Money*)

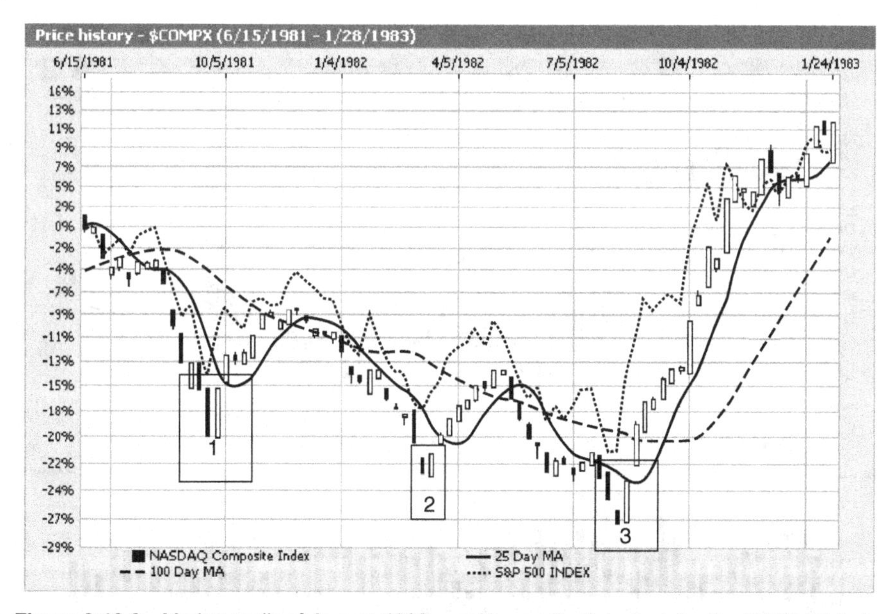

Figure 9.10.6 *Market nadir of August 1982: weekly candlestick chart for the NASDAQ index with comparisons to other indices.*

in terms of weekly candlesticks from June 1981 through January 1983. The 1982 nadir ended a 15-year bear market. At the time, Paul Volker, then chairman of the Federal Reserve Board, was attacking inflation, while Ronald Reagan and Margaret Thatcher were applying highly unpopular economic policies. French President Mitterrand observed that Thatcher combined *the eyes of Caligula with the lips of Marilyn Monroe*. Thatcher's commentary on her friend Ronald Reagan: *Poor dear, there's nothing between his ears*.

Figure 9.10.6 displays a roller coaster ride to ride down to the market nadir the week of 8/16/82 in box 3. The nadir is in terms of a well-defined bullish morning star pattern; see pattern 5 in Table 5.1. The nadir is then followed by a remarkable run of white bodies. The two major relative minima for COMPX nadir—the first occurring just prior to 10/5/81 (box 1) and the second (box 2) in early March 1982—are both defined in terms of bullish engulfing patterns. Note also that the two relative maxima that occur between the three minima are defined in terms of bearish candlestick patterns.

For purposes of smoothing the daily candlestick trends in Figure 9.10.1, Figure 9.10.7 presents the weekly candlestick chart for the NASDAQ Composite Index from 6/30/08 to 7/13/09. Both figures show that the March nadir is associated with bullish indicators—a morning star in the daily chart

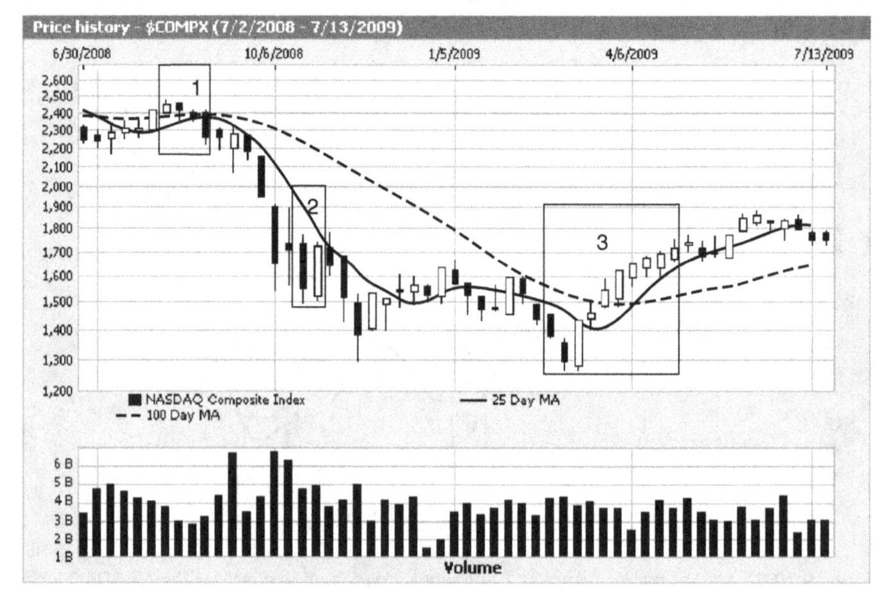

Figure 9.10.7 *Weekly candlestick chart for COMPX from 6/30/08 to 7/13/09. (Source: MSN Money)*

and an engulfing configuration in the weekly chart (see box 2). Relative to daily model forecasts, weekly model forecasts with updates tend to provide more reliable forecasts because of the smoothing effect. A trade-off between modeling smoothed (say, weekly changes) versus less-smoothed (say, daily or intraday changes) time series is that the former tends to provide less profitable rules with less risk, whereas the latter tends to provide more profitable rules with greater risk.

For the weekly candlesticks presented in Figure 9.10.7, model forecasts are now compared with chartist forecasts. Following the bearish candlestick pattern the week of 8/18/08—which is reinforced by bearish bodies in the two weeks that followed (see box 1)—model forecasts are bearish through the week of 11/24/08 (the first white body in box 2). The model prediction that accompanied the bullish piercing pattern on 10/27/08 (box 2) proved to be a false indicator for a market upturn. A model update at that point gave a forecast for a continued downward trend through the next white body. Model forecasts for the plateau period that followed are hit and miss. A bullish morning star pattern defines the March nadir in box 3.

Equation (9.10.1) presents the updated forecasting model based on weekly data through the second dark body in box 3:

$$\begin{aligned} dC(\text{COMPX}, t)^{\wedge} = {} & 0.22_{7.45}[QD\{L(t-1)\}][CB\{R(t-1)\}] \\ & - 72.43_{3.79}[CB\{v(t-1)\}][CB\{R(t-1)\}] \\ & - 0.21_{5.14}[QD\{H(t-1)\}][QD\{cb25(t-1) \\ & - cb100(t-1\}] \\ & - 24.95_{2.75}[LIN\{LW(t-1)\}][CB\{B(t-1)\}]. \end{aligned} \tag{9.10.1}$$

At this point in time, model forecasts again become reliable, which includes correctly forecasting bullish engulfing pattern (the March nadir) in box 3. Moreover, model forecasts confirm the bullish chartist forecasts through the last white body in box 3.

However, for the week that follows those in box 3, the model incorrectly forecasts a downturn. In contrast, the chartist forecast has the downturn occurring two weeks later in terms of the bearish evening star pattern. Both forecasts are in error since a downturn did not materialize. Rather, a plateau period followed around the 1800 price range. When there are inferential disagreements between chartist and model forecasts and/or during plateau periods, recourses are frequent model updates and, in this case, a change from weekly forecasts to daily or intraday forecasts.

Notice that contrary to other modeling exercises, lagged factor scores do not appear in (9.10.1). The reason is that the between and within disequilibria are reflected largely by the disparity variables comprising the interactions—as was conjectured earlier. This type of modeling outcome is more likely to occur when analyses are in terms of smoothed weekly or monthly candlestick data as opposed to daily or intraday candlestick data.

In (9.10.1) the variable $QD\{cb25(t-1) - cb100(t-1)\}$ in the third interaction denotes the quadratic trend in the difference between the 25-day moving average ($cb25$) and the 100-day moving average ($cb100$) from times $t-4$ through $t-1$. For decreasing prices, $QD\{cb25(t-1) - cb100(t-1)\} < 0$, combined with an upward trend in the highs, leads to a positive effect for the third interaction.

The first two interactions involve $CB\{R(t-1)\}$, the cubic trend in the range from times $t-4$ through $t-1$. The positive effect of price increases, measured in terms of $CB\{R(t-1)\} > 0$ and $QD\{L(t-1)\} > 0$, is offset by the negative effect of price volatility. The latter is characterized by excessive price and volume movements, as measured in terms of $CB\{v(t-1)\} > 0$ and $CB\{R(t-1)\} > 0$.

To further illustrate the smoothing effect of weekly candlestick charts (relative to daily charts), Figure 9.10.8 presents weekly price changes in BIDU (see Section 8.4) compared with its competitor Google (GOOG)

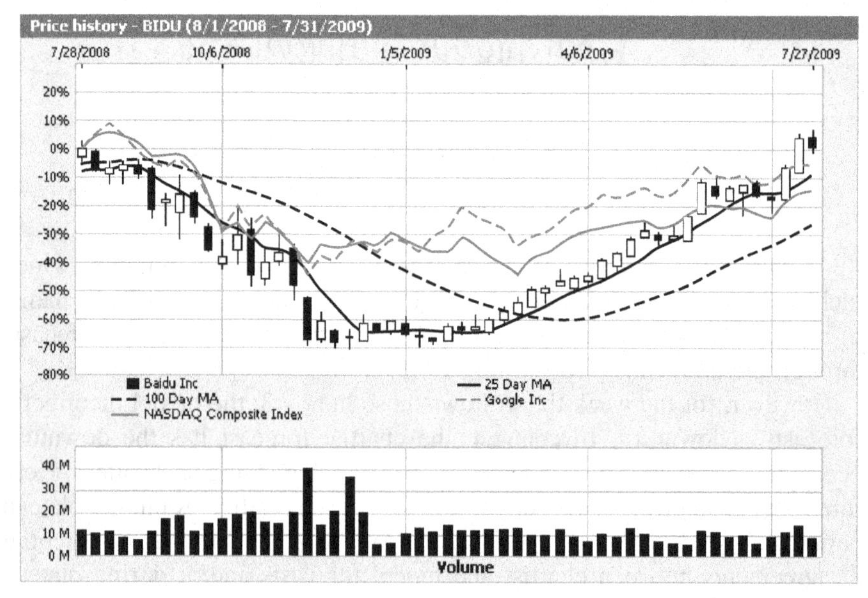

Figure 9.10.8 *Weekly candlestick chart for Baidu (BIDU), including comparisons with Google and the NASDAQ Composite Index. (Source: MSN Money)*

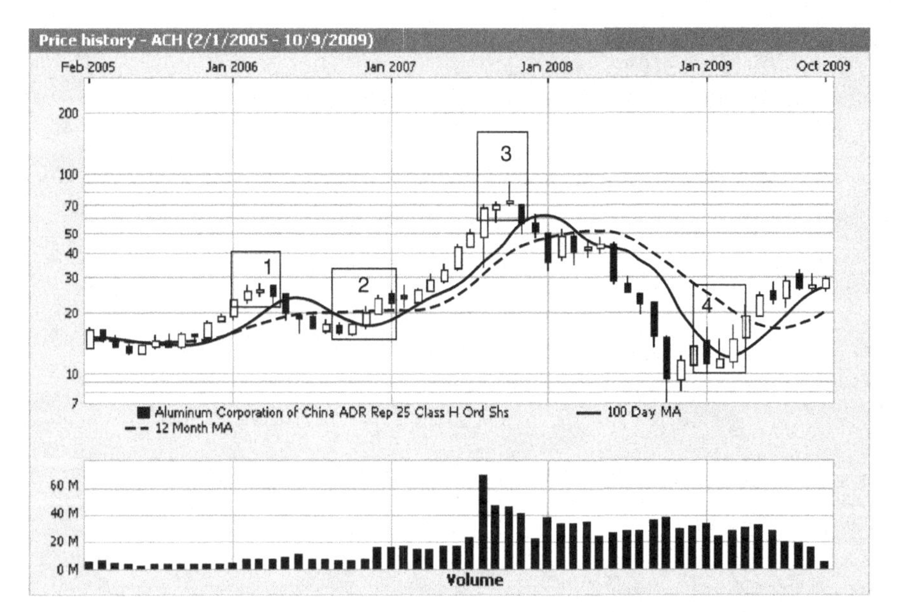

Figure 9.10.9 Monthly candlestick chart for the Aluminum Corporation of China (ACH). (Source: MSN Money)

and the NASDAQ Composite Index (COMPX). During the year under study, BIDU share prices dropped by approximately 70% (far more than the decline for GOOG and COMPX) before returning to their values in July 2008.

There are situations where smoothing in terms of monthly candlesticks may be useful in more conservative active trading. Figure 9.10.9 presents a monthly candlestick chart for the Aluminum Corporation of China. During the February 2005 to October 2009 period, all relative minima (boxes 2 and 4) and maxima (boxes 1 and 3) are identified correctly in terms of bullish and bearish patterns.

In general, the active trader's task is one of scanning a large number of equities on a continuing basis, identifying optimal trading opportunities and weighing the risks. This is a formidable task. James H. Simons, president of Renaissance Technologies, mathematician, and quant investor, is said to be the most successful hedge fund manager ever. Renaissance reportedly spent $600 million on computers and other technology that enable Simons and his army of Ph.D.'s to gather mountains of data on every conceivable securities and futures contract and search for patterns that will tell them what is likely to happen and when (White, 4/28/07). With trading based on insider information on one hand and the market activities of hedge funds such as Renaissance Technologies on the other, the playing field is never level.

10

Modeling Cointegrated Time Series Associated with NBA and NFL Games

10.1 MODELING TRANSITIONS

The transition in modeling simultaneous time series is from financial to sports gambling markets. Financial variables defining $\mathbf{z(t)}$ in (9.4.1) are replaced by the following variables pertaining to game outcomes in the sports gambling markets (SGMs):

$$\mathbf{z_{SGM}(t)} = (D(i,t), GSD(i,t), T(i,t), GST(i,t), WL(i,t), STR(i,t), D(i^*,t^*), \\ GSD(i^*,t^*), T(i^*,t^*), GST(i^*,t^*), WL(i^*,t^*), STR(i^*,t^*))'. \tag{10.1.1}$$

Elements of $\mathbf{z_{SGM}(t)}$ are defined as follows:

$D(i,t) =$ score differential between team i and its opponent, team i^*, in what is game t for team i and game t^* for team i^*; $D(i,t) = -D(i^*,t^*)$

$GSD(i,t) = D(i,t) - \mathrm{LD}(i,t) =$ gambling shock corresponding to the line on the difference, $LD(i,t)$; $GSD(i,t) = -GSD(i^*,t^*)$

Forecasting in Financial and Sports Gambling Markets: Adaptive Drift Modeling, By William S. Mallios

$T(i,t) =$ total points scored by team i and its opponent; $T(i,t) = T(i^*,t^*)$

$GST(i,t) = T(i,t) - LT(i,t) =$ gambling shock corresponding to the line on the total; $GST(i,t) = GST(i^*,t^*)$

$\mathrm{WL}(i,t)$ and $\mathrm{WL}(i^*,t^*) =$ percentage of games won by team i(team i^*) in their previous N games; $N = 10$ for the NBA games and $N = 5$ for NFL games

$STR(i,t)$ and $STR(i^*,t^*) =$ number of consecutive games won or lost by team i(team i^*) prior to game t(game t^*)

In contrast to the assumptions leading to model (9.5.1) for $\mathbf{dz(t)}$, we begin with the assumption that $\mathbf{z_{SGM}(t)}$ can be modeled in terms of the following adaptive process:

$$\mathbf{z_{SGM}(t)} = d\mathbf{z(t)} = \Sigma_{j>0}A_j(t)\mathbf{z_{SGM}(t-j)} + \Sigma_{k\geq 0}B_k(t)\mathbf{w(t-k)} + \boldsymbol{\varepsilon}(t). \tag{10.1.2}$$

$A_j(t)$ and $B_k(t)$ denote, respectively, time-varying coefficient matrices of $\mathbf{z_{SGM}(t-j)}$ and the covariate vector $\mathbf{w(t-k)}$; $\boldsymbol{\varepsilon}\mathbf{(t)}$ denotes the model error.

The vector $\mathbf{w(t-k)}$ includes (but is not limited to) the lines $LD(i,t-k), LT(i,t-k), LD(i^*,t^*-k)$, and $LT(i^*,t^*-k)$ in addition to the following variables:

$H/A(i,t-k)$ and $H/A(i^*,t^*-k) = H/A = 1$ for a home game and $H/A = 0$ for an away game

$D(i,t-p_{i^*})$ and $T(i,t-p_{i^*}) =$ score differential and the total points scored during the previous encounter between the two teams

$GSD(i,t-p_{i^*})$ and $GST(i,t-p_{i^*}) =$ gambling shocks associated with the previous encounter between the two teams

Applications will demonstrate that the time series comprising the 12×1 vector $\mathbf{z_{SGM}(t-k)}$ are cointegrated. As such, an initial task is to estimate $m^* < 12$ nonredundant linear combinations of $\mathbf{z_{SGM}(t-k)}$. Given cointegration, the model component $A_j(t)\mathbf{z_{SGM}(t-j)}$ in (10.1.2) may be written

$$A_j(t)\mathbf{z_{SGM}(t-j)} = C_j(t)D_j(t)\mathbf{z_{SGM}(t-j)} = C_j(t)\mathbf{u_{SGM}(t-j)}, \tag{10.1.3}$$

$$D_j(t)\mathbf{z_{SGM}(t-j)} = \mathbf{u_{SGM}(t-j)}. \tag{10.1.4}$$

$C_j(t)$ is of order $12 \times m^*$, while $D_j(t)$ is of order $m^* \times 12$. The vector $\mathbf{u_{SGM}(t-j)}$ is directly analogous to the vector $\mathbf{u(t-}1)$ in (9.3.5) in that it represents the confounded effects of both within- and between-relation shocks corresponding to time $t-j$.

Substituting the expression for $A_j(t)\mathbf{z_{SGM}(t-j)}$ in (10.1.13) into (10.1.2), we have

$$\mathbf{z_{SGM}(t)} = \Sigma_{j>0}C_j(t)\mathbf{u_{SGM}(t-j)} + \Sigma_{k\geq 0}B_k(t)\mathbf{w(t-k)} + \boldsymbol{\varepsilon}\mathbf{(t)}. \tag{10.1.5}$$

The hth of $m = 12$ equations in (10.1.5) is written

$$z_{\mathrm{SGMh}}(t) = \Sigma_{j>0}\mathbf{c_{hj}(t)}'\mathbf{u_{SGM}(t-j)} + \Sigma_{k\geq 0}\mathbf{b_{hk}(t)}'\mathbf{w(t-k)} + \varepsilon_h(t). \tag{10.1.6}$$

The hth rows of C_j and B_k in (10.1.5) are written $\mathbf{c_{hj}(t)}'$ and $\mathbf{b_{hk}(t)}'$, respectively, and $\boldsymbol{\varepsilon}_h\mathbf{(t)}$ denotes the hth element of $\boldsymbol{\varepsilon}\mathbf{(t)}$.

Analogous to the adaptive drift modeling procedure presented for financial modeling, coefficient drift in (10.1.6) is assumed to be generated in terms of lagged shocks, which include the $\mathbf{u_{SGM}(t-j)}$ [written lexicographically in terms of the vector $\mathbf{u_{SGM}(t-)}$], $GSD(t-1)$ and its earlier lags [written as $\mathbf{GSD(t-)}$], and $GST(t-1)$ and its earlier lags [written $\mathbf{GST(t-)}$]. It should be noted that elements of $\mathbf{u(t-)}$ may largely reflect those of $\mathbf{GSD(t-)}$ and $\mathbf{GST(t-)}$ and that the use of factor analysis scores in estimating $\mathbf{u_{SGM}(t-j)}$ confounds the effects of $\mathbf{u_{SGM}(t-)}, \mathbf{GSD(t-)}$, and $\mathbf{GST(t-)}$.

Models for the vectors $\mathbf{c_{hj}(t)}$ and $\mathbf{b_{hk}(t)}$ are in terms of the following first-order linear regression equations:

$$\mathbf{c_{hj}(t)} = \mathbf{c_{hj}} + FLR(\mathbf{u^*_{SGM}(t-)}, \mathbf{GSD^*(t-)}, \mathbf{GST^*(t-)}; \theta_{\mathbf{chj}}) + \delta_{\mathbf{hj}}\mathbf{(t)}, \tag{10.1.7}$$

$$\mathbf{b_{hk}(t)} = \mathbf{b_{hk}} + FLR(\mathbf{u^*_{SGM}(t-)}, \mathbf{GSD^*(t-)}, \mathbf{GST^*(t-)}; \theta_{\mathbf{bhk}}) + \delta_{\mathbf{hk}}\mathbf{(t)}. \tag{10.1.8}$$

FLR is an acronym for *first-order linear regression* equation in elements of $\mathbf{u^*_{SGM}(t-)}, \mathbf{GSD^*(t-)}$, and $\mathbf{GST^*(t-)}$; regression coefficients are represented by $\theta_{\mathbf{chj}}$ and $\theta_{\mathbf{bhk}}$. As in financial modeling, assumptions imposed on the model errors $\boldsymbol{\delta_{cj}(t)}$ and $\boldsymbol{\delta_{bk}(t)}$ will allow for GARCH-type volatility modeling.

For gradual drift,

$$\begin{aligned}
\mathbf{u}^*_{\mathbf{SGM}}(\mathbf{t-}) &= \mathbf{u}_{\mathbf{SGM}}(\mathbf{t-});\\
\mathbf{GSD}^*(\mathbf{t-}) &= \mathbf{GSD}(\mathbf{t-});\\
\mathbf{GST}^*(\mathbf{t-}) &= \mathbf{GST}(\mathbf{t-}).
\end{aligned} \tag{10.1.9}$$

For abrupt drift where elements of $\mathbf{u}_{\mathbf{SGM}}(\mathbf{t-})$ and/or $\mathbf{GSD}(\mathbf{t-})$ and/or $\mathbf{GST}(\mathbf{t-})$ are sufficiently large in modulus, any or all of the following expressions may hold:

$$\begin{aligned}
\mathbf{u}^*_{\mathbf{SGM}}(\mathbf{t-}) &= \mathbf{u}_{\mathbf{SGM}}(\mathbf{t-}) + H_u \mathbf{v}_{\mathbf{u}}(\mathbf{t-});\\
\mathbf{GSD}^*(\mathbf{t-}) &= \mathbf{GSD}(\mathbf{t-}) + H_{GSD} \mathbf{v}_{\mathbf{GSD}}(\mathbf{t-}),\\
\mathbf{GST}^*(\mathbf{t-}) &= \mathbf{GST}(\mathbf{t-}) + H_{GST} \mathbf{v}_{\mathbf{GST}}(\mathbf{t-}).
\end{aligned} \tag{10.1.10}$$

Coefficients denoted by elements of the matrices H_u, H_{GSD}, and H_{GST} represent respective effects of $\mathbf{v}_{\mathbf{u}}(\mathbf{t-}), \mathbf{v}_{\mathbf{GSD}}(\mathbf{t-})$, and $\mathbf{v}_{\mathbf{GST}}(\mathbf{t-})$, vectors of variables, insignificant in the previous model update, whose direct effects become significant in abrupt drift scenarios.

Substitution of (10.1.17)–(10.1.18) into (10.1.16) results in a second-order reduced regression equation for $z_{SGMh}(t)$, which is written simply

$$z_{SGMh}(t) = f_h[\mathbf{x}_{\mathbf{h}}(\mathbf{t}); \boldsymbol{\psi}_{\mathbf{h}}] + \boldsymbol{\varepsilon}_{Rh}(t). \tag{10.1.11}$$

$\mathbf{x}_{\mathbf{h}}(\mathbf{t})$ includes first- and second-order terms involving $\mathbf{u}^*_{\mathbf{SGM}}(\mathbf{t-}), \mathbf{GSD}^*(\mathbf{t-})$, $\mathbf{GST}^*(\mathbf{t-}), \mathbf{w}(\mathbf{t-}), \mathbf{v}_{\mathbf{u}}(\mathbf{t-}), \mathbf{v}_{\mathbf{GSD}}(\mathbf{t-})$, and $\mathbf{v}_{\mathbf{GST}}(\mathbf{t-})$; $\boldsymbol{\psi}_{\mathbf{h}}$ is the corresponding parameter vector; $\boldsymbol{\varepsilon}_{Rh}(t)$ is the reduced model error, which is subject to volatility modeling. Following estimation of $\mathbf{u}_{\mathbf{SGM}}(\mathbf{t-})$, first-stage model identification or estimation in (10.1.11) follows directly from the stepwise regression as applied in financial modeling in Section 7.3. Based on the squared residuals that correspond to the model errors in (10.1.11), one may then proceed with GARCH-type volatility modeling.

In terms of forecasting outcomes of a game between team i and team i^*, the primary modeling focus is on $D(i,t), D(i^*,t), T(i,t)$, and $\mathrm{T}(i^*,t)$. Forecasts for $D(i,t)$, and $D(i^*,t)$ should, at the very least, not be contradictory in terms of the line; for example, in the 2008 Super Bowl, where the Patriots were 12-point favorites over the Giants, a Giant model forecast of a Giant win by 4 points and a Patriot model forecast of a Patriot win by 1 point are not contradictory in terms of the line; both forecasts indicate that the Giants + 12 is the appropriate bet. Similarly, forecasts for $T(i,t)$ and $T(i^*,t)$ should both be on one side of the line (on the total points scored) or the other.

Forecasts for $GSD(i,t), GSD(i^*,t), GST(i,t)$, and $GST(i^*,t)$ are also highly relevant, for two reasons. First, they provide a check on the validity on forecasts for $D(i,t), D(i^*,t), T(i,t)$, and $T(i^*,t)$; for example, if the forecasts for $D(i,t)$ indicates that team i will beat the line, the forecast for $GSD(i,t)$ should be greater than zero and roughly equal to the difference between the $D(i,t)$ forecast and $LD(i,t)$. Second, the gambling shock forecasts provide alternatives to conventional volatility modeling. Conventionally, volatility modeling associated with, say, $D(i,t)$ is based on GARCH-type modeling of the squared residuals associated with the fitted $D(i,t)$. If the model for $D(i,t)$ is ineffective or biased, the accompanying GARCH model will reflect this bias. Modeling the gambling shock directly is a possible means of avoiding such biases.

In the forthcoming modeling illustrations, the $\mathbf{u_{SGM}(t-j)}$ in (10.1.16) are estimated in terms of factor analysis—as was the case in forecasting price changes in the financial forecasting exercises.

10.2 THE 2007–2008 NEW YORK GIANTS: AS UNEXPECTED AS KATRINA

Reduced model (10.1.11) is applied in forecasting the four New York Giants play-off games in 2007–2008 (the last four NYG games in Figure 1.2.3). The variables under study are $D(i = \text{NYG}, t)$ and $T(i = \text{NYG}, t)$, as defined in (10.1.1).

Modeling begins with the estimation of the $\mathbf{u_{SGM}(t-j)}$ in (10.1.15), defined as the lags of confounded between- and within-relation shocks. As in the financial modeling exercises, factor analysis is applied. The estimated correlation matrix of $\mathbf{z_{SGM}(t)}$ in (10.1.1) is based on five successive years of NYG regular-season and play-off games through the end of the regular 2007–2008 season.

Eigenvectors are extracted from the 12×12 correlation matrix through principal components analysis; see Table 10.2.1. The six vectors associated with eigenroots >0.8 are seen to account for 94.4% of the total variation. These six vectors are then rotated through the varimax procedure to allow for ease in interpretation. The resulting linear combinations are given by the six factors in Table 10.2.2. The 12 row headings in Table 10.2.2 are defined by the variables in (1.1.1). With a slight change in notation, OPP denotes the NY Giant opponent. The factor scores corresponding to each

TABLE 10.2.1 Principal Components Analysis Applied to the Correlation Matrix Associated with NY Giants Team Performance Variables

	Initial Eigenvalues			Extraction Sums of Squared Loadings			Rotation Sums of Squared Loadings		
Component	Total	Percent of Variance	Cumulative %	Total	Percent of Variance	Cumulative %	Total	Percent of Variance	Cumulative %
1	3.042	25.351	25.351	3.042	25.351	25.351	2.065	17.205	17.205
2	2.745	22.874	48.225	2.745	22.874	48.225	1.972	16.430	33.635
3	1.892	15.764	63.989	1.892	15.764	63.989	1.970	16.413	50.048
4	1.779	14.826	78.815	1.779	14.826	78.815	1.913	15.945	65.992
5	1.045	8.710	87.526	1.045	8.710	87.526	1.737	14.473	80.466
6	0.828	6.899	94.424	0.828	6.899	94.424	1.675	13.959	94.424
7	0.266	2.215	96.640						
8	0.201	1.675	98.314						
9	0.087	0.725	99.039						
10	0.053	0.439	99.478						
11	0.036	0.298	99.775						
12	0.027	0.225	100.000						

TABLE 10.2.2 Rotation of the First Six Factors in Table 10.2.1[a]

	Factor					
Variable	u_1	u_2	u_3	u_4	u_5	u_6
D(NYG, *t*−1)	0.924	−0.030	0.042	0.019	0.004	0.332
GSD(NYG, *t*−1)	0.974	0.030	0.012	0.031	−0.024	0.122
T(NYG, *t*−1)	0.016	0.986	0.033	0.072	0.061	−0.014
GST(NYG, *t*−1)	−0.025	0.985	0.009	0.003	0.076	−0.078
WL5(NYG, *t*−1)	0.151	−0.019	0.122	0.001	−0.023	0.944
STR(NYG, *t*−1)	0.480	−0.109	0.116	−0.011	−0.026	0.789
D(OPP, *t**−1)	−0.004	0.005	0.022	0.921	0.300	0.001
GSD(OPP, *t**−1)	0.046	0.069	0.063	0.968	0.112	−0.004
T(OPP, *t**−1)	0.029	0.025	0.980	0.028	0.019	0.140
GST(OPP, *t**−1)	0.033	0.018	0.986	0.054	−0.042	0.064
WL5(OPP, *t*−1)	−0.055	0.076	−0.036	0.099	0.940	0.025
STR(OPP, *t*−1)	0.037	0.066	0.015	0.327	0.858	−0.075

[a] Rotation method: Varimax with Kaiser rotation.

TABLE 10.2.3 New York Giants Model Update Forecasts Versus One-Step-Ahead Forecasts

	D(NYG vs. OPP)^					
NYG Opponent	TB	DAL	GB	NE	*LD*(NYG, *t*)	*D*(NYG, *t*)
At TB	14.6				3.0	10
At DAL	6.7	6.0			−7.0	4
At GB	−6.6	−7.2	−7.4		−7.5	3
NE	5.4	4.8	4.6	4.8	−12.0	3

of the six factors can be shown to define stationary processes and are thus used to estimate the $\mathbf{u}_{\mathrm{SGM}}(\mathbf{t}-\mathbf{j})$.

Table 10.2.3 presents two types of forecasts for NYG play-off games. The type 1 forecast is defined as one-step-ahead forecasting with no further model updates; that is, the model update based on all NYG games played through the end of the 2007–2008 regular season is then used to forecast each successive play-off game with no further model updates. The second type of forecast is defined as one-step-ahead forecasting with a model update following each successive game; that is, following each play-off game the model is updated to allow for changes in coefficients and/or predictors. The type 2 forecast is indicative of the adaptive model drift that accommodates structural changes in team performances on a per game basis.

In Table 10.2.3, the diagonal elements of the 4×4 matrix (comprising the second through the fifth columns) are type 2 forecasts, while lower, off-diagonal elements are type 1 forecasts. Relative to the line on the difference [denoted by $LD(\mathrm{NYG}, t)$] and the actual outcome [$D(\mathrm{NYG}, t)$], all forecasts are correct. Through the Green Bay game, models for the type 1 and type 2 forecasting models are very nearly the same. The type 1 forecasting model is given by

$$D(\text{NYG},t)^{\wedge} = 5.47 - 0.092_{5.73}[GST^*(t-1,\text{NYG})][GST(t-1,\text{OPP})] - 6.558_{3.79}[u_2(t-1)][u_5(t-1)], \quad (10.2.1)$$

where

$$GST^*(t-1,\text{NYG}) = GST(t-1,\text{NYG}) \quad \text{for } GST(t-1,\text{NYG})<0, \\ = 0 \quad \text{for } GST(t-1,\text{NYG}) \geq 0. \quad (10.2.2)$$

The first interaction is defined by two *within-series* GST shocks and the second by two *between- or within-series* shocks; $u_2(t-1)$ and $u_5(t-1)$ denote, respectively, the factor scores corresponding to second and fifth factors (columns) in Table 10.2.2. The factor score for $u_2(t-1)$ refers to shocks relating to T and GST for NYG in their previous game. The factor score for $u_5(t-1)$ refers to shocks relating to their opponent's won–lost percentage and streak. In three of the four NYG play-off games, $u_2(t-1)$ and $u_5(t-1)$ were of opposite sign, whereas for the Patriot game, $u_2(t-1)$ was near zero.

The NYG chart in Figure 1.2.3 provides insights into effects of these interactions. As underdogs in their last three games, NYG won over opponents who had defeated them in their previous encounter during the regular season. Dallas had beaten NYG twice during the regular season, which made it unlikely that they would do so a third time in the second round of the play-offs.

The GB game went into overtime and the weather was frigid.[1] The forecast had NYG losing by 7.4 points, while the official line was virtually the same at 7.5 points; see Figure 10.2.1 for a candlestick chart displaying all GB regular and postseason games. The closeness of the forecast and the line may be indicative of a no-bet situation, which was reinforced by the likelihood of ominous GB weather conditions. An obvious weakness of the forecasting models is that they do not directly reflect effects of weather conditions on game outcomes—unless the bookmaker adjusts the final line to partially reflect anticipated weather condition.

In evaluating model 10.2.1 for effects of MA and bilinear terms, there is evidence of a marginally significant MA($t-2$) effect (illustrating the

[1]In the overtime period, Green Bay quarterback Brett Favre threw a devastating interception that led to the GB loss against the NYG in the second round of the play-offs. Two years later, Minnesota lost to New Orleans in overtime in the second round of the play-offs. Minnesota quarterback Brett Favre threw the final devastating interception.

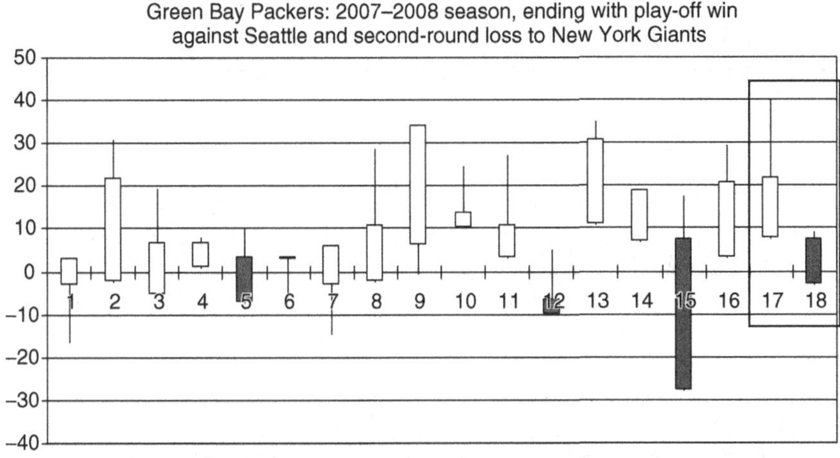

Figure 10.2.1 *Green Bay Packers candlestick chart for the 2007–2008 season.*

nonredundancy of gambling and statistical shocks):

$$NYG\_D(t)^{\wedge} \cong 5.83 + 0.099_{7.27}[GST^{*}(t-1, \text{NYG})][GST(t-1, \text{OPP})] \\ -7.976_{4.24}[u_2(t-1)][u_5(t-1)] - 0.334_{1.77}e_R(t-2). \tag{10.2.3}$$

However, the forecast inferences are the same for equations (10.2.1) and (10.2.2).

The NYG model update just prior to the NE game changed model (10.2.1) slightly, but not the forecast inferences. This model update resulted in

$$NYG\_D(t, \text{OPP} = \text{NE})^{\wedge} = 4.56 + 0.080_{6.27}[GST^{*}(\text{NYG}, t-1)] \\ \times [GST(\text{OPP}, t-1)] - 5.512_{3.52}[u_2(t-1)] \\ \times [u_5(t-1)] + 0.409_{2.42}[LD(\text{NYG}, t)] \\ \times [STREAK(\text{OPP}, t-1)]. \tag{10.2.4}$$

Table 10.2.3 forecasts were confirmed by developing comparable models for each NYG play-off opponent. Such confirmations are critical since each team tends to be unique, so that forecasting models usually differ between teams. In general, ineffective forecasting results from *universal modeling* (where per team databases are pooled in developing a single model to

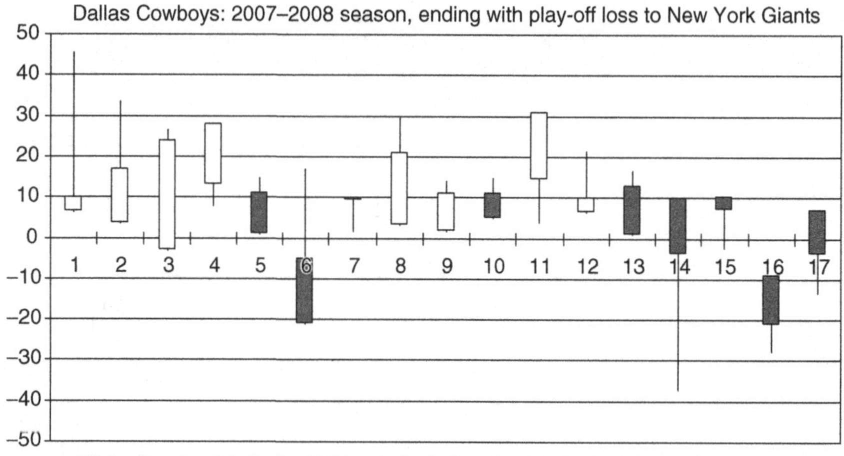

Figure 10.2.2 *Dallas Cowboys candlestick chart for the 2007–2008 season.*

forecast outcomes for all NFL games). When forecasting models for each of two opposing teams yield contradictory predictions relative to the line relative to the line, confirmatory analyses are necessary.

Such contradictory forecasts are illustrated in the Dallas–NYG play-off game. In the latter part of the season Dallas played poorly, in that they failed to beat the line in the four games preceding the NYG game and also lost two of those games; see Figure 10.2.2 for a candlestick chart depicting all regular and postseason Dallas games for the 2007–2008. Against Dallas, the NYG model for $T(\text{NYG}, t)$ forecast a point total less than the line of 47.5 points, while the corresponding model for Dallas forecast a point total of more than 47.5 points. Confirmatory analyses included modeling $GST(\text{NYG}, t)$ and $GST(\text{DALLAS}, t)$—models that directly forecast the gambling shocks corresponding to the total points scored. However, since confirmatory analyses did not resolve the issue, the game was a no-bet situation regarding total points scored.

Forecasts are also suspect when, for an individual team, the forecast based on the model update through the last game differs, in inference, from the forecast based on the previous model update. Such a contradiction occurred in forecasting the total points scored in the NYG vs. GB play-off game, an overtime game that was played under extreme weather conditions. The NYG model update through the Dallas game forecast a point total greater than the line while the model update through the two previous games forecast point totals less than the line. This uncertainty, plus the uncertainty of playing in Green Bay in mid-January, made this a no-bet situation.

Aside from the modeling uncertainties associated with the NYG total point forecasts for the Dallas and GB games, forecasts for the other play-off games against Tampa Bay and NE were correct relative to the line. For the NYG–NE finale, model updates through the end of the regular season, and each play-off game gave total point forecasts that ranged from 34 to 46 points. The line on the total was 53.5 points; 31 points were scored in the game. The NYG model update on total points scored through the GB game is given as follows:

$$\begin{aligned} T(\mathrm{NYG}, t)^{\wedge} = {} & 41.02_{30.1} + 0.067_{7.53}[GST^{**}(\mathrm{NYG}, t-1)][D(\mathrm{OPP}, t-1)] \\ & - 7.94_{4.97}[u_5(t-2)][u_1(t-3)] - 0.055_{5.27}[GST(\mathrm{OPP}, t-1)] \\ & \times [GSD(\mathrm{OPP}, t-2)] - 0.023_{3.59}[GST(\mathrm{OPP}, t-1)] \\ & \times [GST(\mathrm{OPP}, t-2)], \qquad (10.2.5) \end{aligned}$$

where, similar to (10.2.2),

$$\begin{aligned} GST^{**}(\mathrm{NYG}, t-1) &= GST(\mathrm{NYG}, t-1) \quad \text{for } GST(\mathrm{NYG}, t-1) > 0, \\ &= 0 \text{ for } GST(\mathrm{NYG}, t-1) \leq 0. \qquad (10.2.6) \end{aligned}$$

10.3 MISERY FOR THE PATRIOT FAITHFUL

Modeling procedures in Section 10.2 are now applied in forecasting the New England Patriots 2008 play-off games (the last three games in Figure 1.2.2) Results are used to corroborate the NYG forecasts for the finale.

The factor analysis used to estimate cointegrated relations in NYG modeling is reapplied to comparable data for NE. Relative to the results of the principal components analysis for NYG data in Table 10.2.1 (where 94.7% of the variation is explained by the six eigenvectors associated with the six largest eigenroots), the comparable analysis of NE data in Table 10.3.1 show that the first six eigenvectors account for 92.7% of the variation. Rotated eigenvectors for NE data are presented in Table 10.3.2. It is seen that six NE factors are comparable to NYG factors presented in Table 10.2.2. The only difference are in the ordering of factors; for example, for the NYG analysis, the first factor estimates the relation between $D(\mathrm{NYG}, t)$ and $GSD(\mathrm{NYG}, t)$, while for the NE analysis, the fourth factor estimates the relation between $D(\mathrm{NE}, t)$ and $GSD(\mathrm{NE}, t)$.

TABLE 10.3.1 Principal Components Analysis Applied to the Correlation Matrix Associated with New England Patriot Team Performance Variables

	Initial Eigenvalues			Extraction Sums of Squared Loadings			Rotation Sums of Squared Loadings		
Component	Total	Percent of Variance	Cumulative %	Total	Percent of Variance	Cumulative %	Total	Percent of Variance	Cumulative %
1	3.073	25.604	25.604	3.073	25.604	25.604	1.968	16.398	16.398
2	2.440	20.330	45.935	2.440	20.330	45.935	1.930	16.085	32.482
3	1.908	15.900	61.834	1.908	15.900	61.834	1.906	15.882	48.364
4	1.743	14.521	76.355	1.743	14.521	76.355	1.903	15.862	64.226
5	1.081	9.010	85.366	1.081	9.010	85.366	1.851	15.423	79.649
6	0.886	7.382	92.747	0.886	7.382	92.747	1.572	13.098	92.747
7	0.438	3.646	96.393						
8	0.163	1.356	97.749						
9	0.094	0.781	98.530						
10	0.073	0.611	99.141						
11	0.063	0.524	99.665						
12	0.040	0.335	100.000						

TABLE 10.3.2 Rotation of the First Six Factors in Table 10.3.1[a]

	Component					
Variable	1	2	3	4	5	6
NE_LG1DIFF	−0.010	0.022	0.049	0.928	0.039	0.290
NE_LG1SHDIFF	−0.069	0.039	0.024	0.973	0.029	0.089
NE_LG1TOT	0.007	0.980	0.005	0.010	0.028	0.058
NE_LG1SHTOT	−0.002	0.977	−0.004	0.052	0.042	0.067
NE_WL5	0.024	0.046	0.009	0.090	0.160	0.885
NE_STREAK	−0.132	0.082	0.024	0.270	−0.023	0.817
LG1DIFF	0.018	0.019	0.907	0.022	0.346	0.013
LG1SHDIFF	0.075	−0.018	0.964	0.048	0.143	0.017
LG1TOT	0.983	0.001	0.058	−0.048	0.016	−0.076
LG1SHTOT	0.985	0.004	0.031	−0.033	−0.033	−0.021
WL5	−0.043	0.022	0.152	0.009	0.956	0.050
STREAK	0.030	0.060	0.351	0.065	0.874	0.115

[a] Rotation method: Varimax with Kaiser normalization.

The model for $D(\text{NE}, t)$, updated through the last game of the regular season, is given in (10.3.1). The model and resulting forecasts remained largely the same with updates following the first two play-off games.

$$\begin{aligned} D(\text{NE}, t)^\wedge = {} & 3.34 + 0.57_{3.22} LD(\text{NE}, t) \\ & + 0.12_{3.64}[D(\text{NE}, t-1)][STR\_SHD(\text{NE}, t-1)] \\ & + 4.61_{3.83}[u_4(t-2)][u_2(t-3)] - 3.93_{3.07}[u_4(t-2)][u_4(t-3)]. \end{aligned} \tag{10.3.1}$$

The coefficient of $LD(\text{NE}, t)$, the line on the difference for NE, and the effectiveness of the model forecasts indicate that the market was inefficient for the Patriot play-off games. Instead of having a coefficient of $LD(\text{NE}, t)$ close to 1, the analysis produces a coefficient of 0.57. The gambling public clearly misjudged forthcoming NE performances for the play-offs if not at for the entire latter part of the season.

The first interaction is the cross product between $D(\text{NE}, t-1)$ (which was positive for all games but the finale) and $STR\_SHD(\text{NE}, t-1)$, the streak or successive number of positive or negative gambling shocks on the difference going into game t. For each of the three play-off games as well as the three end-of-regular-season games, $GSD(\text{NE}, t) < 0$. The effect of the first interaction is dominant in reducing the value of $D(\text{NE}, t)^\wedge$.

Regarding effects of the factors in Table 10.3.1, the negative effect of the interaction $[u_4(t-2)][u_2(t-3)]$ involves the cross product between factor scores associated, respectively, with the second lag of the fourth factor and the third lag of the second factor. Similarly, $u_4(t-2)$ also interacts with $u_4(t-3)$. For the play-off games, the effects of these shock interactions contributed slightly to $D(\text{NE}, t)^\wedge$.

TABLE 10.3.3 New England Patriots Model Update Forecasts Versus One-Step-Ahead Forecasts

	D(NE vs. OPP)^				
NE Opponent	JAC	SD	NYG	*LD*(NE, *t*)	*D*(NE, *t*)
JAC	9.8			13.5	11
SD	10.1	10.0		14.0	9
NYG	0.8	0.9	0.8	12.0	−3

Table 10.3.3 presents model update forecasts in the same format given for NYG forecasts in Table 10.2.3. Relative to the line, forecasts were correct for all three play-off games. Note that the NE forecasts had the Patriots winning the game against the Giants by about 1 point (well below the line), whereas the NYG forecast had the Giants winning by 4 to 5 points.

For the play-offs, the forecasting model for the total points scored by NE and its opponent, $T(\mathrm{NE}, t)^\wedge$, was incorrect relative to the line for the Jacksonville game but correct for the games against the Chargers and the Giants. Generally, for the 2008 NFL play-off games, TOT forecasts relative to Line(TOT) were less accurate than DIFF forecasts relative to Line(DIFF).

A partial explanation is that TOT values are more volatile than DIFF values. An alternative approach to modeling volatility is to analyze categorized rather than actual values of TOT through alternative procedures such as Bayesian discriminant analysis, Fisher's linear discriminant function, and logistic regression. These procedures are discussed in Chapter 11, which also includes the application of the discriminant analysis that led to the categorized predictions for the NY Giants in Table 1.1.1.

10.4 THE PITTSBURGH STEELERS IN SUPER BOWL 2005

Figure 10.4.1 presents a candlestick chart of regular-season and play-off games for the Pittsburgh Steelers (PIT) during the regular 2005–2006 season and play-off games that concluded with the Super Bowl win over Seattle. PIT began the season by winning seven of its first nine games, but then suffered a major setback when both quarterback Ben Roethlisberger and his backup, Charlie Batch, went down with injuries. With the third-team quarterback as the starter, PIT was upset by Baltimore 16–13. Roethlisberger then returned at which point PIT lost in successive weeks to then undefeated Indianapolis and division rival Cincinnati. PIT then recovered and won the remaining four regular season games to claim the sixth and final seed in the AFC play-offs on the road to winning the

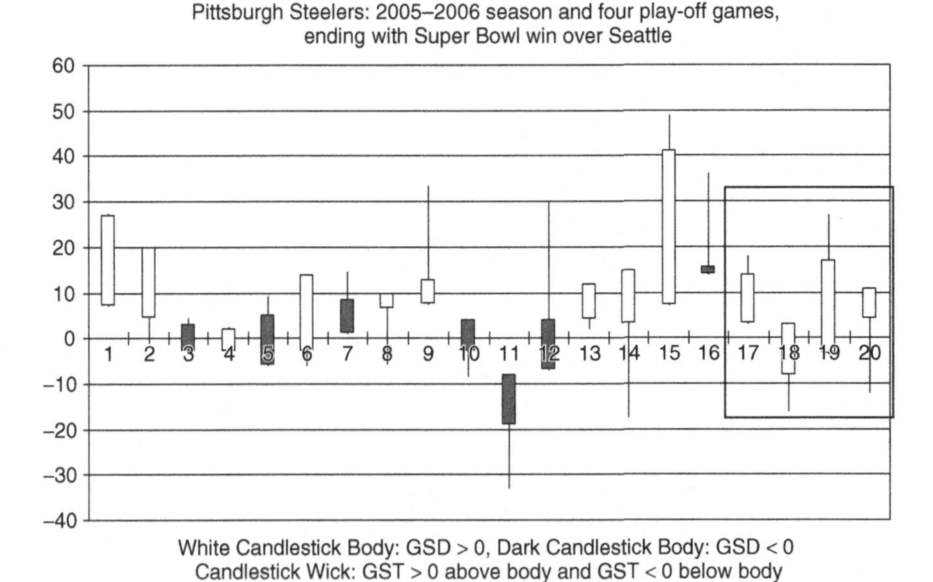

Figure 10.4.1 *Pittsburgh Steelers candlestick chart for the 2005–2006 season.*

2008 Super Bowl. Figure 10.4.1 shows that PIT outperformed the line on the difference in eight of their last nine games. Note also that except for game 10 (when Roethlisberger and his backup were out of the lineup) and game 11 against Indianapolis, a PIT dark body is always followed by a PIT white body.

Tables 10.4.1 and 10.4.2 present results of factor analysis applied to PIT data. These results are roughly the same as those obtained for NYG and NE in Tables 10.2.1 and 10.2.2 and Tables 10.3.1 and 10.3.2, respectively. As in previous analyses for NE and NYG, factor scores corresponding to each of the six factors in Table 10.4.2 define stationary processes and are used to estimate between- and within-relation shocks.

The model update for $D(\text{PIT}, t)$ through the next-to-last regular season game is given in (10.4.1). (Contrary to most other play-off teams, the last game of the regular season was vital to Pittsburgh's hopes of making the play-offs.) With updates after the last regular-season game and throughout the play-offs, the PIT model remained stable. Note that variables relating to total points scored, both by PIT and its opponent in their previous game, dominate the interactions in (10.4.1).

Based on the pre-play-off model update, Table 10.4.3 presents model forecasts against Detroit in the last regular-season games and the four play-off games. Relative to the line, all forecasts are correct.

TABLE 10.4.1 Principal Components Analysis Applied to the Correlation Matrix Associated with Pittsburgh Steelers' Team Performance Variables

	Initial Eigenvalues			Extraction Sums of Squared Loadings			Rotation Sums of Squared Loadings		
Component	Total	Percent of Variance	Cumulative %	Total	Percent of Variance	Cumulative %	Total	Percent of Variance	Cumulative %
1	3.420	28.504	28.504	3.420	28.504	28.504	1.985	16.541	16.541
2	2.639	21.989	50.493	2.639	21.989	50.493	1.970	16.416	32.956
3	2.191	18.255	68.748	2.191	18.255	68.748	1.895	15.795	48.751
4	1.413	11.778	80.526	1.413	11.778	80.526	1.885	15.706	64.457
5	0.872	7.265	87.791	0.872	7.265	87.791	1.855	15.457	79.914
6	0.789	6.574	94.365	0.789	6.574	94.365	1.734	14.450	94.365
7	0.256	2.130	96.495						
8	0.208	1.736	98.231						
9	0.090	0.749	98.980						
10	0.062	0.520	99.500						
11	0.034	0.286	99.786						
12	0.026	0.214	100.000						

TABLE 10.4.2 Rotation of the First Six Factors in Table 10.3.1[a]

	Component					
	u_1	u_2	u_3	u_4	u_5	u_6
D(PIT, *t*−1)	−0.058	−0.189	0.923	−0.080	0.059	0.234
SHD(PIT, *t*−1)	−0.140	−0.175	0.940	−0.023	0.050	0.144
T(PIT, *t*−1)	0.974	−0.111	−0.110	0.032	0.055	0.025
SHT(PIT, *t*−1)	0.976	−0.103	−0.072	0.013	0.033	0.088
*WL*5(PIT, *t*−1)	0.163	0.012	0.144	−0.066	0.062	0.910
STR(PIT, *t*−1)	−0.048	−0.179	0.208	−0.077	−0.064	0.891
D(OPP, *t*−1)	0.018	0.103	−0.044	0.880	0.421	−0.090
SHD(OPP, *t*−1)	0.029	0.097	−0.063	0.970	0.149	−0.073
T(OPP, *t*−1)	−0.128	0.946	−0.211	0.108	0.051	−0.075
SHT(OPP, *t*−1)	−0.100	0.965	−0.145	0.079	0.009	−0.087
*WL*5(OPP, *t*−1)	0.055	−0.008	0.083	0.139	0.937	0.036
STR(OPP, *t*−1)	0.033	0.062	0.020	0.338	0.870	−0.031

[a] Rotation method: Varimax with Kaiser normalization.

TABLE 10.4.3 Pittsburgh Steeler Game Outcomes [Diff(PIT, *t*)] and Model Forecasts [Diff(PIT, *t*)^] Relative to the Line [LineDiff(PIT, *t*)] for the Last Five Games of the 2005–2006 Season

Date	PIT Opp.	Score	LineDiff(PIT, *t*))	Diff(PIT, *t*)	Diff(PIT, *t*)^
1/1/06	DET	W35–21	+15.5[a]	14	5.3
1/8/06	at CIN	W31–17	+3.0	14	8.8
1/15/06	at IND	W21–18	−8.5	3	0.9
1/22/06	at DEN	W34–17	−3.5	17	6.5
2/5/06	SEA[b]	W21–10	+4.0	11	9.7

[a] PIT favored by 15.5 points.
[b] Super Bowl 2006.

$$\begin{aligned} D(\text{PIT},t)^\wedge = {} & 0.21_{4.94}[T(\text{PIT},t-1)][STR\_GST(\text{PIT},t-1)] \\ & - 0.17_{4.98}[u_2(t-1)][TOT(\text{OPP},t-2)] \\ & + 0.04_{3.70}[D\_W(\text{OPP},t-1)][SHTOT(\text{OPP},t-1)] \\ & + 0.29_{2.77}[SHT(t_p)][u_3(t-1)] \\ & + 0.04_{4.16}[TOT(t_p)[WL5(\text{OPP},t-1)] \\ & + 3.06_{2.70}[u_2(t-2)][u_4(t-3)]. \end{aligned} \tag{10.4.1}$$

Regarding the first interaction, the variable $STR\_GST(\mathrm{PIT}, t-1)$ represents the streak or number of consecutive games through game $t-1$ that Pittsburgh's gambling shock on total points scored is greater or less than zero. For the play-offs, GST for PIT alternated between positive and negative values. Regarding the third interaction,

$$\begin{aligned} D\_W(\mathrm{OPP}, t-1) &= D(\mathrm{OPP}, t-1) \quad \text{if } D(\mathrm{OPP}, t-1) > 0, \\ &= 0 \qquad\qquad\qquad\quad \text{if } D(\mathrm{OPP}, t-1) < 0. \end{aligned}$$

The third interaction reflects a motivational factor for PIT in that it has a positive effect on $D(\mathrm{PIT}, t)$ when the PIT opponent had won their previous game and had exceeded the line on total points scored. Similarly, the positive effect of the interaction $[u_2(t-2)][u_4(t-3)]$ has a positive effect when the two lagged shocks reflect positive performance by the PIT opponent in their previous game.

10.5 MIAMI'S FIRST NBA TITLE: 2005–2006

In 2006, the Miami Heat won their first NBA title by defeating the Dallas Mavericks 4–2 in the finals. The Heat's Dwayne Wade was named finals' MVP. This finals matchup featured two teams that never made the finals in the past. The last time this happened was in 1971, when the Milwaukee Bucks met the Baltimore Bullets. The first time this happened was in 1947, when the Chicago Staggs lost to the Philadelphia Warriors. This also happened between the Minneapolis Lakers and the Washington Capitals in 1949 and between the New York Knickerbockers and the Rochester Royals in 1951.

In the Western Conference semifinals, the Mavericks almost gave up a 3–1 series lead to the defending champion San Antonio Spurs but managed to pull out a game 7 overtime win in San Antonio to close out the series. This was just the second time in NBA history that the road team won a game 7 in overtime. The Los Angeles Lakers defeated the Sacramento Kings in the same manner in the 2002 Western Conference finals.

A year earlier, in August 2005, Shaquille O'Neal had signed a five-year extension with the Heat for $100 million. Supporters applauded O'Neal's willingness to take what amounted to a pay cut, and the Heat's decision to secure O'Neal's services for the long term. Critics, however, questioned the wisdom of the move, characterizing it as overpaying an aging and often injured player.

In the second game of the 2005–2006 regular season, O'Neal injured his right ankle and subsequently missed the following 18 games. Many critics

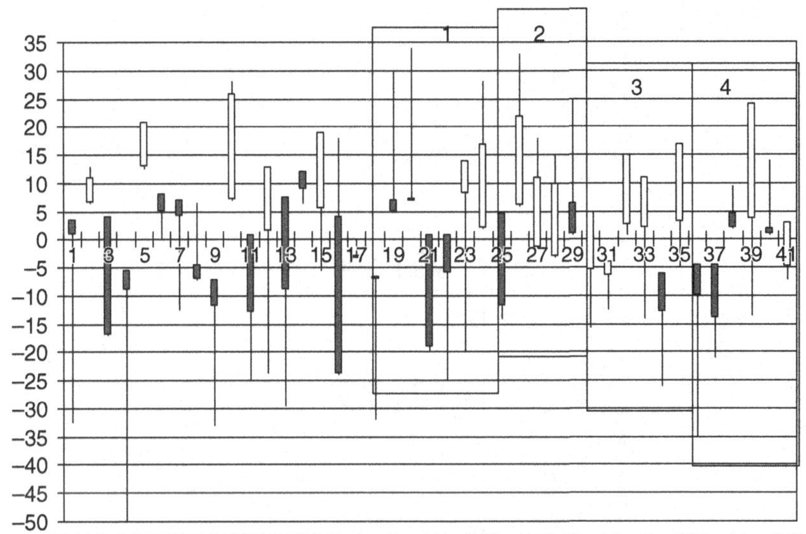

Figure 10.5.1 *Candlestick chart for the Miami Heat for the last 41 games of the 2005–2006 season, including play-off games versus Chicago (won in six games; see box 1), New Jersey (won in five; see box 2), Detroit (won in six; see box 3), and Dallas (won the NBA championship in six; see box 4).*

stated that Miami coach Pat Riley correctly managed O'Neal during the rest of the season, limiting his minutes to a career low. Riley felt that doing so would allow O'Neal to be healthier and fresher come play-off time. Although O'Neal averaged career (or near-career) lows in points, rebounds, and blocks, his view on personal statistics changes: *Stats don't matter. I care about winning, not stats. If I score 0 points and we win, I'm happy. If I score 50, 60 points, break the records, and we lose, I'm pissed off, 'cause I knew I did something wrong. I'll have a hell of a season if I win the championship and average 20 points a game*. During the 2005–2006 regular season, the Heat recorded only a 0.500 record without O'Neal in the lineup (Wikipedia, 2009).

Figure 10.5.1 presents a candlestick chart for Miami for the last 41 games of the season. The dismal Heat performances in the last part of the regular season prior to the play-offs are reflected by the losses and underperformances—the dark bodies—relative to the line. These performances are not indicative of a play-off team, let alone a championship team.

TABLE 10.5.1 Principal Components Analysis Applied to the Correlation Matrix Associated with Miami Heat Team Performance Variables

	Initial Eigenvalues			Extraction Sums of Squared Loadings			Rotation Sums of Squared Loadings		
Component	Total	Percent of Variance	Cumulative %	Total	Percent of Variance	Cumulative %	Total	Percent of Variance	Cumulative %
1	3.080	25.669	25.669	3.080	25.669	25.669	2.046	17.048	17.048
2	2.138	17.815	43.484	2.138	17.815	43.484	1.927	16.061	33.110
3	1.940	16.168	59.652	1.940	16.168	59.652	1.926	16.051	49.160
4	1.553	12.939	72.591	1.553	12.939	72.591	1.706	14.214	63.374
5	1.245	10.378	82.969	1.245	10.378	82.969	1.571	13.093	76.466
6	0.761	6.339	89.309	0.761	6.339	89.309	1.541	12.842	89.309
7	0.443	3.693	93.001						
8	0.374	3.114	96.115						
9	0.221	1.838	97.953						
10	0.101	0.845	98.798						
11	0.090	0.746	99.544						
12	0.055	0.456	100.000						

TABLE 10.5.2 Rotation of the First Six Factors in Table 10.5.1[a]

	Component					
	u_1	u_2	u_3	u_4	u_5	u_6
D(MIA, *t*−1)	0.957	−0.011	−0.099	−0.008	0.058	0.179
SHD(MIA, *t*−1)	0.964	−0.020	−0.106	0.055	−0.091	0.066
T(MIA, *t*−1)	0.021	0.044	−0.036	−0.015	0.865	0.192
SHT(MIA, *t*−1)	−0.038	−0.005	0.048	0.072	0.873	−0.140
*WL*10(MIA, *t*−1)	0.016	−0.154	−0.096	−0.047	−0.082	0.913
STR(MIA, *t*−1)	0.410	0.069	0.032	−0.142	0.199	0.754
D(OPP, *t*−1)	−0.134	0.006	0.883	0.374	0.007	−0.082
SHD(OPP, *t*−1)	−0.094	0.076	0.959	0.113	0.006	−0.020
T(OPP, *t*−1)	−0.034	0.974	0.054	−0.024	0.002	−0.025
SHT(OPP, *t*−1)	0.012	0.969	0.027	0.014	0.038	−0.080
*WL*10(OPP, *t*−1	0.003	−0.041	0.106	0.916	0.041	−0.175
STR(OPP, *t*−1)	0.049	0.041	0.419	0.827	0.023	0.039

[a] Rotation method: Varimax with Kaiser normalization.

Tables 10.5.1 and 10.5.2 present results of factor analysis based on 2.5 years of Miami per game data through the end of the regular 2005–2006 regular season. Note that these results are similar to those for the NFL modeling exercises given in earlier sections of this chapter. Each of the six factors in Table 10.5.2 can be shown to define stationary processes. As such, the corresponding factors scores are used to quantify the confounded effects of between- and within-relation shocks in the Miami modeling exercises.

Model (10.5.1) presents modeling results for $D(\text{MIA}, t)$. The model was updated through the end of the Detroit play-off series and used in forecasting the six games in the final series against Dallas.

$$\begin{aligned} D(\text{MIA}, t)^\wedge = {} & 18.76_{3.82} + 1.38_{3.07} LD(\text{MIA}, t) \\ & - 0.23_{5.09}[D\_W(\text{MIA}, t-1)][STR(\text{MIA}, t-1)] \\ & - 0.11_{3.76}[D\_W(\text{MIA}, t-1)][STR\_W(\text{OPP}, t-1)] \\ & - 0.001_{3.13}[T(\text{OPP}, t-2)][T(\text{OPP}, t-3)] \\ & + 1.83_{3.53}[u_5(\text{MIA}, t-1)][STR\_SHD(\text{MIA}, t-2)] \\ & + 3.77_{3.47}[u_3(\text{MIA}, t-2)][u_1(\text{MIA}, t-3)]. \end{aligned} \qquad (10.5.1)$$

The variable

$$\begin{aligned} D\_W(\text{MIA}, t-1) &= D(\text{MIA}, t-1) && \text{if MIA won their previous game,} \\ &= 0 && \text{if MIA lost their previous game.} \end{aligned}$$

The negative effect of $[D\_W(\text{MIA}, t-1)][STR(\text{MIA}, t-1)]$ on $D(\text{MIA}, t)$ becomes more pronounced the greater their winning streak. Similarly, the

TABLE 10.5.3 Model (10.5.1) Forecasts for the 2007 NBA Championship Series Between Miami (MIA) and Dallas (DAL)

MIA vs. DAL	*LD*(MIA, *t*)	*D*(MIA, *t*)	*D*(MIA, *t*)^	Forecast Relative to *LD*(MIA, *t*)
At DAL-	−4.5[a]	−10	−8.6	Correct
At DAL	−4.5	−14	−10.4	Correct
At MIA	4.5	2	14.4	Incorrect
At MIA	3.5	24	8.3	Correct
At MIA	2.0	1	7.4	Incorrect
At DAL-	5.0	3	0.7	Correct

[a]DAL was favored by 4.5 points.

negative effect of the interaction $[D\_W(\text{MIA}, t-1)]\ [STRw/l(\text{OPP}, t-1)]$ on $D(\text{MIA}, t)$ increases $D(\text{MIA}, t)^\wedge$ if MIA has won their previous game and are meeting an opponent in the midst of a win streak. Conversely, this effect becomes positive if their forthcoming opponent is in the midst of a losing streak. Effects of lagged shocks are in terms of the variables $u_5(\text{MIA}, t-1), u_3(\text{MIA}, t-2)$, and $u_1(\text{MIA}, t-3)$. The interaction $[u_5(\text{MIA}, t-1)][STR\_SHD(\text{MIA}, t-2)]$ has an increasingly positive effect on $D(\text{MIA}, t)^\wedge$ when MIA in a streak of above-average performance (in terms of the line on the difference), particularly in their previous game.

Forecasts based on (10.5.3) are presented in Table 10.5.3. Relative to the game outcomes, all forecasts are correct. However, relative to the line on the difference, four of the six forecasts are correct. A winning margin of 66% is often the norm for these forecasting procedures.

10.6 THE 2006–2007 SAN ANTONIO SPURS: UNEXPECTED TITLISTS

In 2007, the Golden State Warriors qualified for the NBA play-offs for the first time since 1994, the second-longest such streak in league history. In the first round, the Warriors were heavy underdogs and won against the Dallas Mavericks with one of the best records in NBA regular season history. This was after Dallas lost to Miami in the finals a year earlier.

There were expectations of a short series despite the fact that the Warriors, coached by former Dallas coach Don Nelson, had swept the regular season games between the two teams. In addition, the Mavericks failed to play up to the gambling public's expectations during the last weeks of the regular season. Figure 10.6.1 shows that for the 14 games prior to the first-round play-off loss to Golden State, Dallas failed to beat the line on the difference in nine of the games.

Substandard Dallas performances continued with Golden State's game 1 victory in Dallas, behind guard Baron Davis and his frantic style of

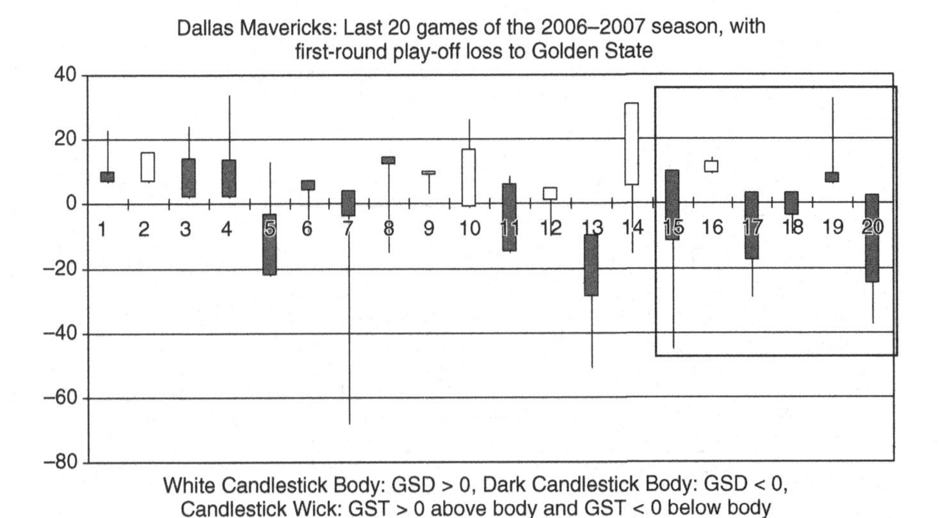

Figure 10.6.1 *Candlestick chart for the 2006–2007 Dallas Mavericks concluding with a first-round play-off loss to Golden State in six games (see the box).*

play. The Mavericks came back to win game 2, to tie the series. However, when the series shifted to Oakland for the next two games, a new X-factor emerged for the Warriors: their home crowd at the Oracle Arena. The Oakland partisans, the highest-paid-attendance crowd for an NBA game in the history of that arena, gave the Warriors a huge lift as they blew out Dallas in game 3, and edged out a close victory in game 4.

As the series shifted back to Dallas, the top-ranked Mavericks found themselves one game from seeing their record-breaking season end prematurely. The Mavericks gave their all and were able to stave off elimination in game 5, but had nothing left in game 6 in Oakland. The Warriors used a third-quarter 18–0 run, sparked by Stephen Jackson's 13 straight points en route to a franchise play-off record seven three-pointers and an unexpected collapse from MVP candidate Dirk Nowitzki (2–13 from the field with 8 points), to finish Dallas. Golden State became the first No. 8 seed to win a best-of-seven series in the first round, in one of the biggest upsets in NBA play-off history.

The Warriors had won their first play-off series since 1991. Ironically, both 2006 NBA finalists (Dallas and Miami) were eliminated in the first round in the 2007 play-offs. The Dallas loss to the Warriors was as unexpected as their loss to Miami in the 2005–2006 finals.

The San Antonio Spurs were to become the unexpected titlists in 2006–2007. Figure 10.6.2 presents the San Antonio candlestick chart for their last 41 games of 2006–2007. The final 20 are play-off games against

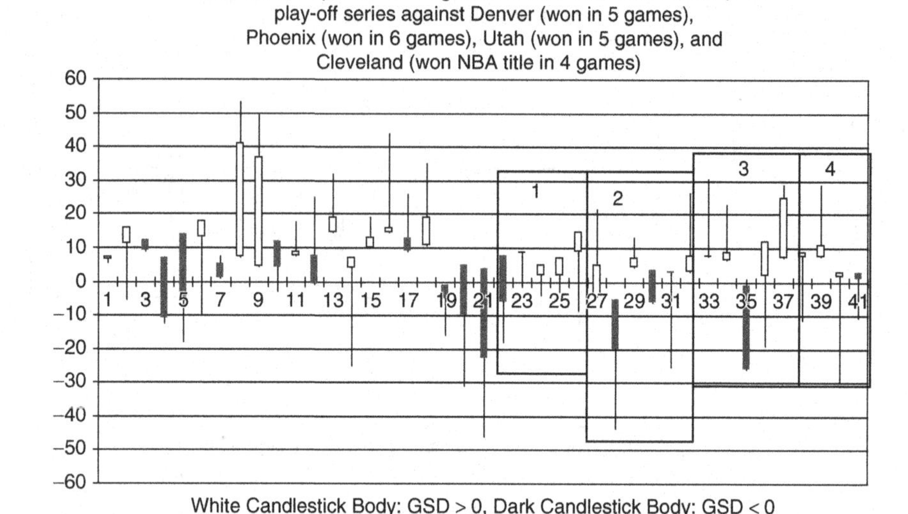

Figure 10.6.2 *Candlestick chart for the 2006–2007 San Antonio Spurs, concluding with play-off games versus Denver (won in five games; see box 1), Phoenix (won in six; see box 2), Utah (won in five; see box 3), and Dallas (won the NBA championship in four; see box 4).*

Denver (won in five; see box 1). Phoenix (won in seven; see box 2), Utah (won in five; see box 3), and the finals against Cleveland (won in four; see box 4). Throughout the play-offs, the Spurs were disciplined and well coached. A dark body play-off loss was always followed by a win or a white body. The Spurs games against their four play-off opponents provided contrasting profiles of team behavior.

With the Dallas loss to Golden State, the stakes of the Phoenix–San Antonio series shot up dramatically. The result was one of the most hotly contested and controversial series in NBA history. The Suns, with a better season record, had home court advantage. Phoenix was led by two-time MVP Steve Nash and the Spurs by three-time finals' MVP Tim Duncan. The Suns quickly lost their home court advantage as the Spurs took a closely contested game 1, a game that saw Nash missing the final minutes due to a bloody gash to his nose. Following game 2, which the Suns won by 20 points, the Suns' star forward Amare Stoudemire accused the Spurs of being a dirty team.

Despite the added scrutiny by the media circles, the Spurs took game 3 in San Antonio with Tim Donaghy, the shamed referee who was later found to fix games, at the helm; see Section 10.7. With the line favoring the Spurs by 3 points, the Suns staged a come-from-behind victory in game 4, to tie the series at 2. However, in the closing minute of game 4 with the Suns leading by 3 points, Nash was shoved into the press table by Robert Horry. This created an altercation that prompted Stoudemire and a teammate to leave the Suns' bench. Although not involved in the altercation, Stoudemire broke an established NBA rule that prohibits players from leaving their bench during altercations. As a result, the NBA suspended Stoudemire for one game, while Horry received a two-game suspension for the flagrant foul, and ejection.

The Suns came into game 5 in San Antonio with the task of beating the Spurs without their star big man. The suspensions created a national outrage as well as a front-court breach for San Antonio to exploit during game 5. The line correctly pegged the Spurs as 3-point favorites, and they won 88–85. Although Stoudemire returned for the Suns for game 6 in San Antonio, the Spurs eliminated them 114–106, to advance to the Western Conference finals against the Utah Jazz. Several NBA pundits opined that the Suns would probably have won the series if the controversial suspensions had not been given to Stoudemire.

For the first time since 1990, neither the No. 1 nor No. 2 seed participated in the Western Conference finals. The series pitted youth against experience as the up-and-coming Utah Jazz faced the seasoned San Antonio Spurs. Coming into the series, the Jazz were not given much of a chance due to their inexperience. Although Carlos Boozer, Deron Williams, and the Jazz were able to hold their own against San Antonio for part of the series, their efforts were not enough.

The first two games of the series, resulting in San Antonio home victories, saw the Spurs blow big first-half leads and the Jazz mount last-gasp rallies that failed. The Jazz, who had been undefeated at home in the postseason, gave their best effort in a 109–83 game 3 rout of the Spurs. However, Jazz fans' euphoria over the team's only series victory gave way to frustration in game 4. The ejections of Utah head coach Jerry Sloan and Jazz guard Derek Fisher energized the Utah home crowd to rain debris onto the court in protest. Game 5 was anticlimatic as the Spurs won 109–84 to advance to the NBA finals against Cleveland. The Spurs then eliminated the overmatched Cavaliers in four games (Wikepedia, 2009).

Tables 10.6.1 and 10.6.2 present results of factor analysis based on 2.5 years of San Antonio game data through the end of the regular 2006–2007 regular season. The six factors for SAN in Table 10.6.2 are somewhat unusual compared with six factors for Miami in Table 10.5.2 and to factor analyses for other NBA games. For example, instead of relations between D and GSD and between T and GST as in Table 10.6.2, the factor analysis results for the Spurs shows, instead, linear relations between D and T, and between GSD and GST. For SAN, the relation between D and

TABLE 10.6.1 Principal Components Analysis Applied to the Correlation Matrix Associated with San Antonio Spurs Team Performance Variables

	Initial Eigenvalues			Extraction Sums of Squared Loadings			Rotation Sums of Squared Loadings		
Component	Total	Percent of Variance	Cumulative %	Total	Percent of Variance	Cumulative %	Total	Percent of Variance	Cumulative %
1	2.892	24.102	24.102	2.892	24.102	24.102	2.162	18.020	18.020
2	2.346	19.552	43.654	2.346	19.552	43.654	2.151	17.923	35.943
3	2.087	17.390	61.043	2.087	17.390	61.043	1.862	15.515	51.458
4	1.430	11.913	72.956	1.430	11.913	72.956	1.807	15.059	66.516
5	1.031	8.591	81.547	1.031	8.591	81.547	1.423	11.858	78.374
6	1.004	8.368	89.915	1.004	8.368	89.915	1.385	11.541	89.915
7	0.422	3.515	93.430						
8	0.329	2.739	96.168						
9	0.200	1.663	97.832						
10	0.119	0.992	98.823						
11	0.082	0.681	99.504						
12	0.059	0.496	100.00						

TABLE 10.6.2 Rotation of the First Six Factors in Table 10.6.1[a]

	Component					
	u_1	u_2	u_3	u_4	u_5	u_6
$D(\text{SAN}, t-1)$	−0.027	0.973	0.047	−0.048	−0.016	0.057
$SHD(\text{SAN}, t-1)$	−0.030	0.057	0.955	0.118	−0.002	0.085
$T(\text{SAN}, t-1)$	−0.034	0.959	0.066	−0.048	−0.036	0.148
$SHT(\text{SAN}, t-1)$	−0.023	0.046	0.961	0.061	0.038	−0.084
$WL10(\text{SAN}, t-1)$	−0.087	0.034	0.007	0.014	−0.022	0.934
$STR(\text{SAN}, t-1)$	−0.022	0.516	−0.015	−0.014	−0.116	0.675
$D(\text{OPP}, t-1)$	0.956	−0.022	−0.037	−0.094	−0.029	−0.068
$SHD(\text{OPP}, t-1)$	−0.114	−0.064	0.108	0.933	−0.031	0.050
$T(\text{OPP}, t-1)$	0.938	−0.049	−0.025	−0.054	0.234	−0.040
$SHT(\text{OPP}, t-1)$	−0.020	−0.028	0.068	0.949	0.030	−0.042
$WL10(\text{OPP}, t-1)$	0.056	−0.065	0.028	−0.004	0.944	−0.064
$STR(\text{OPP}, t-1)$	0.584	0.000	0.012	0.010	0.676	−0.042

[a] Rotation method: Varimax with Kaiser normalization.

T evaluates whether the game as a whole is in or out of balance (as opposed to evaluating whether D is out of balance with respect to GSD), whereas the second relation evaluates whether the two gambling shocks are out of balance. Each of the six factors in Table 10.6.2 can be shown to define stationary processes. As such, the corresponding factors scores are used to quantify the confounded effects of between- and within-relation shocks in the San Antonio modeling exercises.

Except for the Utah series, numerous model updates were required for effectively forecasting $D(\text{SAN}, t)$. The implication is that the Spurs were a team that adjusted their playing style to the particular game situation at hand—to a greater extent relative to the 2006–2007 Miami Heat. For the first series against Denver, the model update through the end of the regular season provided correct forecasts, relative to the line, for the first two games. Due to conflicting forecasts between one-step-ahead forecasts based on an earlier model update and a model update through the last game played, updates were required for the remaining three games of the series—in which case all forecasts were correct relative to the line.

The Phoenix series was a more difficult modeling proposition. The forecast for the first game in Phoenix was correct in picking the winning Spurs relative to the line. However, the forecast also had the underdog Spurs winning the second game—a game that the Spurs lost by 20 points. The latter forecast was counterintuitive, in that it was highly unlikely that the Spurs could win the second game in Phoenix after winning the first away game. While the updated third game forecast was correct, the forecast for the fourth game in San Antonio had the favored Spurs beating the line.

Instead, they lost by 6 points. In all, model forecasts were correct (relative to the line) in three of five games—with $D(\text{SAN}, t) = LD(\text{SAN}, t)$ in game 5.

In the finale against Cleveland, the updated model correctly forecast the San Antonio wins relative to the line in the first two games. However, the forecast for game 3 had the favored Cavaliers beating the line. The Cavaliers lost. The forecast for the final game had the favored Spurs beating the line (3 points), but they won by only 1 point.

Equation 10.6.1, the model update for $D(\text{SAN}, t)$ based on data through the Denver series, gave correct forecasts relative to the line for all six games in the Utah series and did not change with model updates that followed each game of the Utah series. The $D(\text{SAN}, t)$ forecasts for the Utah series are based on the following equation:

$$\begin{aligned} D(\text{SAN}, t)^\wedge = {} & -3.22_{1.82} + 0.008_{6.87}[T(\text{SAN}, t-1)][LD(\text{SAN}, t)] \\ & + 1.13_{6.02}[D\_L(\text{SAN}, t-1)][H\_A(\text{SA}, t)] \\ & + 0.019_{4.68}[SHT(\text{OPP}, t-1)][T(\text{SA}, t_p)] \\ & - 0.018_{4.46}[D(\text{SAN}, t-1)][SHT(\text{OPP}, t-2)] \\ & - 1.86_{3.68}[u_3(\text{SAN}, t-1)][STR\_SHD(\text{SAN}, t-2)]. \end{aligned} \tag{10.6.1}$$

Forecasts based on 10.6.1 are presented in Table 10.6.3. The Spurs were favored in all games except for the third game of the series. All forecasts were correct for this series [relative to the line on $D(\text{SAN}, t)$] as denoted by $LD(\text{SAN}, t)$.

As an adjustment to the line on the difference, the first interaction, $[T(\text{SAN}, t-1)][LD(\text{SAN}, t)]$, indicates that the line should be increased or decreased in proportion to the SAN point total in their previous game.

TABLE 10.6.3 Model (10.6.1) Forecasts for the 2007 NBA Western Conference Championship Series Between San Antonio (SAN) and Utah (UTH)

SAN vs. UTH	*LD*(SAN, *t*)	*D*(SAN, *t*)	*D*(SAN, *t*)^	Forecast Relative to *LD*(SAN, *t*)
At SAN	7.5[a]	8	9.0	Correct
At SAN	6.5	9	19.4	Correct
At UTH	−1.0	−26	−7.6	Correct
At UTH	2.0	12	23.7	Correct
At SAN	7.0	25	12.2	Correct

[a]SAN was favored by 7.5 points.

For example, a low point total for the Spurs in their previous game may possibly indicate a strenuous defensive game, which may have a physical effect on the players in their next game. For the second interaction, $[D\_L(\text{SAN}, t-1)][H\_A(\text{SA}, t)]$, the variable

$$[D\_L(\text{SAN}, t-1)] = D(\text{SAN}, t-1) \qquad \text{when } D(\text{SAN}, t-1) < 0,$$
$$= 0 \qquad \text{when } D(\text{SAN}, t-1) < 0.$$

This interaction has a positive effect when the Spurs are playing at home in game t and lost their previous game. (From Figure 10.6.2 it is seen that a SAN loss was always followed by a SAN win.)

The negative effect of the final interaction $[u_3(\text{SAN}, t-1)]$ $[STR\_SHD(\text{SAN}, t-2)]$ deals with streaks and changes in steaks. The first variable $u_3(\text{SAN}, t-1)$ measures disequilibria between the gambling shock associated with the difference scores and the gambling shock associated with the total points scored in the Spurs' previous game. If, for example, the Spurs had a string of games where they beat the line on the difference through game $t-2([STR\_SHD(\text{SAN}, t-2)] > 0)$ and then showed a negative disequilibrium between the two gambling shocks in game $t-1$, the indication is that Spurs momentum had changed, which in this situation would have a negative effect in their forthcoming game.

10.7 MONITORING NBA REFEREE PERFORMANCES

Modeling assumptions for MLB, NBA, and NFL game outcomes are premised on the assumption of a level playing field (i.e., outcomes are not subject to fixes). The case of Tim Donaghy may suggest otherwise. Recall that Donaghy was the NBA referee who made calls affecting the point spreads on games after betting on those games; see Section 1.2. Such revelations make it imperative that procedures be established and/or improved for monitoring individual referee performances in the manner of quality control, as used in engineering and manufacturing to ensure that product or services are designed and produced to meet or exceed customer requirements and demands.

To illustrate the process of monitoring behavior in simplest terms, we present an analogous example of detecting deviant performance in law enforcement. In the case of *United States* v. *Barajas* (Cr.S-93-495-WBS), a California Highway Patrol officer (Officer Smith) ticketed Barajas on Interstate Highway 5. Smith was subsequently charged with racial

TABLE 10.7.1 Number of Tickets Issued by Officer Smith (S) and Other Officers (OF) to Hispanic (H) and Non-Hispanic (NH) Drivers[a]

	H	NH	Row Total
S	104	106	210
OF	154	810	964
Column total	258	916	1174

[a]Summary statistics:

- Percent of tickets issued by Smith: 17.9%.
- Expected percentage of all tickets issued by Smith under the assumption that all six officers issued the same number of tickets: 16.7%.
- Percent Hispanic among those ticketed by Smith: 49.5%.
- Percent Hispanic among those ticketed by the five other officers: 16.0%.

profiling by ticketing drivers with surnames identified as definite or probable Hispanic (H). Table 10.7.1 presents data on the numbers of H drivers and non-Hispanic (NH) drivers ticketed by Officer Smith (S) and the five other officers (OF) who most frequently worked the same beat during the January–March period of 1994. The table includes summary statistics.

Let $P_H(S)$ and $P_H(OF)$ denote, respectively, the proportions of H among those ticketed by S and OF over the period in question. The standardized normal test statistic that tests the hypothesis that $P_H(S) = P_H(OF)$ versus the alternative that $P_H(S) > P_H(OF)$ indicates the probability that $P_H(S) = P_H(OF)$ is true is less than one time in a trillion. Given the evidence of racial profiling by Officer Smith, the case against Barajas was dismissed.

There are analogies between monitoring and detecting the aberrant behaviors of Officer Smith and Referee Donaghy. R. J. Bell, president of the sports betting site Pregame.com, was interviewed by ESPN sports commentator Wayne Drehs following disclosure of the Donaghy indictment. In the two seasons in which the FBI was investigating Donaghy for allegedly fixing games, Bell found that NBA teams scored more points than the Las Vegas total line on total points scored in 57% of games when Donaghy was part of the officiating crew. With a league average of 49 to 51%, Bell concluded that the odds of such an occurrence are 19 to 1. When Bell analyzed the numbers from the two seasons prior to the two in question, he found that Donaghy-officiated games beat the line on total points just 44% of time. Bell concluded that the odds of an increase from 44% to 55% are about 1 in 1000.

ESPN.com's own research into Donaghy's last two seasons supported Bell's claims. *In the 66 games Donaghy refereed in the 2005–06 season, the two teams in his games combined to score an average of 196.8 points. The average over/under, according to* BoDog.com, *was 186.6, a difference of almost 10 points. In 2006–07, Donaghy refereed 73 games. In those contests, the two teams combined to score 201.37 points and the average over/under was 187.9 points, a difference of more than 13 points per game*. In response to the ESPN.com research, Bell added: *Vegas is too good for that to happen. The standard range should be somewhere around five or six, maybe. Not 10 or 13*.

Bell's research on Donaghy's officiating went further. *At the start of the 2007 calendar year, there were 10 straight games in which Donaghy was part of the officiating crew and the point spread moved a point and a half or more before tip-off, indicating big money had been wagered on the game. In those 10 contests, according to Bell, the big money won all 10 times. "They say follow the money, right?" Bell said. "Well, when the money is right 10 straight times, something is going on. To me, that's the gavel clicking down." ... Just as interesting are the numbers from April 15 to the postseason. During that stretch, there were eight games in which Donaghy was part of the officiating crew and the line moved more than a point and a half before the tip, Bell said. And in those games, including over/under bets and win/loss wagers, the big money was just 2–7. "It means one of two things," Bell said. "Perhaps in the playoffs, they felt too much scrutiny and they weren't trying to do anything and the results are just random. Or perhaps there was some sort of turnabout with the individual in question and he went the other way"* (Drehs, 4/24/07).

Based on the statistical findings in the Donaghy case,[2] it would appear that additional monitoring techniques are appropriate—techniques in terms of candlestick graphics and alternative probability assessments. Figure 10.7.1 presents a candlestick chart for 40 consecutive NBA games in which a hypothetical referee, referee X, was part of the officiating crew.

[2]*Donaghy wrote a book about corruption in NBA officiating while serving time in a federal prison for being the most corrupt NBA official ever. ... While publisher Random House will reportedly not publish* Blowing the Whistle: The Culture of Fraud in the NBA *because of liability concerns*, [the Web site] deadspin.com *printed what it calls excerpts of the book. ... The NBA denies it threatened any legal action against Random House in an effort to stop the book. ... It's far more likely the publisher pulled back when its senior legal team got a look at the completed manuscript that lacks corroboration for the most serious allegations. ... Whether or not Donaghy's allegations are true, most of them are believable. Not only to anyone who has watched a game, but the league's own rank-and-file players and coaches. ... Donaghy admits stars get preferential treatment, some refs have it in for some players and coaches, and a losing home team is likely to get a favorable whistle to make it competitive. ... By understanding the dynamics of intra-league relationships and referee tendencies during his 13 years with the NBA, Donaghy writes he was able to gamble successfully on the outcome* (Wetzel, 10/29/09).

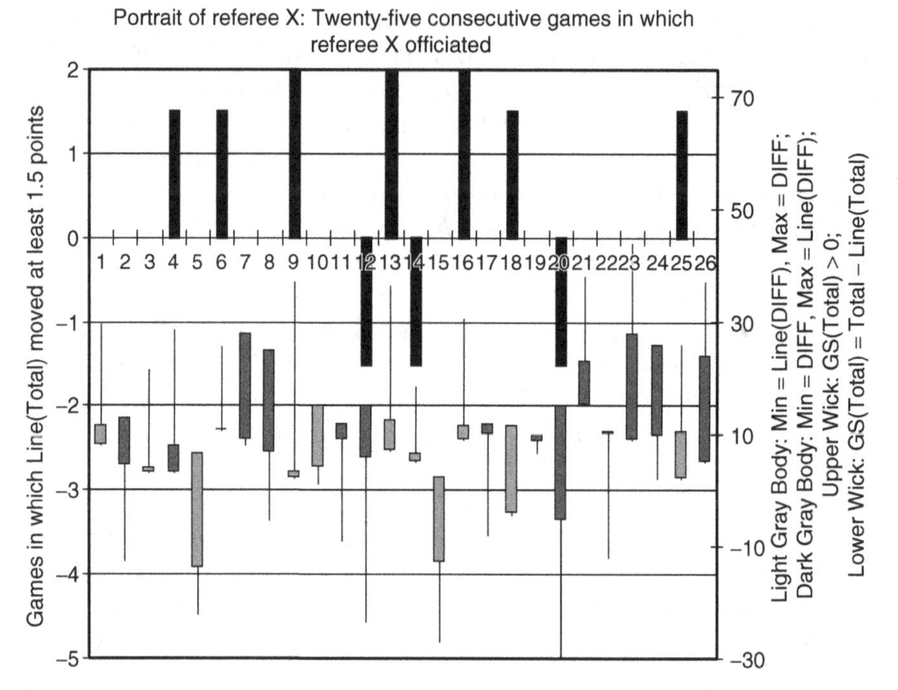

Figure 10.7.1 *Hypothetical candlestick chart for consecutive officiating performances by referee X..*

It is seen that the line on total points scored (*LTOT*) moved at least 1.5 points in 10 of the 26 games. Moreover, in all but one of these 10 games, the value of corresponding gambling shock ($GST = TOT - LTOT$) was in agreement with the direction of the movement of *LTOT*; the only disagreement occurred in game 14. Based on these figures, the question at hand is whether there is evidence of irregularities in referee X's officiating performance in terms of (1) the number of games in which the line moved at least 1.5 points and (2) the number of times the *LTOT* and *GST* were in agreement.

Similar to the example in Table 10.7.1, the answer depends on performances by other established NBA referees during the same period of time. Suppose first that we examine the likelihood that, out of 26 games involving referee X, 14 would have $GST \leq 0$ and 12 with $GST > 0$. Typically, the norm for established referees is quantified in terms of the binomial distribution:

$$f(x_a; n, p_a) = [n_a!(n - n_a)!/n!]p_a^{na}(1 - p_a)^{n-na}, \qquad (10.7.1)$$

where

$$n_a = \text{number of games out of } n = 26 \text{ for which } GST > 0$$
$$n - n_a = \text{number for which } GST \leq 0$$
$$p_a = \text{probability that } GST > 0 \text{ for any given game}$$

For referee X, the question is whether $P(n_a \geq 14)$ is sufficiently small, where the estimate of p_a based on the average of p_a for the other established referees. If the estimated value of p_a is, say, 0.50, then, for referee X, $P(n_a \geq 14) \approx 0.35$ does not warrant undue suspicion.

A difficulty with the distributional assumption in (10.8.1) is that p_a may vary sufficiently between the other referees so as to invalidate the binomial assumption. To address this shortcoming, we assume that p_a varies according to the beta distribution with parameters $\alpha > 0$ and $\gamma > 0$:

$$g(p_a; \alpha, \gamma) = p_a^{\alpha-1}(1 - p_a)^{\gamma-1}/\beta(\alpha, \gamma), \tag{10.7.2}$$
$$\beta(\alpha, \gamma) = [\Gamma(\alpha)\Gamma(\gamma)]/[\Gamma(\alpha) + \Gamma(\gamma)],$$
$$E(p_a) = \alpha/(\alpha + \gamma),$$
$$\text{variance}(p_a) = \alpha\gamma/(\alpha + \gamma)^2(\alpha + \gamma + 1),$$

where $\Gamma(\cdot)$ denotes a gamma function. Multiplying (10.8.1) by (10.8.2) and integrating out p_a over the range [0, 1], we have the binomial-beta probability density:

$$h(n_a; n, \alpha, \gamma) = [n_a!(n - n_a)!/n!][\beta(n_a + \alpha, n - n_a + \gamma)/\beta(\alpha, \gamma)], \tag{10.7.3}$$

$$E(n_a) = n\alpha/(\alpha + \gamma),$$
$$\text{variance}(n_a) = n^2\alpha\gamma/(\alpha + \gamma)^2(\alpha + \gamma + 1).$$

J.G. Skellam (1948) published the results in 10.7.3 in the late 1940s. Relative to the binomial distribution, 10.7.3 will usually provide more reliable expectations and assessments of probabilities in these types of applications.

Next consider the assessment of referee X performance—as depicted in Figure 10.7.1—in terms of:

n_1: the number of times in the last 26 games that the line moved at least 1.5 points and the associated GST was in agreement with the line movement

n_2: the number of time the line moved at least 1.5 points but the associated *GST* was not in agreement

$n_3 = n - n_1 - n_2$: the number of times the line did not move more at least 1.5 points

Similar to the binomial assumption in (10.8.1), the trinomial distribution

$$f(n_1, n_2; n_3, p_1, p_2) = [n_1!n_2!(n_3)!/n!]\, p_1^{n1} p_2^{n2} (1 - p_1 - p_2)^{n3}, \tag{10.7.4}$$

where $n = n_1 + n_2 + n_3$, is typically applied in applying obtaining expectations that determine whether the probability $P(n_1 \geq 9, n_2 < 2)$ is in accord with the average probability associated with the other referees during the same time period.

If we assume that $p_1 = 0.15$ and $p_2 = 0.5$ for the other referees during the same 26-game time span, the corresponding chi-squared goodness-of-fit test is highly significant. The implication is that the near-zero value $P(n_1 \geq 9, n_2 < 2;\ p_1 = 0.15, p_2 = 0.5)$ based on (10.7.4) should arouse suspicions regarding the performance of referee X and his officiating crew.

Analogous to the replacement of (10.7.1) with (10.7.3) in determining outcome expectations, it is likely that p_1 and p_2 for the other referee crews are not constant from game to game, but instead, have a tendency to vary between games. If so, p_1 and p_2 in (10.8.3) are likely to follow the multivariate beta distribution:

$$\begin{aligned} g(p_1, p_2; \alpha_1, \alpha_2, \alpha_3) &= p_1^{\alpha-11} p_2^{\alpha-12} (1 - p_1 - p_2)^{\alpha-13} / \beta(\alpha_1, \alpha_2, \alpha_3), \\ \beta(\alpha_1, \alpha_2, \alpha_3) &= [\Gamma(\alpha_1)\Gamma(\alpha_2)\Gamma(\alpha_1)]/[\Gamma(\alpha_1) + \Gamma(\alpha_2) + \Gamma(\alpha_3)]. \end{aligned} \tag{10.7.5}$$

Multiplying (10.8.3) by (10.8.4) and performing appropriate integrations, we have

$$\begin{aligned} h(n_1, n_2; n, \alpha_1, \alpha_2, \alpha_3) &= [n_1!n_2!(n_3)!/n!][\beta(n_1 + \alpha_1, n_2 + \alpha_2, n_3 + \alpha_3)/ \\ &\quad \times \beta(\alpha_1, \alpha_2, \alpha_3)], \\ E(n_i/n) &= \alpha_i/(\alpha_1 + \alpha_2 + \alpha_3), \\ \text{variance}(n_i/n) &= \alpha_i(\alpha_1 + \alpha_2 + \alpha_3 - \alpha_i)(n + \alpha_1 + \alpha_2 + \alpha_3)^2/ \\ &\quad \times (1 + \alpha_1 + \alpha_2 + \alpha_3 - \alpha_i)n, \\ \text{covariance}(n_i/n, n_{i*}/n) &= \alpha_i \alpha_{i*}(\alpha_1 + \alpha_2 + \alpha_3 + n)/(\alpha_1^2 + \alpha_2^2 + \alpha_3^2) \\ &\quad \times (\alpha_1 + \alpha_2 + \alpha_3 + 1)n, \end{aligned} \tag{10.7.6}$$

where $i^* \neq i = 1, 2, 3$. See Mallios (1989) for applications of the multinomial–multivariate beta distribution.

Expectations based on the compound distribution in (10.7.5) tend to be more realistic than those based on the trinomial distribution in (10.7.4) since they allow for acceptable between-game variations in judging NBA referee performances.

11

Categorical Forecasting

11.1 FISHER'S DISCRIMINANT FUNCTION

Effective categorical forecasts have at least three attributes: They provide an added means of risk assessment; they are useful in spread betting; and they serve to confirm point forecasts, especially regarding the variables $D(i,t), GSD(i,t), T(i,t)$, and $GST(i,t)$. Table 1.1.1 presents an illustration of categorical forecasts for the Giants–Patriots matchup in the 2008 Super Bowl, where odds and probabilities are given for each of three categories $D(\text{NYG},t) : D(\text{NYG},t) < -7, 7 \leq D(\text{NYG},t) \leq 7$, and $D(\text{NYG},t) > 7$. Table 1.1.2 presents analogous forecasts for weekly changes in Microsoft share prices. In this chapter we discusses methodologies that underlie these examples.

Three procedures are commonly employed in categorical modeling (Afifi, 1996): (1) Bayesian discriminant analysis, (2) logistic regression, and (3) Fisher's discriminant function. Suppose, for example, that $GSD(i,t)$ is categorized into mutually exclusive and exhaustive categories. For two categories, we would logically choose

$$GSD(i,t) > 0 \quad \text{and} \quad GSD(i,t) \leq 0. \tag{11.1.1}$$

Modeling attaches probabilities to each of the two outcomes, which tells us which side of the line to bet on and the associated risk. For three categories,

Forecasting in Financial and Sports Gambling Markets: Adaptive Drift Modeling, By William S. Mallios

we might choose

$$GSD(i,t) \le -2, \quad -2 < GSD(i,t) < 2, \quad \text{and} \quad GSD(i,t) \ge 2. \tag{11.1.2}$$

A sufficiently large probability associated with the middle category in (11.1.2) may indicate a no-bet situation in the sense that game is *too close to call relative to the line*. Or, for spread bets on $D(i,t)$ (see Section 8.3), where L_S and U_S denote the lower and upper limits of the spread, $D(i,t)$ would be categorized according to

$$D(i,t) < L_S, \quad L_S \le D(i,t) \le U_S, \quad \text{and} \quad D(i,t) > U_S. \tag{11.1.3}$$

A sufficiently large probability associated with one end category or the other would indicate a favorable bet.

Fisher's discriminant function applies to two categories, as in (11.1.1). Let

$$\begin{aligned} CGSD(i,t) &= 1 \qquad \text{when } GSD(i,t) > 0, \\ &= 0 \qquad \text{when } GSD(i,t) \le 0. \end{aligned} \tag{11.1.4}$$

Substituting $GSD(i,t)$ for $z_{SGMh}(t)$ in (10.1.20) and then replacing $GSD(i,t)$ with $CGSD(i,t)$, we have

$$CGSD(i,t) = f_h[\mathbf{x_h(t)}; \boldsymbol{\psi_h}] + \varepsilon_{RCGSD}(t). \tag{11.1.5}$$

In forthcoming discussions, the subscript h in $f_h[\mathbf{x_h(t)}; \boldsymbol{\psi_h}]$ will be understood to correspond to the categorized variable under discussion—in this case $CGSD(i,t)$.

Model identification and estimation in (11.1.5) is identical to that for $z_{SGMh}(t)$ in (10.1.20) with one exception. One may choose to impose Lagrangian constraints in the estimation procedure to assure that $-1 \le CGSD(i,t)^\wedge \le 1$; $CGSD(i,t)^\wedge$ denotes the estimate of $CGSD(i,t)$. Without the Lagrangian constraints, values of $CGSD(i,t)^\wedge$ may, on occasion, lie outside the interval [−1, 1].

Let

$$CGSD(i,t) = CGSD(i,t)^\wedge + e_{RCGSD}(t), \tag{11.1.6}$$

where $e_{RCGSD}(t)$ denotes the residual corresponding to reduced model error $\varepsilon_{RCGSD}(t)$. Risk may be assessed in terms of forecasts for time-varying variability. A variance forecast, denoted by $e^2_{RCGSD}(t)^\wedge$, is available through GARCH-type modeling.

Without reference to the variance forecast, $CGSD(i,t)^\wedge$ provides one measure of the degree of certainty regarding the wager [i.e., the greater (less) the departure of $CGSD(i,t)^\wedge$ from 0.5, the greater (less) the certainty]. To incorporate $e^2_{RCGSD}(t)^\wedge$ in the decision-making process, we assume that $CGSD(i,t)$ follows the beta distribution in (10.8.2). The mean forecast $CGSD(i,t)^\wedge$ and variance forecast $e^2_{RCGSD}(t)^\wedge$ are then equated to the first two moments of the beta distribution [as given in (10.7.2)]:

$$CGSD(i,t)^\wedge = \alpha/(\alpha+\gamma) \quad \text{and} \quad e^2_{RCGSD}(t)^\wedge = \alpha\gamma/(\alpha+\gamma)^2(\alpha+\gamma+1). \tag{11.1.7}$$

Methods of moment estimators are obtained by solving for α and γ. This solution can be used either as a final estimate or as an initial estimate in obtaining maximum likelihood estimates of α and γ. Substituting the estimate $(\alpha^\wedge, \gamma^\wedge)$ for (α, γ) in the beta distribution and integrating $g(p_a; \alpha^\wedge, \gamma^\wedge)$ in (10.7.2) over the range [0, 0.5], we obtain the probability that $CGSD(i,t)$ exceeds 0.5:

$$P(GSD(i,t) \geq 0.5; \alpha^\wedge, \gamma^\wedge).^1 \tag{11.1.8}$$

Sufficiently large or small values of both $CGSD(i,t)^\wedge$ and (11.1.8) would determine which side of the line to bet on and the associated risk.

In the 2006 Super Bowl, the Steelers were 4-point favorites over Seattle and won by 11 points; see Table 10.4.3. The model forecast for $D(PIT,t)$ was 9.7 points, a correct prediction relative to the line. As a confirmatory analysis, values of $D(PIT,t)$ were categorized according to (11.1.4) and modeled according to Fisher's discriminant function. The result was $GSD(\text{PIT}, t = \text{2006 Super Bowl})^\wedge = 0.704$.

GARCH modeling of the squared residuals corresponding to $GSD(\text{PIT}, t = \text{2006 Super Bowl})$ led to the forecast $e^2_{RCGSD}(\text{PIT}, t = \text{2006 Super Bowl})^{2\wedge} = 0.0199$. As in (11.1.7), the mean and variance forecasts are equated to the expected value and variance of the beta distribution. Solving for α and γ, we obtain the following method of moments estimates: $\alpha^\wedge = 9.5$ and $\gamma^\wedge = 4.0$. For the 50% point of the beta distribution, $P(GSD > 0.5; \alpha^\wedge = 9.5, \gamma^\wedge = 4.0) = 0.714$. Since the point and categorical forecasts reinforced one another, the game represented a highly favorable scenario for betting on the Steelers relative to the line.

[1]Tables of percentage points of the beta distribution are given in the *CRC Handbook of Tables for Statistics and Probability*, Second Edition, W.R. Beyer, Editor, The Chemical Rubber Company, Cleveland, Ohio, 1974. Tables are also referenced on numerous Web sites dealing with the beta distribution.

11.2 BAYESIAN DISCRIMINANT ANALYSIS

Suppose that $z_{SGMh}(t)$ in (10.1.11) is categorized into T mutually exclusive and exhaustive categories, with C_{hk} denoting the kth category. The objective of Bayesian discriminant analysis is to assign a probability to C_{hk} based on the observed value of $\mathbf{x_h(t)}$ in (10.1.11). From Bayes' theorem, we have

$$P(C_{hk};\mathbf{x_h(t)}) = P(C_{hk})P(\mathbf{x_h(t)};C_{hk})/[\Sigma_{k^*}P(C_{hk^*})P(\mathbf{x_h(t)};C_{hk^*})], \tag{11.2.1}$$

where $k, k^* = 1,\ldots,T. P(C_{hk};\mathbf{x_h(t)})$ is the posterior probability that $GSD(i,t)$ belongs to C_{hk} after observing $\mathbf{x_h(t)}$. $P(C_{hk})$ is the prior probability that $GSD(i,t)$ belongs to C_{hk} [or the probability that $GSD(i,t)$ belongs to C_{hk} prior to observing $\mathbf{x_h(t)}$]. $P(\mathbf{x_h(t)};C_{hk})$ is the probability that the observed value of $\mathbf{x_h(t)}$ is known to have been drawn from category C_{hk}. The prior probabilities may be based either on subjective judgment or on past empirical distributions (see Section 7.4). For uniform priors, $P(C_{hk}) = 1/T$.

$P(\mathbf{x_h(t)};C_{hk})$ may represent any multivariate distribution for $\mathbf{x_h(t)}$. In software packages such as SPSS, $P(\mathbf{x_h(t)};C_{hk})$ is represented by a multivariate normal distribution: $\mathbf{x_h(t)}$ **:Normal** $[\boldsymbol{\mu}_{\mathbf{hk}}\mathbf{(t)}, V_{hk}(t)]$; that is,

$$\begin{aligned} &f[\mathbf{x_h(t)};\boldsymbol{\mu}_{\mathbf{hk}}\mathbf{(t)},V_{hk}(t)] \\ &\quad = (2\pi)^{-p/2}|V_{hk}(t)|^{-p/2}\exp[-(\mathbf{x_h(t)}-\boldsymbol{\mu}_{\mathbf{hk}}\mathbf{(t)})'V_{hk}(t)^{-1} \\ &\qquad \times[(\mathbf{x_h(t)}-\boldsymbol{\mu}_{\mathbf{hk}}\mathbf{(t)})/2]. \end{aligned} \tag{11.2.2}$$

[Applications by Mallios (1989) utilize alternative distributional assumptions for $\mathbf{x_h(t)}$.] If $V_{hk}(t) = V_h(t)$ for all k, (11.2.1) becomes

$$\begin{aligned} &P(C_{hk};\mathbf{x_h(t)}) \\ &\quad = P(C_{hk})\exp\{\lambda_{hk} + \boldsymbol{\lambda}'_{\mathbf{hk}}\mathbf{x_h(t)}/[\Sigma_{k^*}P(C_{hk^*})\exp\{\lambda_{hk^*} + \boldsymbol{\lambda}'_{\mathbf{hk^*}}\mathbf{x_h(t)}\}], \end{aligned} \tag{11.2.3}$$

where

$$\lambda_{hk} = -\boldsymbol{\mu}_{\mathbf{hk}}\mathbf{(t)}'\boldsymbol{\mu}_{\mathbf{hk}}\mathbf{(t)}/2 \quad \text{and} \quad \boldsymbol{\lambda}'_{\mathbf{hk}} = \boldsymbol{\mu}_{\mathbf{hk}}\mathbf{(t)}'V_h(t). \tag{11.2.4}$$

If $V_{hk}(t) \neq V_h(t)$, (11.2.1) becomes

$$P(C_{hk};\mathbf{x_h(t)}) = \frac{P(C_{hk})\lambda_{0hk}\exp\{\lambda_{hk} + \boldsymbol{\lambda}'_{\mathbf{hk}}\mathbf{x_h(t)} + \mathbf{x_h(t)}'\Gamma_{hk}\mathbf{x_h(t)}}{\Sigma_{k^*}P(C_{hk^*})\lambda_{0hk^*}\exp\{\lambda_{hk^*} + \boldsymbol{\lambda}'_{\mathbf{hk^*}}\mathbf{x_h(t)} + \mathbf{x_h(t)}'\Gamma_{hk^*}\mathbf{x_h(t)}\}}, \tag{11.2.5}$$

where

$$\lambda_{0hk*} = |V_{hk}(t)|^{-p/2} \quad \text{and} \quad \Gamma_{hk*} = [V_{hk*}(\mathbf{t})]^{-1}. \qquad (11.2.6)$$

In (11.2.2), the quantity in the exponent,

$$D_{hk}^2 = (\mathbf{x_h(t)} - \boldsymbol{\mu}_{\mathbf{hk}}(\mathbf{t}))' V_{hk}(t)^{-1} (\mathbf{x_h(t)} - \boldsymbol{\mu}_{\mathbf{hk}}(\mathbf{t})), \qquad (11.2.7)$$

is known as the *Mahalanobis D^2 distance* (Afifi, 1996). D_{hk}^2 is a weighted measure of the distance between $\mathbf{x_h(t)}$ and $\boldsymbol{\mu}_{\mathbf{hk}}(\mathbf{t})$, the centroid of the normal distribution of $\mathbf{x_h(t)}$ that corresponds to category C_{hk}. If $\mathbf{x_h(t)}$ is closer to $\boldsymbol{\mu}_{\mathbf{hk}}(\mathbf{t})$ than to $\boldsymbol{\mu}_{\mathbf{hk*}}(\mathbf{t})$—as measured in terms of D_{hk}^2 and D_{hk*}^2—$\mathbf{x_h(t)}$ is more likely to belong to C_{hk} than to C_{hk*}. In terms of forecasting, the observed value of $\mathbf{x_h(t)}$ in (11.2.7) is converted into the Mahalanobis distances that determine probabilities that each category is likely to occur.

Under (11.2.5), the modeling procedure requires two initial steps, the first being the estimation of population parameters and the second being the selection of variables in $\mathbf{x_h(t)^*}$ from those in $\mathbf{x_h(t)}$:

1. The parameters $\boldsymbol{\mu}_{\mathbf{hk}}(\mathbf{t})$ and $V_{hk}(t)$ are usually unknown and replaced by the per category sample estimates: say, $\boldsymbol{\mu}_{\mathbf{hk}}(\mathbf{t})^\wedge$ and $V_{hk}(t)^\wedge$.
2. The vector $\mathbf{x_h(t)}$ in (10.1.11) contains a large number of possible predictor variables that have been identified through the adaptive drift modeling procedure for the continuous variable $z_{SGMh}(t)$. Since $z_{SGMh}(t)$ is now categorized according to the $C_{hk}, \mathbf{x_h(t)}^*$ must be selected from $\mathbf{x_h(t)}$ in order to forecast $P(C_{hk}; \mathbf{x_h(t)}^*)$.

As such, the estimation of $P(C_{hk}; \mathbf{x_h(t)}^*)$ is preceded by the adaptive drift modeling procedure for $z_{SGMh}(t)$ that first identifies $\mathbf{x_h(t)}$. The scanning procedure associated with Bayesian discriminant analysis—which differs from that used in stepwise regression analysis—is applied in selecting the $\mathbf{x_h(t)}^*$. The $\mathbf{x_h(t)}^*$ selected for $P(C_{hk}; \mathbf{x_h(t)}^*)$ may or may not be the same as the $\mathbf{x_h(t)}^*$ selected in forecasting the continuous variable $z_{SGMh}(t)$. Statistical software such as SPSS include a stepwise scanning algorithm for choosing elements of $\mathbf{x_h(t)}^*$ and estimating $P(C_{hk}; \mathbf{x_h(t)}^*)$ under both (11.2.3) and (11.2.5). When the normality assumption is invalid (or lacks robustness under moderate assumption violations), logistic regression analysis provides an alternative modeling approach, as discussed in Section 11.3.

The algorithm for choosing the subset $\mathbf{x_h(t)}^*$ from $\mathbf{x_h(t)}$ is based on the application analysis of variance in the first step and then a series of analyses of covariance in subsequent steps. For this algorithm, the T categories are considered as T treatments in an analysis of variance in a completely randomized design model. In the first step, each variable in $\mathbf{x_h(t)}$ is regressed on the T treatments. That variable in $\mathbf{x_h(t)}$ for which the variation among treatments is most significant, say $x_{h1}(t)^*$, is chosen for inclusion in $\mathbf{x_h(t)}^*$. The second step repeats the first step in selecting $x_{h2}(t)^*$ after adjusting for $x_{h1}(t)^*$ as a covariable. The third step repeats the previous step in selecting $x_{h3}(t)^*$ after adjusting for the covariables $x_{h1}(t)^*$ and $x_{h2}(t)^*$. This process continues until no remaining variable in $\mathbf{x_h(t)}$ accounts for significant variation among treatments. This algorithm for selecting $\mathbf{x_h(t)}^*$, which is termed the *variable selection stage*, precedes and is distinct from the estimation stage where parameters in (11.2.3) and (11.2.5) are estimated.

Once $\mathbf{x_h(t)}^*$ is selected, the $P(C_{hk}; \mathbf{x_h(t)}^*)$ are estimated after $\boldsymbol{\mu}_{\mathbf{hk}}\mathbf{(t)}^*$ and $V_{hk}(t)^*$ are replaced by sample estimates. The normality assumption thus becomes a deciding factor in estimating parameters in $P(C_{hk}; \mathbf{x_h(t)}^*)$. The Box (1949) statistic is used to test the hypothesis $H_0 : V_{hk}(t) = V_h(t)$ for all k versus $H_a : V_{hk}(t) \neq V_h(t)$.

Regarding categorical forecasts for NBA and NFL games, as well as those for equity price changes, the equality of variance–covariance matrices under H_o is usually rejected. This is to be expected since the extreme categories tend to display greater variability than the middle categories. Moreover, off-diagonal elements of the $V_{hk}(t)$ may vary when comparing categories corresponding to winning performances with those corresponding to losing performances. The extent of differences between the $V_{hk}(t)$ tend to be team specific.

Forecasts for the three categories in Table 1.1.1 are based on the application of (11.2.5). The Box test rejects $H_0 : V_{hk}(t) = V_h(t)$ for $k = 1, 2, 3$. The stepwise algorithm for selecting $\mathbf{x_h(t)}^*$ identified the three variables on the right-hand sides of equation (10.2.4) along with $GSD(\text{NYG}, t-1)$ and $[u_6(t-10)][STREAK(\text{NYG}, t-1)]$. The Mahalanobis D^2 distance is presented for each of the three categories in Table 11.2.1.

Based on Table 11.2.1 results, the expected winning margin (EWM) for the Giants is obtained by multiplying the midpoint (or another representative value) of each category by the probability associated with that

TABLE 11.2.1 Results of Bayesian Discriminant Analysis: Categorical Forecasts for the Giants vs. the Patriots in Super Bowl 2008

DIFF(NYG, t^*)[a]	D^2	Probability	Odds
<-7	1.77	0.334	\$2.01 to 1
$[-7, 7]$	2.37	0.106	\$13.3 to 1
>7	0.43	0.560	\$0.79 to 1

[a] $*t =$ Super Bowl 2008 vs. NE.

category. The midpoint of [−7, 7] is zero. However, for the six games that the NYG won or lost by 7 or fewer points, the average value is 3.50. For the outer categories, the associated midpoints are taken as the average number of points that accompanied NYG losses and wins by more than 7 points. During the 2007–2008 season, the Giants lost by an average of 15.8 points in the five losses by more than 7 points and won by an average of 15 points in the six wins by more than 7 points. As such,

$$\begin{aligned} EWM\,(\text{NYG}, t = \ & \text{Super Bowl 2008 vs. NE}) = 0.334(-15.8) + 0.116(0) \\ & + 0.560(15) = 3.4 \text{ points} \end{aligned}$$

or

$$\begin{aligned} EWM\,(\text{NYG}, t = \ & \text{Super Bowl 2008 vs. NE}) = 0.334(-15.8) + 0.116(3.5) \\ & + 0.560(15) = 3.5 \text{ points.} \end{aligned}$$

Relative to the line, these categorical forecasts confirmed the Table 10.2.3 point forecast of a NYG win by 4.8 points, which is based on equation (10.2.4).

11.3 LOGISTIC REGRESSION ANALYSIS

For periods of market inefficiency, Bayesian discriminant analysis usually provides effective categorical forecasts results when nonconstant dispersion matrices are imposed under the normality assumption. When the multivariate normality assumption is invalid, logistic regression provides both an alternative and a companion modeling approach.

If uniform priors are assumed in (11.2.3), then

$$P(C_{hk}; \mathbf{x_h(t)}) = \exp\{\lambda_{hk} + \boldsymbol{\lambda}'_{\mathbf{hk}}\mathbf{x_h}(\mathbf{t})/[\Sigma_{k^*} \exp\{\lambda_{hk^*} + \boldsymbol{\lambda}'_{\mathbf{hk^*}}\mathbf{x_h}(\mathbf{t})\}] \tag{11.3.1}$$

defines a system of T equations. Without the prespecified relationships between the regression parameters $(\lambda_h, \boldsymbol{\lambda}_{\mathbf{hk}})$ in terms of the parameters of the multivariate normal distribution [as given in (11.2.4) or some other prespecified distribution], the equations defined by (11.3.1) are underidentified; that is, there is more than one solution to the $(\lambda_{hk}, \boldsymbol{\lambda}_{\mathbf{hk}})$ that leads to the same probabilities.

To make the system in (11.3.1) identifiable in the absence of a distributional assumption for $\mathbf{x_h(t)}$, the value of $(\lambda_{hk}, \boldsymbol{\lambda}_{\mathbf{hk}})$ for one of the equations is set to zero. It is arbitrary which of the $(\lambda_{hk}, \boldsymbol{\lambda}_{\mathbf{hk}})$ is equated to zero since they each yield the same probabilities. Suppose that we set to zero $(\lambda_{h1}, \boldsymbol{\lambda}_{\mathbf{h1}})$, the regression parameters corresponding to the first category (termed the *reference category*). Then

$$P(C_{h1}; \mathbf{x_h(t)}) = 1/[1 + \Sigma_j \exp\{\lambda_{hj} + \boldsymbol{\lambda}'_{\mathbf{hj}}\mathbf{x_h(t)}\}],$$
$$P(C_{hj}; \mathbf{x_h(t)}) = \exp\{\lambda_{hj} + \boldsymbol{\lambda}'_{\mathbf{hj}}\mathbf{x_h(t)}/[1 + \Sigma_j \exp\{\lambda_{hj} + \boldsymbol{\lambda}'_{\mathbf{hj}}\mathbf{x_h(t)}\}], \tag{11.3.2}$$

where $j = 2, \ldots, T$. These equations define the multinomial logistic regression model and lead to the following probability ratios relative to the reference category:

$$P(C_{hj}; \mathbf{x_h(t)})/P(C_{h1}; \mathbf{x_h(t)}) = \exp\{\lambda_{hj} + \boldsymbol{\lambda}'_{\mathbf{hj}}\mathbf{x_h(t)}\}. \tag{11.3.3}$$

The SPSS stepwise procedure for selecting $\mathbf{x_h(t)}^*$ from $\mathbf{x_h(t)}$ is analogous to that for stepwise regression except that it is in terms of nonlinear rather than linear estimation The last step of the variable selection stage provides estimates of the $P(C_{hj}; \mathbf{x_h(t)}^*)/P(C_{h1}; \mathbf{x_h(t)}^*$ which allow the estimation of $P(C_{h1}; \mathbf{x_h(t)}^*$ and the $P(C_{hj}; \mathbf{x_h(t)}^*$.

11.4 ALLOCATING BETTING MONIES IN THE SPORTS GAMBLING MARKETS

A bet on one side of the line on a game carries the risk of a total loss, whereas a bet on a financial equity in terms of a long or short position carries the risk of a partial loss. How, then, should bettors allocate available gaming monies to avoid ruin?

For team i in their forthcoming game t, the forecast $P(GSD(i,t))^\wedge = 0.75$ portends a highly favorable bet. However, given the gambler's available gaming monies, the 25% chance of losing has a direct bearing on how much to bet. Kelly (1956) betting provides one solution to the allocation problem. For bets with two outcomes, one involving losing the entire amount bet and the other involving winning the amount wagered (after a 10% betting commission fee) multiplied by the payoff odds, the Kelly formula is given by

$$F_R = (O_D P - Q)/O_D. \tag{11.4.1}$$

F_R denotes the fraction of the monies available for betting; O_D denotes the odds to \$1 that a particular team will win, P denotes the true probability of winning, and $Q = 1 - P$. For even-money bets where the line $O_D = 1$,

$$F_{\mathrm{R}} = 2P - 1. \tag{11.4.2}$$

An intuitive derivation of (11.4.2) is as follows. If the gambler bets $(2P-1)M$, where M denotes the available betting monies, a winning bet results in a gain of $2PM$ (ignoring the betting commission). For a losing bet, the loss is $2(1 - P)M$. If the gambler makes N bets under the same conditions and wins N^* of these bets, the gambler will have won

$$2^N \mathrm{P}^{N*}(1 - P)^{N-N*}M, \tag{11.4.3}$$

assuming independence between bets. If a second gambler bets $(2P-1+P_0)M$ for some positive or negative P_0, the gain is $(2P + P_0)M$ for a win and $(2(1 - P) - P_0)M$ for a loss. For the same number of wins and losses as in (11.4.3), the second gambler will have

$$(2P + P_0)^{N^*}(2(1 - P) - P_0)^{N-N^*}M. \tag{11.4.4}$$

The derivative of (11.4.4) with respect to P_0 implies that $P_0 = 2[(N^*/N) - P)$. But since $\lim_{N\to\infty}(N^*/N) = P$, the gambler's final gain is maximized under (11.4.2).

For the case where $P(GSD(i,t))^\wedge = 0.75$ is the estimate of $P, F_R = 2(0.75) - 1 = 0.5$ (i.e., one should bet 50% of their available betting monies). As opposed to single events such as the Super Bowl and World Series games, there are usually a number of bets to be made on concurrent games at a given time. During the NFL regular season, there are lines on DIFF(i,t) for as many as 16 games. Of this number, a majority will likely have favorable modeling forecasts for either the home or visiting teams. And if one adds the lines on TOT(i,t), the number of favorable bets may double. For simultaneous bets on concurrent games, outcomes are, for the most part, independent. The same cannot be said for comparable bets on equities in financial markets.

Suppose, for example, that there are favorable betting scenarios for five NBA games scheduled for the same day. In (11.1.8), let $P(GSD(i,t) \geq 0.5; \alpha^\wedge, \gamma^\wedge)$, be abbreviated by $P(i)$ and suppose that the five favorable betting scenarios are: $P(1) = 0.75, P(2) = 0.7, P(3) = 0.65, P(4) = 0.60$, and $P(5) = 0.55$. Applying (11.4.2) to each scenario, we have, respectively, $F_R(1) = 0.5, F_R(2) = 0.4, F_R(3) = 0.3, F_R(2) = 0.2$, and $F_R(5) = 0.1$. This means that 50% of the gambling money would be allocated to the most

TABLE 11.4.1 Allocation of $1000 to Five Favorable Betting Scenarios Under Kelly Betting

Betting Scenario	$Fr = 2P$-1	Proportion of $1 000 Bet	Amount Bet
$P(1) = 0.75$	0.50	0.500	$500
$P(2) = 0.70$	0.40	0.200	200
$P(3) = 0.65$	0.30	0.090	90
$P(4) = 0.60$	0.20	0.042	42
$P(5) = 0.55$	0.10	0.017	16.80
Total		0.849	$848.80

favorable bet. Forty percent of the remaining 50% [or $(0.5)(0.4) = 20\%$] would be allocated to the second most favorable bet. Thirty percent of the remaining 30% [or $(0.3)(0.3) = 9\%$] would be allocated to the third most favorable bet. Twenty percent of the remaining 21% [or $(0.2)(0.21) = 4.2\%$] would be allocated to the fourth most favorable bet. Finally, 10% of the 16.8% [or $(0.10)(0.168) = 1.68\%$] would be allocated to the fifth most favorable bet. Table 11.4.1 summarizes Kelly betting for this scenario when $1000 is available for betting. Of the $1000, $151.80 is not bet.

12

Financial/Mathematical Illiteracy and Adolescent Problem Gambling

12.1 THE CALL FOR FINANCIAL/MATHEMATICAL LITERACY IN 21st-CENTURY AMERICA

Financial and mathematical literacy underlies effective functioning of the nation's economy and markets.

> *... access to wealth should be the hope of every American and financial literacy is an essential tool to make that hope a reality.* (Former Treasury Secretary Paul O'Neill, 2002)

> *... for an increasingly complex financial system to function effectively market participants must make the type of informed judgments that promote their own well-being and foster the most efficient allocation of capital.* (Alan Greenspan, 2002)

A report by the Federal Deposit Insurance Corporation issued 12/2/09 reported that 60 million Americans live without a bank account or use nonbank operations such as pawn shops and payday lenders to handle their finances. *The FDIC survey revealed vast racial disparities in access to financial services. Almost 22% of black households had no bank account compared with 3.3% for white households. The report could increase political pressure on banks to do more for their communities after unprecedented government efforts to bail out the sector* (O'Conner and Guerrera, 12/3/09).

Forecasting in Financial and Sports Gambling Markets: Adaptive Drift Modeling, By William S. Mallios

Numeracy underlies our daily actions and personal finance. *How do you split a lunch bill three ways? How do you mentally estimate discounts, tips, and sale prices? How do you compare credit card offers with different interest rates for different periods of time? How do you choose between insurance plans or finance plans for buying a house?*

The term *innumeracy* was coined to describe an endemic deficiency: *a lack of grasp of numbers*. Innumeracy and financial illiteracy are inextricably related and may be described as ongoing epidemics based on common viruses. If so, what are the origins?

> *American high schools are obsolete. Until we design them to meet the needs of the 21st century, we will keep limiting, even ruining, the lives of millions of Americans.* (Bill Gates)

American high schools were designed in the early 20th century and are currently accomplishing the goals of that outdated era. When secondary education became the norm for American youth, its goal was to separate students who could manage classic theoretical education from those who could not, while fostering enough literacy to feed the thriving industrial base with productive workers. From 1900 to 1950, roughly 30% of the students made it through these schools; the remainder dropped out and went into the many skilled and semiskilled jobs of that time. After 1950, these jobs were gradually diminished or outsourced and the need for an educated, skilled workforce increased dramatically.[1]

Unfortunately, the need for rigorous education in a global economy has not been translated into action. As such, we are faced with transforming an outdated system into one that meets globalization's needs. The transformation of higher education is equally challenging. For centuries, particularly in Europe, where the Industrial Revolution started, education usually followed economic growth. Now higher education is preceding that of the economy, and higher education precedes globalization.[2]

[1]Paraphrased from *American Schools can be Saved by Business* by S. Weill, Chairman of Citigroup, and J. Ferrandino, President of the National Academy Foundation, *Financial Times*, 6/1/05.

[2]*China has experienced the strongest growth in scientific research over the past three decades of any country*. . .*and is on course to overtake the U.S. by 2020*. . . .*Three main factors are driving Chinese research. First is the government's enormous investment*. . .*at all levels of the system from schools to postgraduate research. Second is the organized flow of basic science to commercial applications. Third is the efficient and flexible way in which China is tapping the expertise of its extensive scientific diaspora in North America and Europe, tempting back mid-career scientists with deals that allow them to spend part of the year working in the west and part in China* (Cookson, 1/26/10).

To reinvent secondary education, *schools within schools* have been created through public–private partnerships with the business and professional worlds. The reason such schools are said to provide viable and sustainable solutions is that they embody what reformers characterize as the new "3 R's": rigor, relevance, and relationships. The new "3 R's" are intended to ensure that the content is challenging and provides skills to compete in the 21st century and that students are connected to and interact with adults—mentors, role models, and teachers—who are interested in their success. Although there have been claims of success, the effectiveness of such programs remains uncertain. Longitudinal studies, if designed properly, should provide answers.

12.2 DATA, INFORMATION, AND THE INFORMATION AGE

Information, defined as the communication of knowledge or intelligence, is contrasted with *data*, defined as well or ill-defined, objective or subjective measurements from which information can be extracted. Under these definitions, it is a misnomer to refer to data as information since information must be extracted from data if knowledge is to be communicated. Without a means of extracting such information—such as through mathematical reasoning—data are useless. The following statement illustrates this misnomer.

> *More information has been produced in the last 30 years than in the previous 5000. A weekday edition of the* New York Times *has more information than the average 17th-century man or woman would have come across in an entire lifetime.*

If *data* were substituted for *information*, this quotation would reflect the fact that we live in an unprecedented *Age of Data Acquisition* and, perhaps more precisely, an *Age of Data Oversaturation*. Indeed, if data acquisition processes were to cease henceforth, it would take decades to analyze and extract information from current databases. The Age of Data Acquisition bodes ill for societies where limited mathematical reasoning and financial illiteracy are the norm, especially in complex, heterogeneous societies such as ours.

The reality is that financial illiteracy is far more the rule than the exception. Such illiteracy has carried over from primary to secondary to higher education and beyond. Consider, for example, results of a study that examined the existing state of financial and statistical literacy among graduating California State University (CSU) business students. (The CSU system, with 450,000 students, is the world's largest system of higher education.) Schools and colleges of business within the CSU system collaborated to create an exit examination for graduating seniors. The examination, termed the Business Assessment Test (BAT), was given for the first time in the spring of 2005 to 2454 graduating seniors on 13 campuses that host business programs. A total of 80 questions spanned content in eight business disciplines.

Table 12.2.1 summarizes the BAT scores. Average scores, in terms of the percentage of questions answered correctly in each of the eight disciplines, are given both system-wide and for the 213 graduating seniors at the Craig School of Busines, CSU Fresno. The substandard test scores in finance and statistics can be attributed to the low state of financial and mathematical literacy among graduating CSU business students—a condition that is not limited to graduating business school students.

Since there were faculty controversies regarding the selected questions that comprised the BAT, particularly in statistics and finance, the writer administered an anonymous financial literacy test (described in detail in Section 12.3) to 250 business students at CSU Fresno in the spring of 2006. The students were mostly upper division. All had completed prerequisite courses in mathematics and were currently enrolled in statistics. Many had completed or were concurrently enrolled in economics, business law, and finance. Results of selected questions are indicative of the graduating seniors' abilities in finance.

- 78% did not know that bond prices and interest rates moved in opposite directions.
- 79% did not know the meaning of selling short.
- 72% did not know the meaning of a bear market.

TABLE 12.2.1 Results of the 2005 CSU Business Assessment Test[a]

Business Discipline	Average Scores System-wide	Average Scores for CSU Fresno
Management	51.96	51.64
Accountancy	48.60	47.70
Business law	49.92	47.89
Statistics	*41.62*	*34.74*
Finance	*38.11*	*36.31*
Economics	46.23	44.03
Marketing	55.88	53.05
Management information systems	63.14	61.40

[a]Average percent of questions answered correctly within each of eight business disciplines. Sample size: system-wide, 2454; CSU Fresno, 213.

- 74% answered the following question incorrectly: *Suppose you deposit $1000 in a savings account at an interest rate of 5% compounded annually. What will be the total worth of your savings in two years?*

These students were also asked to express their views on American society through the following choices:

- *Despite its shortcomings, American society offers equal opportunity—a fair chance for advancement.* (44%)
- *For many, the window of opportunity in America is limited.* (29%)
- *Though some get ahead through merit, the gospel of equal opportunity is basically a sham.* (13%)
- *No opinion.* (14%)

Analyses showed that low test scores were associated with *apparent societal disenchantment*, which is reflected by answers to the last question. An implication is that we must begin to secure our intellectual health in finance and mathematics and it must begin early in a student's intellectual training. Otherwise, we doomed to the consequences of past and current realities.

> *Markets need a fresh supply of losers just as builders of the ancient pyramids of Egypt needed a fresh supply of slaves.* (Alexander Elder)

12.3 THE COMPANION EPIDEMIC OF ADOLESCENT PROBLEM GAMBLING

The epidemics of financial and mathematical illiteracy cannot be divorced from the emerging epidemic of adolescent problem gambling. As early as 1991, *Time* magazine estimated that of the nearly 8 million compulsive gamblers in the United States, fully 1 million were teenagers. Based on the exponential growth of casino, Internet, and TV poker gambling over the past two decades, World Poker tour officials estimate that 100 million people in the United States play poker—up from 50 million 18 months ago (Growth Sets the Stage for Addiction, Web MD feature, 1/24/05) It is understandable that the science of gaming focuses on finding the most effective means of manipulating impulses so that gamblers' losses are maximized (*U.S. News & World Report*, 5/23/05).

These trends have spurred considerable research on the mental health effects of gambling on the general public, especially since the mid-1990s when studies conducted by the American Psychiatric Association (APA) found little research on pathological gambling due to the limited availability of survey data (*APA Diagnostic and Statistical Manual*, 4th edition, 1994).

In 1997, the Harvard Medical School Division of Addictions conducted a meta-analysis of 120 gambling studies in the United States and Canada to better understand the complexity and prevalence of disordered gambling (*Estimating the Prevalence of Disordered Gambling Behavior in the United States and Canada: A Meta-analysis*, Harvard Medical School Division on Addictions, 1997). Two of the main research questions were, first, to identify the percentage of the population that could be classified according to three levels—level 1: no evidence of problem gambling, level 2: problem gambling, and level 3: pathological gambling (addiction)—and second, to estimate the magnitude of the problem in subpopulations of the general public, such as youths, college students, and adults.

The Harvard study found that just under 2% of the adult population displayed level 3 addiction, and that 4% of youths and almost 5% of college students displayed level 3 pathological addiction. Of greater concern to the researchers was that 4% of the adult population displayed level 2 profiles, while almost 9.5% of the youth and college students displayed level 2 profiles. This finding disturbed the researchers since they felt that given the exponential growth of gambling venues, the percentages of youth and college students moving from level 2 to level 3 would show drastic increases over the next 10 to 15 years. They noted that *numbers gambling is so common that many people do not even consider lottery playing to be gambling* (p. 74).

The study found very limited regional or socioeconomic data available to identify the number or specific subpopulations that might display level 2 or level 3 tendencies of gambling addiction. Without this information, local health professionals cannot generate the necessary public support to commit staff and program funding to address this mental health issue appropriately. Local health providers are thus limited in their ability to assist teachers and parents who identify those needing assistance. Also, there are no protocols for referrals for treatment for those displaying level 2 and 3 tendencies.

12.4 RESULTS OF A PILOT STUDY ON ADOLESCENT PROBLEM GAMBLING AND FINANCIAL/MATHEMATICAL LITERACY

The financial literacy questionnaire, summarized briefly in Section 12.2, was adminsitered along with the South Oaks Gambling Screen Revised for

Adolescents (SOGS-RA). SOGS-RA is a well-know, validated instrument that is comprised of questions 16–35 in Table 12.4.1. [3] However, only responses to questions 24–35 are used to classify respondents according to level 1, 2, or 3 gambling profiles. Classification of individuals into level 2 or 3 is based on the number of *yes* answers to questions 25–36 and answers of *every time* or *most of the time* to question 24 (Winters, 1993). The purpose of the Table 12.4.1 questionnaire was to evaluate the level of basic financial literacy, to estimate the prevalence of adolescent problem gambling and to estimate relations, if any, between financial literacy and the propensity to gamble.

Table 12.4.1 presents the scored results of the SOGS_RA portion of the Table 12.4.1 questionnaire. Also included are results from an earlier SOGS_RA survey conducted in Oregon (Carlson et al., 1998). The Oregon results are similar to comparable studies conducted elsewhere in the United States and Canada. The disturbing CSUF results are comparable to those from a study conducted on institutionalized adolescents in Louisiana. It appears that the CSUF sample is a reflection of ethnicities, socioeconomic backgrounds, and casino gambling environments that differ from the sample in the Oregon study. Over 60% of the CSUF respondents stated that they first gambled for money during high school or before. One implication is that proposed intervention programs aimed at controlling problem gambling must occur at the high school level or before.

Table 12.4.3 presents the three gambling profile percentages for each of the three levels of financial literacy. The low level is defined by four or fewer correct answers to questions 1–13, the middle level by five to nine correct answers, and the high level by nine or more correct answers. The significant interaction between rows and columns indicates that, on average, the severity of gambling problems tends to increase (decrease) as financial literacy scores decrease (increase).

Inferences based on Table 12.4.3 results become more complicated once ethnicity is considered. Table 12.4.4 presents the trend in gambling profiles within each of the five ethnic groups. The Black/African-American group shows an unusually disproportionate percentage of problem/pathological gamblers. The corresponding percentage for the

[3]The *CSU/University of California Integrated Second Year Readiness Test* is a *diagnostic test of topics needed for success in a second-year integrated mathematics course*. It was designed to test undergraduates' preparedness for the required course in business statistics. The test consists of 45 multiple-choice questions covering topics in basic algebra and geometry, topics that are normally presented in the first two years in high schools with college prep options. Student performances on this test, administered by the writer to business students over the course of eight semesters, directly reflected those of the financial literacy questionnaire in Table 12.4.1.

TABLE 12.4.1 Financial Literacy Questionnaire and the South Oaks Gambling Screen: Revised for Adolescence

---

INSTRUCTIONS: You will be answering this questionnaire anonymously. Do not put your name or self-identifying information on the answer (scantron) form. For all questions, select one and only one of the choices given.

Answer questions 1–3 to the best of your ability—in the sense that you should not guess but, rather, choose the option "don't know the answer" when there is doubt in your mind regarding the correct answer.

1. When you purchase **stocks** issued by a large corporation, you:
   A. are lending money to the corporation
   B. are exercising effective control of the corporation's management (79%)
   C. have a legal claim on the corporation's earnings
   D. are buying shares of the corporation's ownership
   E. don't know answer to question (8%)
2. When you purchase **bonds** issued by a large corporation, you:
   A. are lending money to the corporation (61%)
   B. are exercising effective control of the corporation's management
   C. have a legal claim on the corporation's earnings
   D. are buying shares of the corporation's ownership
   E. don't know answer to question (61%)
3. Which of the following is true regarding the **price of bonds**?
   A. bond prices increase as interest rates increase
   B. bond prices are directly related to stock prices
   C. bond prices decrease as interest rates increase (22%)
   D. bond prices are directly related to the rate of unemployment
   E. don't know answer to question (19%)
4. Which of the following is true about **sales taxes**?
   A. sales taxes are regulated by the federal government
   B. the national sales tax is currently at 6.5%
   C. individuals with very low incomes are not required to pay sales taxes
   D. they are taxes that increase consumer costs (41%)
   E. don't know answer to question (11%)
5. For a savings account at a bank, which of the following is correct regarding **interest earned**?
   A. earnings from savings accounts may not be taxed
   B. sales tax may be charged on interest earned
   C. interest cannot be earned unless one is at least 18 years old
   D. income tax may be charged on the earned interest if one's income is sufficiently high (29%)
   E. don't know answer to question (37%)
6. Which group would have the greatest difficulty during periods of **high inflation**?
   A. working couples
   B. illegal aliens
   C. retired people living on fixed retirement income (67%)
   D. individuals whose savings exceed 10% of their gross income
   E. don't know answer to question (18%)

TABLE 12.4.1 ***(Continued)***

7. Which of the following types of investment would best protect the purchasing power of a family's savings in the event of a sudden **increase in inflation**?
   A. a twenty-five-year corporate bond
   B. a house financed with an adjustable interest rate
   C. a house financed with a fixed interest rate (47%)
   D. renting an apartment instead of owning a condo
   E. don't know answer to question (35%)
8. If the quantity of available **housing in a community** is less than the quantity of housing demanded, the average price of housing:
   A. will remain approximately the same
   B. will fall to clear the market
   C. will rise to clear the market (67%)
   D. will be unaffected by the supply and demand for housing
   E. don't know answer to question (10%)
9. In a **bear market**, prices of securities generally:
   A. increase
   B. usually remain approximately the same
   C. decrease (12%)
   D. stabilize with increased volatility
   E. don't know answer to question (72%)
10. A nation's **productivity** is measured in terms of:
   A. the total value of all final goods and services produced (50%)
   B. the average dollar (or dollar equivalent) output per hour of work
   C. the total amount invested in its equity capital markets
   D. the difference between gross revenues and expenditures
   E. don't know answer to question (22%)
11. If you "**sell short**" the stock issues of a particular corporation, you are:
   A. liquidating all the stock that you own in the particular corporation
   B. selling the stock only for the short term
   C. selling the stock at the current price and repurchasing the stock at a future date (21%)
   D. trading corporate warrants for stock shares
   E. don't know answer to question (58%)
12. When **government expenditures** exceeds its receipts during a specified period of time:
   A. the national debt tends to be unchanged
   B. there is a budget surplus
   C. the balance of payments between imports and exports decreases
   D. there is a budget deficit (78%)
   E. don't know answer to question (13%)
13. Suppose you deposit $100 in a savings account at an **interest rate** of 5% compounded annually. What will be the total worth of your savings in two years?
   A. $105.00
   B. $110.00
   C. $110.25 (26%)

(continued overleaf)

TABLE 12.4.1 (*Continued*)

D. $111.25
E. don't know answer to question (42%)

Please answer questions 14–45 honestly by selecting on your answer form one and only one choice for each of the following questions. It is emphasized that your answers to these questions cannot be identified with your name.

14. How sure do you feel about your ability to manage your own finances?
    A. Not sure at all — I wish I knew a lot more about money management (13%)
    B. Not too sure — I wish I knew more about money management (19%)
    C. Somewhat sure — I understand most of what I'll need to know (48%)
    D. Very sure — I understand money management very well (20%)
15. Which of the following choices best reflects your view of American society?
    A. Despite its shortcomings, American society offers equal opportunity — a fair chance for advancement. (44%)
    B. For many, the window of opportunity in America is limited. (29%)
    C. Although some get ahead through merit, the gospel of equal opportunity is basically a sham. (13%)
    D. No opinion. (14%)
16. Have you ever played the numbers or bet on lotteries?
    A. Not at all (46%)
    B. Less than once a week (50%)
    C. Once a week or more (4%)
17. Have you ever played cards for money?
    A. Not at all (31%)
    B. Less than once a week (57%)
    C. Once a week or more (12%)
18. Have you ever bet on sporting events?
    A. Not at all (49%)
    B. Less than once a week (47%)
    C. Once a week or more (4%)
19. Have you ever gone to casinos (legally or otherwise)
    A. Not at all (16%)
    B. Less than once a week (77%)
    C. Once a week or more (7%)
20. Have you ever watched the poker tournaments on television?
    A. Not at all (25%)
    B. Less than once a week (57%)
    C. Once a week or more (18%)
21. Have the television poker tournaments ever motivated you to play cards for money?
    A. Not at all (64%)
    B. Less than once a week (27%)
    C. Once a week or more (9%)

TABLE 12.4.1 (*Continued*)

22. Have you ever participated in any type of Internet gambling (such as poker, sports betting, etc.)?
    A. Not at all (77%)
    B. Less than once a week (17%)
    C. Once a week or more (6%)
23. What is the largest amount of money you have ever gambled with on any one day?
    A. Never gambled (13%)
    B. Less than $10 (18%)
    C. More than $10 but less that $100 (44%)
    D. More than $100 but less than $1 000 (23%)
    E. More than $1 000 (2%)
24. How often have you gone back another day to try and win back money that you lost gambling?
    A. Every time (2%)
    B. Most of the time (9%)
    C. Some of the time (24%)
    D. Never (55%)
    E. I have never gambled (10%)
25. When you were betting, have you ever told others you were winning money when you weren't?
    A. Yes (8%) B. No (92%)
26. Has your betting money ever caused any problems for you, such as arguments with family and friends, or problems at school or work?
    A. Yes (13%) B. No (87%)
27. Have you ever gambled more than you had planned to?
    A. Yes (43%) B. No (57%)
28. Has anyone criticized your betting or told you that you had a gambling problem, whether you thought it true or not?
    A. Yes (12%) B. No (89%)
29. Have you ever felt bad about the amount of money you bet or about what happens when you bet money?
    A. Yes (37%) B. No (63%)
30. Have you ever felt like you would like to stop betting, but didn't think you could?
    A. Yes (11%) B. No (89%)
31. Have you ever hidden from family or friends any betting slips, IOUs, lottery tickets, money that you won, or any signs of gambling?
    A. Yes (9%) B. No (91%)
32. Have you had money arguments with family or friends that centered on gambling?
    A. Yes (9%) B. No (91%)
33. Have you borrowed money to bet and not paid it back?
    A. Yes (5%) B. No (95%)
34. Have you ever skipped or been absent from school or work due to betting activities?
    A. Yes (8%) B. No (92%)

(*continued overleaf*)

TABLE 12.4.1 (*Continued*)

35. Have you borrowed money or stolen something in order to bet or to cover gambling activities?
    A. Yes (6%) B. No (94%)
36. To what extent do you think that answers to questions 16–35 give a valid assessment of gambling by college students?
    A. Not at all (7%)
    B. To a lesser extent (22%)
    C. Somewhere between a lesser extent and a greater extent (42%)
    D. To a greater extent (14%)
    E. No opinion (15%)
37. What percent of your friends gamble for money?
    A. Less than 5% (44%)
    B. 6% to 15% (18%)
    C. 16% to 33% (17%)
    D. 34% to 50% (11%)
    E. More than 50% (10%)
38. What percent of your friends do you think have a gambling problem?
    A. Less than 3% (76%)
    B. 3% to 6% (8%)
    C. 7% to 9% (9%)
    D. 9% to 12% (5%)
    E. More than 12% (2%)
39. At what age did you first gamble for money?
    A. Never gambled (14%)
    B. Before high school (23%)
    C. During high school (37%)
    D. During college (24%)
    E. After college (2%)
40. What is your gender?
    A. Female (36%) B. Male (64%)
41. You are currently enrolled as a:
    A. Freshman (8%)
    B. Sophomore (28%)
    C. Junior (53%)
    D. Senior (11%)
    E. Other (0%)
42. What is the highest level of schooling that your mother completed?
    A. Did not complete high school (12%)
    B. Completed high school and did not attend college (20%)
    C. Completed some college (36%)
    D. College graduate or beyond (26%)
    E. Don't know (6%)

TABLE 12.4.1 (*Continued*)

43. What is the highest level of schooling that your father completed?
    A. Did not complete high school (15%)
    B. Completed high school and did not attend college (20%)
    C. Completed some college (25%)
    D. College graduate or beyond (33%)
    E. Don't know (7%)
44. How do you describe yourself?
    A. Asian or Asian-American (16%)
    B. Black or African-American (6%)
    C. Hispanic or Latino (27%)
    D. White or Caucasian (44%)
    E. Other (7%)
45. How would you describe your current employment status?
    A. I work full time throughout the year (15%)
    B. I work part time throughout the year (59%)
    C. I work full or part time in the summer and don't work during the school year (11%)
    D. I never have been formally employed outside the home (7%)
    E. Other (8%)

TABLE 12.4.2 CSU Fresno Pilot Survey: Results of SOGS-RA Testing

Gambling Level	CSU Fresno (%)	Oregon Study (%)
Level 1 (no apparent problem)	61.6	84.7
Level 2 (problem gambler)	20.5	11.2
Level 3 (pathological gambler)	17.9	4.1

TABLE 12.4.3 CSU Fresno Survey Results: Gambling Profile Levels with Each Financial Literacy Level

	Gambling Profile		
Financial Literacy Level	Level 1 (%)	Level 2 (%)	Level 3 (%)
Low	30	30	40
Medium	40	37	23
High	43	39	18

Asian/Asian-American group is also significantly higher than those for the Hispanic/Latino–White/Caucasian–Other groups. The peculiarity is that the financial literacy scores for the Asians/Asian-Americans were significantly higher than those for all other groups. This result would appear to contradict the results in Table 12.4.3.

A possible explanation is that relative to the other ethnic groups, Asians/Asian Americans have a propensity to both gamble and to excel in finance and mathematics. *Two huge casinos in Connecticut—Foxwoods and*

TABLE 12.4.4 CSUF Fresno Survey Results: Gambling Profile Levels within Ethnicity

Ethnicity	Gambling Profile	
	Level 1 (%)	≥ Level 2 (%)
Asian/Asian-American	57.6	42.4
Black/African-American	33.3	66.7
Hispanic/Latino	65.5	34.5
White/Caucasian	64.3	33.7
Other	63.3	36.7

Mohegan Sun—send more than 100 buses every day to pick up customers in predominantly Asian neighborhoods of Boston and New York.... In a Washington Post *article titled "Casinos Are Aggressively Courting Asian Americans,"...Foxwoods Resort Casino (the world's biggest in terms of gambling floor space) estimates that at least a third of its customers are Asian.... If financial advisors knew as much about Asians as the casinos do, they'd have a much better chance of attracting them as clients. Casino dealers know not to touch Chinese customers on the shoulder—a sign of bad luck. They don't say the number 4, which sounds like the word for death. ("Nine" also sounds to the Japanese like the word for pain.) At Pai Gow and baccarat tables, which have numbered seats, Foxwoods has even omitted the No. 4 seat.... How does a propensity for gambling jibe with the tendency to save? Surely people don't accumulate money just for the pleasure of risking it on games of chance.... The underlying motivation may have a lot to do with joss, which has connotations of both "luck" and "fate." It's an important factor in the lives of many Chinese, whose philosophy is often a blend of Christianity and Buddhism/Taoism. Gambling is a way of inviting good joss, which can make a person wealthy in a heartbeat* (Luck and the Chinese, Investment Advisor, 8/1/08).

As an initial modeling exercise, Fisher's discriminant analysis (see Section 11.1) was applied. The dependent variable is defined as follows:

$$\begin{aligned} LEV &= 0 \quad \text{if gambling profile} = \text{level 1} \\ &= 1 \quad \text{if gambling profile} = \text{level 2 or level 3.} \end{aligned} \tag{12.4.1}$$

The independent variables to be scanned for significance in a linear model include the total number of correct answere to financial literacy questions 1–13, 16–23, 37–40 and 42–45 and all two-factor interactions between these variables. Answers to questions 42–45 were first converted to dummy variables. Through stepwise linear regression, LEV was predicted in terms of the following six variables (with significance at the 0.05 level or below and $R^2 = 0.61$):

1. Question 22: The greater the participation in Internet gambling, the greater the likelihood of $LEV = 1$.
2. Question 38: The greater the percentage of friends that have gambling problems, the greater the likelihood of $LEV = 1$.
3. Question 20 (gender); females are less likely to be problem gamblers that males.
4. Financial literacy score (FLS): The greater the FLS, the less the likelihood of $LEV = 1$.
5. The interaction FLS $\times$ question 23: The greater the FLS and the greater the amount gambled on any day, the greater the likelihood of $LEV = 1$.
6. The interaction FLS $\times$ question 16: The greater the FLS and the greater the frequency of playing the numbers or betting on lotteries, the greater the likelihood of $LEV = 1$.

The relatively large level 2 percentage for the Asian/Asian-American group in Table 11.4.4 is explained in terms of the last three variables in the list above. While the Asian/Asian-American respondents had higher financial literacy scores relative to the other ethnicities, they also gambled in greater amounts and played the numbers games and lotteries with greater frequencies.

Results of this study form the basis for working conjectures that should be used in designing follow-on studies and proposing remedial interventions. Major conjectures/recommendations are as follows:

1. There is an emerging epidemic of adolescent problem gambling in the Central California Valley.
2. The emerging problem gambling epidemic is related directly to the nationwide epidemic of financial and mathematical illiteracy.
3. Intervention programs are necessary to address both problem gambling and financial/mathematical illiteracy—with particular emphasis on adolescents.
4. Beginning immediately at the high school level, culturally competent intervention programs are necessary in at least three areas:
   a. Identifying adolescent problem gamblers and referring them to effective treatment programs
   b. Establishing new, effective treatment programs whenever necessary and evaluating the effectiveness of all programs
   c. Developing instruction in financial/gambling literacy that is tailored for each semester of high school

5. The candlestick charts in finance and sports provide an exceptionally effective means of promoting mathematical/financial literacy among adolescents, especially when such instruction includes competitive games.

These recommendations parallel results in a previous study by the writer (Mallios, 1989) that led to a court-sponsored program that identified likely recedivists among convicted drunk drivers and referred such recidivists to effective treatment programs.

13

The Influenza Futures Markets

13.1 MARKETS FOR EXPERT INFORMATION RETRIEVAL

The modeling transition is to seasonal influenza. The data under study are in terms of (1) total deaths and (2) combined total deaths due to influenza and pneumonia in the United States for each of 406 consecutive weeks starting with the first week of 1996 and ending with the 51st week of 2003. Figure 13.1.1 presents graphs of $z_1(t)$ and $z_2(t)-2$, where

$$z_1(t) = \log_e(\text{deaths in the United States from influenza and pneumonia during week } t), \quad (13.1.1)$$

$$z_2(t) = \log_e(\text{total deaths in the United States during week } t).$$

The forthcoming analyses would have been more germane had they been applied to a system of time series that included variables such as the weekly incidence of seasonal flu, primary influenza viral pneumonia, secondary influenza due to bacterial infection, and the H1N1 virus. However, as of this writing, it appears that such a database has not been fully developed.

Avian and swine flu have been viewed as commodities in the futures market for purposes of expert information retrieval. The Iowa Electronic Market (IEM) has operated influenza futures markets since 2004. The influenza market is a closed, invitation-only market with participants limited to medical professionals and scientists in the health care fields who

---

Forecasting in Financial and Sports Gambling Markets: Adaptive Drift Modeling, By William S. Mallios

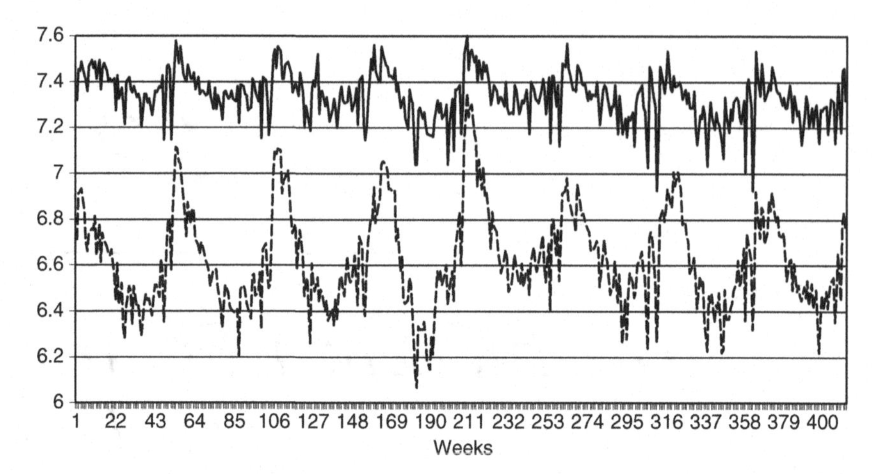

Figure 13.1.1 *Weekly U.S. trends of deaths due to influenza and pneumonia (DIP) and total deaths (DT) for each of 406 consecutive weeks starting with the first week of 1996 and ending with the 51st week of 2003. The top and bottom trends denote, respectively,* $\log_e(DT) - 2$ *and* $\log_e(DIP)$.

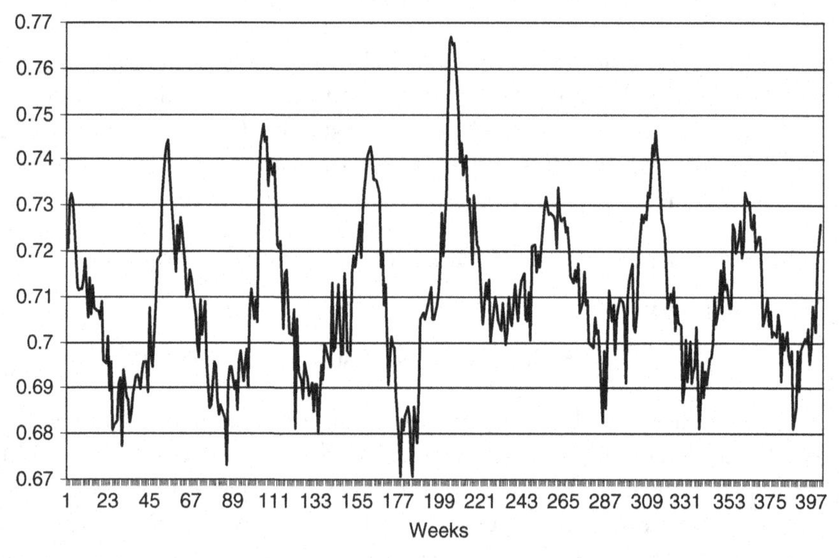

Figure 13.1.2 *Weekly trends in Figure 13.1.1 presented in terms of the ratio* $R = \log_e(DPI)/\log_e(DT)$.

have some knowledge of the virus's development. In 2009, IEM expanded to include the H1N1 virus for purposes of building practitioner-level predictions of how the virus would spread, its severity, and its duration (FluPrediction@Uiowa.edu).

As of October 2009, IEM predicted a 70% chance that there will not be enough H1N1 vaccine available to inoculate the target of 50 million Americans until at least December 2009. It also predicted a more than 90% chance that more than half of the U.S. flu cases this season will be swine flu (Weitzman, 10/29/09). Figure 13.1.2, an alternative version of Figure 13.1.1, presents the ratio $R = \log_e(\text{DPI})/\log_e(\text{DT})$ over the same time period.

13.2 ADAPTIVE SEASONAL TIME SERIES MODELING

Seasonal time series modeling can be approached in terms of either ARMA-type processes (Box and Jenkins, 1976) or periodic regression (Bliss, 1958; Mallios, 1989). For the seasonal, weekly data in Figure 13.1.1, the observations for a particular week of the year are related not only to the previous week's observations, but also to the observations of the comparable week in the preceding year. As such, differencing such as

$$dz_h(t) = [z_h(t) - z_h(t-52)] - [z_h(t-1) - z_h(t-53)] \qquad (13.2.1)$$

is employed to achieve stationarity in ARMA-type processes when coefficients are non-time varying. With time-varying coefficients, the modeling procedure for (13.2.1) follows that given for (9.5.2).

A periodic regression model with a single harmonic is given by

$$\begin{aligned} z_h(t) &= a_0 + a_1 \cos(mt - \theta) + \Sigma_{i>0}\, c_i u(t-i) + \varepsilon_h(t) \\ &= a_0 + b_1 \cos(mt) + b_2 \sin(mt) + \Sigma_{i>0} c_i u(t-i) + \varepsilon_h(t), \end{aligned} \qquad (13.2.2)$$

where $h = 1, 2$. The first expression for $z_h(t)$ in (13.2.2) is linearized through the trigonometric identity $\cos(U - V) = \cos U \cos V + \sin U \sin V$; a_1 is the semiamplitude, θ is the phase angle (or the time in angular measure of the maximum); $b_1 = a_1 \cos\theta$ and $b_2 = a_1 \sin(\theta)$. The constant $m = 2\pi/52$ converts the weeks (t) in a single cycle to angular measure in radians. When the two time series are cointegrated, there is a single cointegration vector; the c_i represent the effects of the $u(t-i)$, the lagged between-relation shocks. When the $u(t-i)$ do not adequately reflect both between- and within-relation shocks, effects of moving average variables can be recovered in a subsequent stage of modeling.

Let $b_1{}^\wedge$ and $b_2{}^\wedge$ denote OLS estimates of b_1 and b_2. Then the amplitude is estimated by $a_1{}^\wedge = (b_1^{2\wedge} + b_2^{2\wedge})^{1/2}$ since $(b_1^{2\wedge} + b_1^{2\wedge}) = a_1^{2\wedge} \cos\theta^2 + a_1^{2\wedge} \sin\theta^2 = a_1^{2\wedge}$; the phase angle θ is estimated by

$\theta\wedge = \arctan(b_2\wedge/b_1\wedge)$. Adaptive periodic regression,[1] as applied to the $z_h(t)$ in (13.1.1), is expressed in terms of

$$z_h(t) = a_0(t) + a_1(t)\cos mt \cos(\theta(t)) + a_1(t)\sin mt \sin(\theta(t)) + \Sigma_{i>0}c_i(t)u(t-i) + \varepsilon_h(t). \quad (13.2.3)$$

The parameters $a_1(t)$ and $\theta(t)$ allow the semiamplitude and phase angle to vary from one seasonal cycle to the next. For the model error, it is assumed that

$$\boldsymbol{\varepsilon}(\mathbf{t}) = \varepsilon_h(\mathrm{t}) : [0, \Sigma_\varepsilon(t)], \quad (13.2.4)$$

where allowances are made for possible time-varying volatility.

Exploratory applications of the two modeling approaches—model (13.2.3) and model (9.5.2) with $dz_h(t)$ given by (13.2.1)—to the data in Figure 13.1.1 resulted in somewhat more effective forecasts for the periodic regression approach. As such, the current analysis will be limited to model (13.2.3). However, in application, where frequent model updates are advisable, forecasts based on both modeling options should be compared for effectiveness.

Each of the time-varying parameters in (13.2.3) is assumed generated in terms of a first-order linear regression process in elements of $\mathbf{u^*(t-)}$, a vector containing relevant lags of the variable $u^*(t)$; that is,

$$\omega(t) = \omega + \mathbf{w}'_\omega\mathbf{u}^*(\mathbf{t-}) + \delta_\omega(t), \quad (13.2.5)$$

where $\omega = a_0,\ a_1, \theta$ and c_i; for example, $\theta(t) = \theta + \mathbf{w}'_\theta\mathbf{u^*(t-)} + \delta_\theta(t)$; the $\delta_\omega(t)$ denote model errors; the $\mathbf{w}_\omega$ are regression vectors representing the effects of $\mathbf{u}^*(\mathbf{t-})$ on $\omega(t)$. When the $u(t-i)$ adequately reflect both between- and within-relation shocks, $\mathbf{u}^*(\mathbf{t-}) = \mathbf{u(t-)}$, where $\mathbf{u(t-)}$ is defined to contain relevant lags of $u(t)$. When the $u(t-i)$ do not adequately reflect within-relation shocks, $\mathbf{u}^*(\mathbf{t-}) = (\mathbf{u(t-)}, \boldsymbol{\varepsilon}(\mathbf{t-}))'$, where $\boldsymbol{\varepsilon}(\mathbf{t-})$ is defined to contain all relevant moving average variables.

For

$$\theta(t) = \theta + \mathbf{w}'_\theta\mathbf{u}^*(\mathbf{t-}) + \delta_\theta(t) \quad (13.2.6)$$

[1]Model (13.2.2) is easily generalized to include higher harmonics (Bliss, 1958). Such generalizations are appropriate when they can be given physical interpretation. As reported in the news media in November 2009, health officials were quoted as saying that H1N1 may peak two to three times within 2009–2010. If so, additional harmonics should be considered in (13.2.2).

in (13.2.5), $\theta(t)$ is approximated by $\theta + \mathbf{w}'_\theta\mathbf{u}^*(\mathbf{t}-)$, whereupon $\cos(\theta + \mathbf{w}'_\theta\mathbf{u}^*(\mathbf{t}-))$ and $\sin(\theta + \mathbf{w}'_\theta\mathbf{u}^*(\mathbf{t}-))$ are approximated by the following second-order series expansions:

$$\cos[\theta + \mathbf{w}'_\theta\mathbf{u}^*(\mathbf{t}-)] \approx \cos\theta - \mathbf{w}'_\theta\mathbf{u}^*(\mathbf{t}-)\sin\theta - 0.5[\mathbf{w}'_\theta\mathbf{u}^*(\mathbf{t}-)]^2\cos\theta,$$
$$\sin[\theta + \mathbf{w}'_\theta\mathbf{u}^*(\mathbf{t}\text{-})] \approx \sin\theta - \mathbf{w}'_\theta\mathbf{u}^*(\mathbf{t}-)\cos\theta - 0.5[\mathbf{w}'_\theta\mathbf{u}^*(\mathbf{t}-)]^2\sin\theta. \tag{13.2.7}$$

Substitution of expressions on the right-hand sides of (13.2.5) and (13.2.7) for the corresponding time-varying coefficients in (13.2.3) results in a linear, higher-order reduced equation for $z_h(t)$ that includes first-order effects of the variables $\cos(mt), \sin(mt), \mathbf{u}^*(\mathbf{t}-)$, and higher-order effects of interactions. This reduced equation is written in abbreviated notation as

$$z_{\mathrm{Rh}}(t) = f[\cos mt, \sin mt, \mathbf{u(t-)},\mathbf{v(t-)}; \Xi] + \varepsilon_{\mathrm{Rh}}(t), \tag{13.2.8}$$

where Ξ represents the parameter vector and $\varepsilon_{Rh}(t)$ the reduced model error. Equation (13.2.8) is used as a starting point for identifying an effective forecasting model.

When $\mathbf{u(t-)}$ does not adequately reflect the moving average effects [as denoted by the vector $\boldsymbol{\varepsilon}(\mathbf{t}-)$], it is first necessary to go through the process of identifying significant predictors in model (13.2.8) in order to obtain an initial estimate of elements of $\boldsymbol{\varepsilon}(\mathbf{t}-)$. The next step is to identify MA and bilinear variables to be included in the model. Following the latter identification, nonlinear estimation is applied in obtaining an updated forecasting model.

13.3 FORECASTING WEEKLY INFLUENZA AND PNEUMONIA DEATHS

Modeling is initiated by showing that the two series in (13.1.1) are cointegrated. Following the procedure in (9.1.2)–(9.1.5), we fit the regression of $z_{\mathrm{DIP}}(t) = z_1(t)$ on $z_{\mathrm{DT}}(t) = z_2(t)$ and obtain

$$z_{\mathrm{DIP}}(t) = -8.48_{15.2} + 1.62_{27.1} z_{\mathrm{DT}}(t) + u_{\mathrm{DIP}}(t), \qquad R^2 = 0.640, \tag{13.3.1}$$

where $u_{\mathrm{DIP}}(t)$ denotes the residual. Following (9.1.5), $du_{\mathrm{DIP}}(t) = u_{\mathrm{DIP}}(t) - u_{\mathrm{DIP}}(t-1)$ is regressed on $u_{\mathrm{DIP}}(t-1)$ and lags of $du_{\mathrm{DIP}}(t)$ with the following result:

$$\begin{aligned} du_{\text{DIP}}(t)^{\wedge} = & -0.16_{3.62}u_{\text{DIP}}(t-1) - 0.41_{6.78}du_{\text{DIP}}(t-1) \\ & -0.25_{4.50}du_{\text{DIP}}(t-2) + 0.11_{2.10}du_{\text{DIP}}(t-49) \\ & +0.23_{4.13}du_{\text{DIP}}(t-52) + 0.13_{2.33}du_{\text{DIP}}(t-53). \end{aligned} \qquad (13.3.2)$$

Since H_0 in (9.1.4) is rejected, the two time series are cointegrated. As such, the $u(t-i)$ in (13.2.3) are estimated by the $u_{\text{DIP}}(t-i)$ as given by (13.3.1). When the analysis in (13.3.1)–(13.3.2) is performed on different segments of the data in Figure 13.1.1, the two time series remain cointegrated, although the coefficient estimates in (13.3.1) usually change to a lesser extent.

For real-time forecasting with model updates occurring on a weekly basis, stepwise regression (see Section 7.2) is applied for purposes of identifying significant predictors for $z_{Rh}(t) = z_{R\text{DIP}}(t)$ in (13.2.10). As mentioned earlier, once significant explanatory variables in model (13.2.10) are tentatively identified, the next stage is to recover, if necessary, information from significant moving average terms.

For the data in Figure 13.1.1, the following analysis is in terms of a model update through the third week of 2002 (or week 316 in Figure 13.1.1). This update precedes the peak period for $z_{\text{DIP}}(t)$ during 2002. Weighted stepwise linear regression (see Section 7.2) is applied to (3.2.10) to identify significant predictors. The weights are in terms of the weeks in Figure 13.1.1; that is, the more recent the week, the greater the weight of the observation.

The resulting prediction equation for $z_{\text{DIP}}(t)$ is given as follows:

$$\begin{aligned} z_{\text{DIP}}(t)^{\wedge} = & \ 6.61 + 0.08_{5.26}\cos mt + 0.10_{6.44}\sin mt + 0.21_{2.09}u(t-2) \\ & +0.37_{4.14}u(t-3) + 0.23_{2.53}u(t-4) \\ & -0.26_{3.00}u(t-49) + 0.27_{3.09}u(t-53) \\ & +1.14_{2.18}[u(t-1)][u(t-2)] - 1.59_{2.96}[u(t-3)][u(t-4)] \\ & +2.16_{3.57}[u(t-3)][u(t-52)] \\ & -0.88_{2.70}[u(t-52)]^2 + 3.13_{4.47}[u(t-1)][u(t-2)](\cos mt) \\ & -1.48_{2.44}[u(t-1)][u(t-50)](\cos mt); \end{aligned} \qquad (13.3.3)$$

$R^2 = 0.77$. For this particular model update, the effects of moving average variables are, for the most part, reflected in terms of the lagged $u(t)$ and their interactions.

Based on model (13.3.3), the one-step-ahead forecasts in Table 13.3.1 are seen to be reasonably close to the observed values. These results are

TABLE 13.3.1 One-Step-Ahead Weekly Forecasts Based on a Model Update Through the Second Week of 2002[a]

Week	$z_{DIP}(t)$	$z_{DIP}(t)^\wedge$
2	6.85	6.90
3	6.79	6.82
4	6.86	6.79
5	6.93	6.85
6	6.87	6.84
7	6.85	6.81
8	7.01	6.92
9	6.95	6.98
10	7.01	7.00
11	6.97	7.01
12	6.93	6.98
13	6.78	6.93
14	6.79	6.84

[a]Forecasts of $\log_e$(total deaths due to influenza and pneumonia) $= z_{DIP}(t)$ are given for the 3rd through 14th weeks of 2002; model forecasts are denoted by $z_{DIP}(t)^\wedge$.

presented only for purposes of illustration. When forthcoming peak periods are officially monitored for purposes of forecasting the severity of influenza, model updates should be performed on a weekly basis. For model updates in scenarios such as Figure 13.1.1, the first harmonic (i.e., $\cos mt$ and $\sin mt$) tends to be the dominant predictor and explains approximately 60% of the total variation. However, the coefficients and the other explanatory variables are subject to change. Model updates are also imperative for purposes of evaluating the need for including higher harmonics in the model.

Regarding time-varying volatility, GARCH-type modeling associated with model update in (13.3.3) results in the following forecast for $e^2_{\mathrm{DIP}}(t)$:

$$e^2_{\mathrm{DIP}}(t)^\wedge = 0.009_{5.72} + 0.269_{4.06} e^2_{\mathrm{DIP}}(t-1). \qquad (13.3.4)$$

Model forecasts based on (13.3.3) and (13.3.4) can be used to heuristically identify and define the level of an epidemic. If $z_{\mathrm{DIP}}(t)^\wedge$ and $e^2_{\mathrm{DIP}}(t)^\wedge$ are used as estimates of the mean and variance of, say, the normal probability distribution (NPD), three epidemic levels may be defined:

- Level 1 epidemic: $z_{\mathrm{DIP}}(t)$ exceeds the upper 10% level of the NPD.
- Level 2 epidemic: $z_{\mathrm{DIP}}(t)$ exceeds the upper 5% level of the NPD.
- Level 3 epidemic: $z_{\mathrm{DIP}}(t)$ exceeds the upper 1% level of the NPD.

References

Afifi A, Clark V. *Computer Aided Multivariate Analysis*. Chapman & Hall, London, 1996.

Anderson A. S.E.C. moves to ban edge held by fast traders. *New York Times*, 9/18/09.

Atwood M. *Payback*: *Debt and the Shadow Side of Wealth*. Bloomsbury, New York, 2008.

Authers J. Hedge funds poised for harsh phase of evolution. *Financial Times*, 10/18/08.

Authers J. Like baseball stats, bond risks are hard to fathom. *Financial Times*, 5/23/09.

Authers J. The short view. *Financial Times*, 7/16/09.

Authers J. Return of the banks' golden age is an illusion. *Financial Times*, 7/18/09.

Authers J. The short view. *Financial Times*, 10/27/09.

Bachelier L. Theory of speculation. In *Random Character of the Stock Market Prices*, Cootner PH, ed. MIT Press, Cambridge, MA, 1964.

Blas J. The short view. *Financial Times*, 8/28/09.

Blas J. Sugar crisis worsens as Indonesia tender fails. *Financial Times*, 1/22/10.

Bliss CI. *Periodic Regression in Biology and Climatology*. No. 615. Connecticut Agricultural Experiment Station, 1958.

Blitz R. A better hand. *Financial Times*, 2/4/09.

Bollerslev T. Generalized autoregressive heteroskedasticity. *Journal of Econometrics*, 31: 307–327, 1986.

Bouchaud J. Economics needs a scientific revolution. *Nature*, 455, Oct. 2008.

Bowe C. Ghost of Shoeless Joe still haunts US sport. *Financial Times*, 7/24/06.

Box GEP. A generalized distribution theory for a class of likelihood criteria. *Biometrika* 36: 823–834, 1949.

Box GEP, Jenkins GM. *Time Series Analysis*. Holden-Day, San Francisco, 1976.

Brenner R, Terk I. Gambling has been falsely demonized for centuries. *Financial Times*, 10/5/06.

Brittan S. The overwhelming case for paying stealth taxes. *Financial Times*, 11/25/99.

Brittan S. How the budget hole developed. *Financial Times*, 7/24/09.

Brooks D. Holding back, letting go. *Wall Street Journal*, 6/8/99.

Brooks D. The great seduction. *New York Times*, 6/10/08.

Burkhart L. NBA commissioner on legalized gambling. Onlinecasinosphere.com, 12/11/09.

Burnham T. *Mean Markets and Lizard Brains: How to Profit from the New Science of Irrationality*. Wiley, Hoboken, NJ, 2005.

Carlson M et al. *Adolescent Gambling in Oregon: A Report to the Oregon Gambling Treatment Addiction Foundation*. 1998.

Chung J. Systemic failure turns yet more ire on SEC. *Financial Times*, 12/15/08.

Chung J. Ex-Lehman executive charged over insider trading scheme. *Financial Times*, 12/15/08.

Chung J. Insider trading crackdown. *Financial Times*, 10/17/09.

Clauset A., Shalizi CR, Newman MEJ. Power-law distributions in empirical data. *SIAM Review*, 51(4): 661–703, 2009.

Coates J. Traders would do well to track their hormones. *Financial Times*, 4/15/08.

Coates J. Alpha males must trade on more than machismo. *Financial Times*, 11/25/09.

Coles S. *An Introduction to Statistical Modeling of Extreme Values*. Springer-Verlag, New York, 2001.

Cookson C. Testosterone warning to traders. *Financial Times*, 4/15/08.

Cookson C. China set for global lead in scientific research. *Financial Times*, 1/26/10.

Cowling C. *Populists, Plunger, and Progressives*. Princeton University Press, Princeton, NJ, 1965.

Daneshku S, Done K. Fines call over EARD deals. *Financial Times*, 7/29/07.

DeGrauwe P. Warring economists are carried along by the crowd. *Financial Times*, 7/22/09.

De Luca C. The ghost of Shoeless Joe: Say it ain't so. cdeluca@ suntimes.com, 7/24/09.

Dobra JL, Cargill TF, Meyers RA. Efficient markets for wagers: the case of professional basketball wagering. In *Sportsmetrics*, Goff B, Tollison R, eds. Texas A&M University Press, College Station, TX, 1990.

Drehs W. Numbers indicate unlikely outcomes in games Donaghy officiated. ESPN, 4/24/07.

Engle R. Autoregressive conditional heteroskedasticity with estimates of the variance of inflationary expectations. *Econometrica*, 50: 987–1007, 1982.

Engle R, Granger CWJ. Cointegration and error correction representation. *Econometrica*, 55: 251–276, 1987.

Farrell G. Trader revealed as main Galleon witness. *Financial Times*, 2/3/10.

Ferguson N. A memo to market dinosaurs: It really is Darwinian out there. *Financial Times*, 12/13/07.

Fergusen N. *The Ascent of Money: A Financial History of the World*. Allen Lane, New York, 2008.

Fiedler I, Rock J. Skill vs. luck: Let's get it right. *Gaming Law Review and Economics*, 13(1): 50–57, 2009.

Forrester J. Counterintuitive behavior of social systems. *Technology Review*, 73(3): 52–68, 1971.

Friedman B. Overmighty finance levies a tithe on growth. *Financial Times*, 8/27/09.

Gangahar A. Last-minute swings suggest return of *quants*. *Financial Times*, 3/30/09.

Garrahan M, Edgecliffe-Johnson A. Disney in $4 bn bid to snare Marvel's superstars. *Financial Times*, 9/1/09.

Gelsi S. XOM buys XTO. *MarketWatch*, 12/14/09.

Gorton G, Rouwenhorst G. *Facts and Fantasies About Commodity Futures*. Yale ICF Working Paper. Yale University, New Haven, CT, 2004.

Gould SJ. *Hen's Teeth and Horse's Toes*. Norton, New York, 1984.

Gould SJ, Eldridge N. Punctuated equilibria: the tempo and mode of evolution reconsidered. *Paleobiology*, 3: 115–151, 1977.

Granger CWJ. Testing for causality. *Journal of Economic Dynamics and Control*, 2: 329–352, 1980.

Granger CWJ. Cointegrated variables and error correction models. Discussion paper. UC San Diego, San Diego, CA, 1983.

Granger CWJ, Anderson AP. An *Introduction to Bilinear Time Series Models*. Vandenhoeck & Ruprecht, Gottingen, Germany, 1978.

Granger CWJ, Morgenstern O. *Predictability of Stock Prices*. Heath Lexington, Lexington, MA, 1970.

Groz M. Residual games: a new deal for players. *Proceedings of the 13th International Conference on Gambling and Risk Taking*. Institute for the Study of Gambling and Commercial Gaming, University of Nevada, Reno, NV, 2006.

Hall B. France steps up drive to sell morality to markets. *Financial Times*, 1/3/09.

Harper C. Goldman Sachs $100 million trading days reach record. Bloomberg.com, 8/5/09.

Hollinger P. Furor as report on EADS hits internet. *Financial Times*, 4/9/08.

IAV. For a new culture thrift: confronting the debt culture. www.newthrift.org, 2008.

Isaacson M. Cub fans too high or too low. *Chicago Tribune*, 10/7/08.

Jackson T. Frankly, my dear, that record has gone with the wind. *Financial Times*, 1/11/10.

Jenkins GM, Alavi AS. Some aspects of modeling and forecasting multivariate time series. *Journal of Time Series*, 2: 1–47, 1981.

Johansen S. *Likelihood-Based Inference in Cointegrated Vector Auto-Regressive Modeling*. Oxford University Press, Oxford, UK, 1995.

Jones S. Old strategies lift hedge funds. *Financial Times*, 8/13/09.

Kahneman D, Tversky A. An analysis of decisions under risk. *Econometrica*, 47: 263–291, 1979.

Kaufman H. *Henry Kaufman on Money and Markets: A Wall Street Memoire*. McGraw-Hill, New York, 2001.

Kay J. Blown off course by butterflies. *Financial Times*, 11/26/08.

Kay J. How bank bonuses let us all down. *Financial Times*, 2/25/09.

Kearns J. 3Com option trades may have been more than luck. Bloomberg.com, 11/12/09.

Keen S. $A carry trade in SMH, age. *Debtwatch*, 10/27/09.

Kelly JL. A new interpretation of information rate. *Bell System Technical Journal*, 35: 917–926, 1956.

Kim C, Nelson CR. The time varying parameter model for modeling changing conditional variance: the case of the Lucas hypothesis. *Journal of Business and Economics Statistics*, 7: 433–449, 1989.

Kirchgaessner S. Christian right joins US battle against betting. *Financial Times*, 8/14/07.

Kuper S, Blitz R. Ugly truth behind beautiful game. *Financial Times*, 11/21/09.

Lawrence S. State lottery offers new game show. *Fresno Bee*, 1/16/09.

Lewis M. *Moneyball: The Art of Winning an Unfair Game*. Norton, New York, 2003.

Lo A. The adaptive market hypothesis: market efficiency from an evolutionary perspective. *Journal of Portfolio Management*, 30: 15–29, 2004.

Lo A. Reconciling efficient markets with behavioral finance: the adaptive market hypothesis. *Journal of Investment Consulting*, 7: 21–44, 2005.

Mackay C. *Extraordinary Popular Delusions and the Madness of Crowds*. 1841.

Mackenzie M, Chung J. High frequency trading under scrutiny at watchdog's meeting. *Financial Times*, 1/13/10.

MacKinnon JG. Critical values for cointegration tests. In *Long-Run Economic Relationships in Cointegration*, Engle RF, Granger CWJ, eds. Oxford University Press, Oxford, UK 1991, pp. 267–276.

Malkiel BG. *A Random Walk Down Wall Street*. Norton, New York, 1985.

Mallios WS. *Statistical Modeling: Applications in Contemporary Issues*. Iowa State University Press, Ames, IA, 1989.

Mallios WS. *The Analysis of Sports Forecasting: Modeling Parallels Between Sports Gambling and Financial Markets*. Kluwer, Norwell, MA, 2000.

Mandlebrot BB, Hudson RL. *The (Mis)Behavior of Markets: A Fractal View of Risk, Ruin and Rewar*. Profile Books, London, 2004.

Mandlebrot BB, Taleb N. A focus on the exceptions that prove the rule. *Financial Times*, 3/24/06.

Martin R. Managers must be judged on the real score. *Financial Times*, 5/12/09.

Marwah K, ed. *Selected Papers of Lawrence Klein: Theoretical Reflections and Econometric Applications*. Singapore: World Scientific, 1997.

Masters B. US ruling exposes insider trading divide. *Financial Times*, 10/12/09.

McTeer B. In praise of an economic revolutionary. *Wall Street Journal*, 7/5/01.

Mehlman M. *Bioethics: Performance Enhancing Drugs in Sports*. www.TheDoctorWillSeeYouNow.com, 2005.

Meyer G. Gurus who sparked commodities rush are betting on the long term. *Financial Times*, 2/10/10.

Milne R. Hedge funds take pounding as Porsche drives for control of VW. *Financial Times*, 10/28/08.

Milne R, Burgess K. Hedge funds left reeling as VW races on. *Financial Times*, 10/29/08.

Muth JF. Rational expectations and the theory of price movement. In *Rational Expectations and Economic Practice*, Lucas R, Sargent T, eds. University of Minnesota Press, Minneapolis, MN, 1981.

Nisson S. *Japanese Candlestick Charting Techniques*. New York Institute of Finance, New York, 1991.

Nuttall N. Executive quits in Galleon case. *Financial Times*, 11/3/09.

O'Conner S, Guerrera F. Report reveals vast banking "inderclass" among US adults. *Financial Times*, 12/3/09.

Prowse M. Greedy bosses, lying politicians and cheating teachers. *Financial Times*, 6/16/02.

Quenouille MH. *The Analysis of Multiple Time Series*. Hafner, New York, 1957.

Reed J. *Ten Days That Shook the World*. Penguin Classics, New York, 1919.

Richards J. How markets attacked the Greek piñata. *Financial Times*, 2/12/10.

Rose IN. The politics behind the Internet gaming bills. *Proceedings of the 14th International Conference on Gambling and Risk Taking*. Institute for the Study of Gambling and Commercial Gaming, University of Nevada, Reno, NV, 2009.

Santa Clara P. Letters to the Editor. *Financial Times*, 11/5/08.

Schneider W. Analytical uses of Kalman filtering in economics: a survey. *Statistical Papers*, 29: 3–33, 1988.

Schwarz A, Rashbaum W. NBA referee is the focus of a federal inquiry. *New York Times*, 7/21/07.

Sharpe W. The Sharpe ratio. *Journal of Portfolio Management*, 21(1): 49–58, 1994.

Simon M. On human action. *Philosophical Books*, 25: 38–41, 1984.

Skellam J. A probability distribution derived from the binomial distribution by regarding the probability of success as variable between sets of trials. *Journal of the Royal Statistical Society*, Series B, 10: 257–261, 1948.

Smith C. *Adam Smith's Political Philosophy: The Invisible Hand and Spontaneous Order*. Routledge Studies in Social and Political Thought. Routledge, New York, 2006.

Smith V. *Bargaining and Market Behavior: Essays in Experimental Economics*, collected works. Cambridge University Press, Cambridge, UK: 2000.

Snopes.com. September 11 PutCall, 12/11/05.

Soros G. Do not ignore the need for financial reform. *Financial Times*, 10/26/09.

Spengler O. *The Decline of the West*. Knopf, New York, 1932, p. 49.

SPSS. Version 16.0.2. Apr. 2008.

Tait A, Blitz R. Top EU court deals blow to online gambling. *Financial Times*, 9/9/09.

Taleb N. *Fooled by Randomness: The Hidden Role of Chance in Life and in the Markets*. Random House, New York, 2005.

Taleb N. *The Black Swan: The Impact of the Highly Improbable*. Random House, New York, 2007.

Taleb N, Triana P. Bystanders to this financial crime were many. *Financial Times*, 12/8/08.

Tett G. Credit crunch causes analysts to rethink rational market theory. *Financial Times*, 6/16/09.

Tett G, Davies P, Van Duyn A. A new formula? Complex finance contemplates a more fettered future. *Financial Times*, 10/10/08.

The Economist. The inefficiency of markets: slaves to some defunct economist. 6/11/09.

Thompson J, Hollinger P. Provence resists Sarkozy's attempt to highjack Camus. *Financial Times*, 12/2/09.

Thorp E. *Beat the Dealer*. Random House, New York, 1966.

Tversky D, Kahneman D, Slovic P. *Judgment under Uncertainty: Heuristics and Biases*. Cambridge University Press, Cambridge, UK, 1981.

Tyler C. "Profit from it" at Dow Jones. Optionetics.com, 5/3/07.

Wadhwani S. How efficient market theory gave rise to policy mistakes. *Financial Times*, 12/17/08.

Weitzman H. Taking a gamble on swine flu. *Financial Times*, 10/29/09.

West M, Harrison J. *Bayesian Forecasting and Dynamic Models*, 2nd ed. Springer-Verlag, New York, 1997.

Wetzel D. 7 hours, 49 minutes ago. Yahoo! 10/29/09.

White R. The billion dollar boffin. *Financial Times*, 4/28/07.

Winters KC, Stinchfield RD, Fulkerson J. Towards development of an adolescent problem severity scale. *Journal of Gambling Studies*, 9: 63–84, 1993.

Wolfers J. Point shaving: corruption in NCAA basketball. *American Economic Review*, May 2006.

Wright S. Correlation and causation. *Journal of Agricultural Research* 20: 557–585, 1921.

Zellner A. *An Introduction to Bayesian Inference in Econometrics*. Wiley, New York, 1971.

Zezima K. Sweet dreams in hard times add to lottery sales. *New York Times*, 9/13/08.

Index

Forecasting in Financial and Sports Gambling Markets: Adaptive Drift Modeling, By William S. Mallios

Printed in the United States
By Bookmasters